Elm

Frontispiece. English Elm and megalith at Avebury (Wi), by Michael Young.[1] The association between elms and prehistoric sites is one of the main themes explored in this book.

ELM

R. H. RICHENS

*Director of the Commonwealth Bureau of Plant Breeding
and Genetics, Cambridge 1964–1979*

*Secretary of the International Commission for
the Nomenclature of Cultivated Plants 1968–1979*

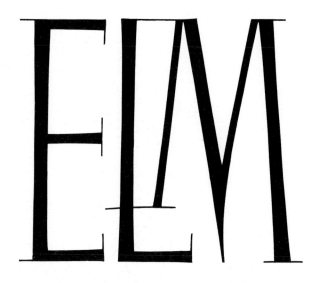

CAMBRIDGE UNIVERSITY PRESS

Cambridge

London   New York   New Rochelle

Melbourne   Sydney

Published by the Press Syndicate of the University of Cambridge
The Pitt Building, Trumpington Street, Cambridge CB2 1RP
32 East 57th Street, New York, NY 10022, USA
296 Beaconsfield Parade, Middle Park, Melbourne 3206, Australia

© Cambridge University Press 1983

First published 1983

Printed in Great Britain at the Alden Press, Oxford

Library of Congress catalogue card number: 82–17690

British Library Cataloguing in Publication Data

Richens, R. H.
Elm.
1. Elm – Great Britain
I. Title
583'.962   QK495.U4

ISBN 0 521 24916 3

Contents

Illustrations

Acknowledgements

Permission to reproduce one stanza of Olney Hymns has been given by Sir John Betjeman.

The author and publisher would like to thank the following for permission to reproduce illustrations:

Frontispiece, Mr Michael Young and *The Times*; 2, *East Anglian Magazine*; 10, Princes Risborough Laboratory; 18, Nationalmuseet, København; 19, National Museum of Antiquities of Scotland; 29, *Kansas Science Bulletin*; 30, E. J. Brill; 31, *Journal of Helminthology*; 38, École Nationale Supérieure Agronomique, Montpellier; 40*a* and *d*, Mr H. H. Keifer; 40*b* and *c*, *Acta Zoologica Hungarica*; 43, Phaidon; 47, GC & HTJ; 48, Miss Stella Ross-Craig and Bell & Hyman; 50, GC & HTJ; 52, Mr Anthony Day; 53, Tate Gallery; 54, *Berkshire Archaeological Journal*; 55, *Edgar Allen News*; 56, Fishmongers' Company and Courtauld Institute of Art; 57, Provost and Fellows of King's College, Cambridge; 58, Mr Henry Moore and Lund Humphries; 59, Tate Gallery; 60 and 61, Design Council; 62, Trustees of the Sir John Soanes Museum; 63, *Country Life*; 64, Cheltenham Art Gallery and Museum; 65, Tunbridge Wells Public Library; 67, *Cambridge Evening News*; 71, Victoria and Albert Museum; 76, Greater London Council Print Collection; 78, *Country Life*; 79, Royal Albert Memorial Museum, Exeter; 81, Edward Arnold; 82, British Museum; 83, Ashmolean Museum; 84, F. Lewis; 87, British Museum; 88, Arthur Rackham Estate and Hodder & Stoughton; 89, Fitzwilliam Museum, Cambridge; 90, Ashmolean Museum; 94 and 96, Tate Gallery; 97, Trustees of the National Gallery, London; 98 and 99, Tate Gallery; 100, National Trust; 101 and 102, Tate Gallery; 103, Mr Robin Tanner and Robin Garton; 106, British Museum; 107, F. Lewis; 108, Hutchinson; 109, Tate Gallery; 110, Bristol Art Gallery; 111, Ipswich Museums and Galleries; 112, Ernest Benn; 113, Mr Rowland Hilder; 114, Paul Nash Trust and Tate Gallery; 115, Miss Joan Hassall; 118, Beverley Borough Council; 119, City of York Art Gallery; 121, Stafford Art Gallery; 124, British Museum; 125, National Trust and Courtauld Institute of Art; 126, Macmillan, London and Basingstoke; 127, Whitworth Art Gallery, University of Manchester; 128, British Museum; 132, Castle Museum, Nottingham; 133, Harris Museum and Art Gallery, Preston; 136, Cambridge University Aerial Photography Collection; 142, F. Lewis; 143, British Museum; 144, Borough of Haringey; 145, London Transport; 148, Ashmolean Museum; 149, Marquess of Salisbury; 150, Paul Nash Trust, Tate Gallery and Lund Humphries; 151, Mr John Piper.

Abbreviations

Av	Avon	La	Lancashire
Bd	Bedfordshire	Le	Leicestershire
Br	Berkshire	LF	leaf formula (defined in chapter 2, note 26)
Bu	Buckinghamshire	Li	Lincolnshire
Ca	Cambridgeshire	Me	Merseyside
Ch	Cheshire	Nb	Northumberland
Cl	Cleveland	Nf	Norfolk
Co	Cornwall	Ng	Nottinghamshire
Cu	Cumbria	Np	Northamptonshire
Db	Derbyshire	NY	North Yorkshire
Do	Dorset	Ox	Oxfordshire
Du	Durham	ph.	photograph
Dv	Devon	repr.	reproduction
ES	East Sussex	Sa	Salop
Ex	Essex	Sf	Suffolk
Gl	Gloucestershire	So	Somerset
GL	Greater London	Sr	Surrey
GM	Greater Manchester	St	Staffordshire
Ha	Hampshire	SY	South Yorkshire
He	Hertfordshire	TW	Tyne and Wear
Hu	Humberside	Wa	Warwickshire
HW	Hereford and Worcester	Wi	Wiltshire
illus.	illustration	WM	West Midlands
IW	Isle of Wight	WS	West Sussex
Ke	Kent	WY	West Yorkshire

Preface

Elm has been a major feature of the English landscape for many centuries. Its recent disappearance from so many parts of the country through the ravages of the second cycle of Dutch elm disease has been one of the most cataclysmic changes this landscape has ever suffered. The object in writing this book is to present, while still possible, a comprehensive account of this tree.

The main experimental material utilized has been an all-England biometrical collection of elm leaves. This was started in the early 1950s and completed in 1976, just ahead of the latest outbreak of Dutch elm disease. It has been supplemented for comparative purposes by similar though smaller biometrical collections of elm leaves from the Continent, in particular France and Spain. Since over this period, my duties at the Commonwealth Bureau of Plant Breeding and Genetics, Cambridge, over which I was latterly Director, have claimed most of my time, the task of collecting material, investigating it and assembling elucidatory data has been restricted to such interstices of free time as this way of life affords. The opportunity for sustained writing came with my retirement, when a room was put at my disposal in the Department of Applied Biology, Cambridge.

There is much hypothesis in the book, based on a conceptual model described in chapter 3. If the reader seeks certainty, he must go elsewhere. There is little point, however, in presenting the enormous mass of data here assembled without some unifying concept. The boundaries between fact and theory are clearly drawn. If the hypotheses presented here on the basis of the facts to hand are unacceptable to the critical reader, he has the material for constructing his own alternative.

An author in another field, but one with some overlap with the present study, has written: 'It is sometimes worth while for a historian deliberately to press a new line of enquiry beyond the point of proof, to suggest ways in which old material may perhaps serve new purposes, and to leave to the future rather than the present the judgment between a searchlight and a will-o'-the-wisp'.[2] This has been the outlook of the present investigation. The decision to publish the results so far obtained has been strengthened by awareness that the field work on which this book rests is unlikely to be possible again in the present century or perhaps ever.

In a project spread over so many years it is not possible to acknowledge individually every colleague and friend who has helped by contributing material or information, by discussing implications, or by deflating hypotheses. An author covering as wide a field as here would be foolish to suppose his touch is equally sure everywhere. I have been saved from many pitfalls by experts in specialist fields into which I have strayed. I ought, however, to mention Mr H. M. Heybroek, Miss O. Holbek, Mr N. Janjić, Professor A. I. Kuptsov and Mr P. Popovski who have provided me with elm samples from parts of Europe and elsewhere to which I have not penetrated. I owe a particular debt to Mr J. N. R. Jeffers, who has collaborated with me in the multivariate analysis of my collections.

My thanks are also due to the many librarians, archivists and curators of art galleries who have assisted this investigation. I would like to record a particular note of appreciation of the facilities of the University Library, Cambridge. It is doubtful whether the extensive range of published material utilized in this book could have been assembled without the ease of access to it that this library affords.

For reading and commenting on parts of this book in draft, and for assisting in translation problems, I am particularly beholden to Dr C. E. Blacker, Mr D. R. Goddard, Dr J. F. P. Hopkins, Professor M. B. Loraine and Dr G. J. Walker.

The expenses of the later stages of research and of publication were assisted by grants from the Institute of Terrestrial Ecology and the National Environment Research Council.

Finally, and most particularly, I gladly record

my thanks for the way in which my family have helped and tolerated so long a preoccupation. In the days before inflation, my children took frequent advantage of a standing offer of 1d for any elm citation I had not previously encountered. Latterly, they contributed material from their foreign travels and my son, Dr J. E. Richens, has driven me over many miles of Continental terrain. My debt of gratitude to my wife, for constant companionship and help, including the typing of a difficult manuscript, exceeds all others.

R. H. Richens

Department of Applied Biology
Cambridge
1982

I

Introduction

As indicated in the preface, the scope of this book is the elm in England.[1] Of the various boundaries demarcating England from Wales and Scotland, the oldest and most stable of all have been utilized, those of the relevant medieval dioceses. The one elm topic for which no attempt has been made to provide comprehensive cover is Dutch elm disease. This has given rise to so voluminous an output of articles that their inclusion would have seriously unbalanced the book. Key publications are listed in chapter 6.

Since the investigations described have been very much a multidisciplinary exercise, their presentation has involved difficulty. The jargons appropriate to the various particular aspects of the study are unlikely to be read comfortably by those not habituated to them. To obtain a generally intelligible text, much of the succinctness that technical language affords has had to be sacrificed.

It is also hoped that the book will simultaneously provide a readable account of the elm and serve as a reference book for those seeking further information on particular points. The two functions are not easy to combine. What has been done has been to relegate a great deal to the end notes of each chapter and to assemble much of the local detail in the three penultimate chapters.

The first 11 chapters hopefully constitute a moderately readable text, bowdlerized of such jargon as has proved feasible. However, most readers will probably find heavy patches, and it is recommended that the table of contents be scanned first, and the chapter covering the ground most familiar to the reader be selected for initial reading. Having read this, chapters 2–11 can be tackled with maximum confidence. So, a botanist would be encouraged to start with chapter 2, a philologist with chapter 4 and anyone whose prime interest is literary with chapter 10. The sequence of chapters follows the general drift of the argument, and so, in the nature of the case, opens with botanical considerations.

A two-tier referencing system has been used. References in full in the end notes are to publications in which elm is not of primary concern. References directing the reader to the general bibliography are of articles primarily on elm or on organisms exclusively associated with it.

The severity of the last Dutch elm disease epidemic has created a problem as to what tense to use. For reasons that should emerge, the present tense is used uniformly for general statements about elm. The past tense is used uniformly for individual trees, which even if standing at the time of writing may have gone by the time this book appears in print.

The classification of the elms is treated in detail in chapter 7, but in order to be able to discuss topics dealt with in chapters 3–6, the principal kinds are introduced in anticipation.

There are some 30 species of elm, of which two are accepted as occurring in England. The first of these is the native species *Ulmus glabra*,[2] for which the current vernacular name is Wych Elm.[3] Except where

1,120, 122

Fig. 1. Wych Elm at Moor Court (HW), by Worthington Smith.[3]

Latin names need to be used, they are avoided in this book. This is partly to obtain a more easily read text. It is partly a matter of principle. The vernacular names of cultivated plants and other plants of special interest to man are usually more stable than the Latin names. This is the case with the elms. Throughout this book standardized vernacular names for the principal kinds of elm are used. These are always spelt with initial capitals. They are to be understood in the sense given to them in this introduction. They correspond exactly with the Latin names here associated with them. Occasionally, it will be necessary to refer to historical material in which vernacular names are used in a different sense. Such names are printed in lower-case italics. For example, *wych elm* in Essex in Tudor times referred to a variety of the second species to be considered.

Wych Elm occurs mainly in northern England and along the Welsh border. It is a tree of spreading habit. The seed is central in the fruit. The leaves are 44 large and with relatively short stalks.

The second species is the introduced *U. minor*.[4] There is no generally accepted vernacular name for

Fig. 2. Narrow-leaved Elm at Swaffham Prior (Ca), by A. Sinclair.[5]

Fig. 3. English Elm in Surrey, by Edmund Evans after Samuel Read.[6]

SUNDAY MORNING.

FROM "THE LUMP OF GOLD," BY CHARLES MACKAY.   DRAWN BY SAMUEL READ.

Embower'd amid the Surrey Hills
  The quiet village lay,
Two rows of ancient cottages
  Beside the public way,
A modest church, with ivied tower,
  And spire with mosses grey.

Beneath the elm's o'erarching boughs
  The little children ran,
The self-same shadows fleck'd the sward
  In days of good Queen Anne;

And then, as now, the children sang
  Beneath its branches tall;
They grew, they loved, they sinned, they died,—
  The tree outlived them all;—
—But still the human flow'rets grew,
  And still the children play'd,
And ne'er the tree lack'd youthful feet
  To frolic in its shade,
The ploughboy's whistle in the spring,
  Or chant of happy maid.

* * * * * * *

'Twas Sunday morn, and Parson Vale,
  Beloved of high and low,
With smiles for all men's happiness,
  And heart for every woe,
Walked meekly to the parish-church
  With hair as white as snow;
Walked meekly to the parish-church
  Amid his daughters three;
There were more angels at his side
  Than mortal eyes could see:
The four were seven—for with them went
  Faith, Hope, and Charity.

this. It will be referred to as Field Elm, on the analogy of the Continental vernacular names *orme champêtre* and *feldulme*. The seed is displaced towards the apex of the fruit. The leaves are usually smaller than in the Wych Elm. The leaf stalks, relative to the blade, are longer.

Fig. 4. Cornish Elm at Coldrenick (Co).[7]

The Field Elm is a highly variable species, of which five varieties are accepted for present purposes. The first is *U. minor* var. *minor*, which has no vernacular name. It is called here the Narrow-leaved Elm.[5] Its canopy is open compared with the next kind. It is the characteristic elm of East Anglia, northern Essex, parts of the east Midlands and eastern Kent. The leaves are almost always narrower relative to their length than those of the English Elm, mentioned next. They are often light green in colour. *2, 152* *45*

The most frequent elm in England is *U. minor* var. *vulgaris*. It has been commonly called *U. procera* by English botanists. It is generally given the vernacular name English Elm[6] and this usage is followed here. It is the elm of the southern Midlands and further south. Its dense rounded canopy and dark orbicular leaves at once identify it. *3, 63, 108* *46*

The third variety is *U. minor* var. *cornubiensis*. It is generally called the Cornish Elm,[7] as here. It is the principal elm of Cornwall and western Devon. It is very closely related to the Narrow-leaved Elm. The straight trunk and narrow silhouette are distinctive. *4, 127*

U. minor var. *lockii* is the fourth variety. The Latin name renders its vernacular name, Lock's Elm,[8] which is used here in preference. Systematic botanists in recent years have called this elm *U. plotii*. As will appear later, this name is inappropriate. It has a very *5*

Fig. 5. Lock's Elm at Laxton (Np), by Mary Neal.[8]

scattered distribution in the north Midlands. Its habit is characteristic. The silhouette is narrow and the leading branch droops to one side.

Lastly, there is *U. minor* var. *sarniensis*. This is the characteristic Field Elm of Guernsey, hence the 6 vernacular name adopted here of Guernsey Elm.[9] Unlike the previous four varieties, which are of ancient introduction, the status of the Guernsey Elm in England is that of a recent, nineteenth century, horticultural introduction.

Wych Elm and Field Elm readily hybridize. The resultant first-generation hybrids are given the general Latin designation *U.* × *hollandica*. The same name is also given to the sexual progeny of the first-generation hybrids. Since this is the only cross between different elm species relevant to most of this book, the elms resulting from it will be simply referred to as Hybrid Elms.

These are a highly variable assemblage. First-generation hybrids usually have large leaves, as in the Wych Elm parent, and a relatively long leaf stalk, as in the Field Elm parent. Two Hybrid Elms have been widely planted. The first, *U.* × *hollandica* nm. *hollandica*, 7 is known in England as the Dutch Elm,[10] the term used

here. It is not what goes under the corresponding vernacular name, *hollandse iep*, in the Netherlands. The former elm was introduced into England in the seventeenth century. It is recognizable by its large, coarse leaves and open, scraggy habit.   49

The second hybrid, *U.* × *hollandica* nm. *vegeta*, is called the Huntingdon Elm[11] in English, the latter name being preferred here. It originated near Huntingdon in the late eighteenth century. It has large, smooth leaves. The main branches spread out 50 fanwise.

In chapters 12–14, the whole of England is divided into elmscape regions. In those prefixed by Z, as Z1, no elm is frequent. In regions prefixed by G, as G7, Wych Elm is common, in those prefixed by V, as V2, English Elm is common, while in those prefixed by M, as M3, Field Elm other than English Elm is abundant. The frequent occurrence of Hybrid Elms is indicated by prefixed H. If more than one kind of elm is frequent, double or triple prefixed letters are used, as, for example, GV13, where both Wych Elm and English Elm are common, or GMV1, where three kinds of elm coexist at high frequency. The basis of this arrangement is the all-England biometrical collection mentioned in the preface. This was drawn from the elms of the closes around ancient villages. It does not cover more recent amenity plantings, as in parks and along roads.

Fig. 6. Guernsey Elm, by Nigel Luckhurst.[9]

Fig. 7. Dutch Elm, by Nigel Luckhurst.[10]

It is convenient to list here some conventions used throughout the book. Iron Age is used solely for the pre-Roman Iron Age. The Middle Ages or Medieval Period are terms used only for the centuries following the Norman Conquest. The term place name is used inclusively to cover not only habitation sites and natural features, but also minor named entities such as fields and closes. Europe and the Continent refer to the mainland excluding the British Isles.

Frequent allusion is made to hypothetical linguistic roots, such as *lem-*, the postulated Celtic root for elm. These are prefixed, as customary, by an asterisk. Linguistic roots are treated here as no more than codes for the presumed ancestral term in a language group or family. Since there is no intention of reproducing the various speculations by linguists as to the exact form the ancestral term took, its spelling has been simplified for easy reading.

Fig. 8. Huntingdon Elm, along Gonville Place, Cambridge, by Nigel Luckhurst.[11]

2
Botany

Till recently, oak, ash and elm were the principal trees of the English landscape. The first two are indigenous. Some elm populations are also native. Most, however, were introduced by man at times and for reasons which are investigated in this book. The association between man and elm has been long, continuous and remarkably complex. It will be a recurrent theme in what follows.

Salient characters

Elm is the best known representative of the family Ulmaceae, named after *Ulmus*, the botanical Latin name of elm. The family consists exclusively of trees and shrubs. Its members have undivided leaves which are inserted alternately along the stem. Stipular scales are present at the base of the leaves when they first emerge. The flowers are inconspicuous and usually borne in clusters. The floral envelope is a perianth. This is divided into segments at the base of which the stamens arise, usually one stamen opposite each perianth segment. The ovary has two styles and contains a single ovule. Chemically, the family is distinguished by lack of diagnostic substances.

There are two large families related to the Ulmaceae, the Moraceae and the Urticaceae. The Moraceae, which include the mulberries and figs, differ from the Ulmaceae in the presence of a milky juice. The Urticaceae, which include the nettles, are mostly herbaceous. The Ulmaceae consist of two tribes, the Ulmeae and the Celtidae.[1] The latter are mainly tropical or subtropical, but the genus *Celtis* is present in warm temperate regions. The distinguishing char-

acter of the Celtidae is their ovoid seed. They can usually be recognized also by the leaf. Occasionally this has a single main vein as in the Ulmeae. Generally, however, there are three main veins diverging from the leaf base. This is never the case in the Ulmeae. The Celtidae will not be considered further.

The Ulmeae, which include the elms, are with the exception of one genus a north temperate tribe. They occur both in the Old World and the New. The exceptional genus is *Holoptelea*. This has one species in peninsular India and another in western Africa. The genus most closely related to that of the elms is *Zelkova*. Today, this occurs in Crete, the Caucasus and the Far East. In the geological past, it was present also in the mainland of Europe and in North America. Of the other genera, *Hemiptelea* and *Pteroceltis* are now confined to China. *Hemiptelea* was present in Europe till the climatic decline preceding the last cycle of glaciations. *Planera* is confined to North America. The Ulmeae differ from the Celtidae in having flat seeds. The leaves always have a single main vein.

The elms are all placed in the genus *Ulmus*.[2] This is the genus most numerous in species and most widely distributed of all the genera in the tribe Ulmeae. It is essentially a north temperate assemblage though three species extend into the tropics. It occurs throughout Europe as far north as Scotland, southern Finland and north central Russia. Though absent from most of Siberia, it is present in central Asia, Turkey, Lebanon, Israel, Iran, Afghanistan and the Himalayas. In the Far East, it is widespread in China, Korea and Japan. On the Asian mainland it extends northward to the Soviet Far East. The southern Kurile Islands are its northern limit in the islands of the Far East. In south-east Asia it extends through Malaya to Sumatera, Sulawesi and Flores. The only elm of long standing in Africa is in northern Algeria. In North America, elms are only native in the eastern states. One species occurs in Mexico, Central America and Colombia.

The deciduous leaf provides two characters that together distinguish the elms from all other members of the Ulmeae and from almost all other trees. The first is its bilateral asymmetry, which can be very pronounced. The second is the doubly toothed leaf margin. The principal teeth are themselves toothed. The fruit (seed in loose parlance) is variable in structure but never succulent as in the related genus *Zelkova*. It is conspicuously winged in most species, including all that occur in Europe.

50
46
49

The elms are all trees, eventually with furrowed bark. The roots spread out underground to a distance up to twice the height of the tree.[3]

The elm leaf is extremely variable. The range of variation is described in the next section. Features that are universal are the way in which the leaves are folded lengthwise in the bud[4] and their rapid decay when shed in autumn.[5] While there are differences in the number of secondary veins branching off the main vein and in their frequency of forking, the pattern of the fine veins hardly varies and is a valuable means of identifying fossil elm leaf fragments. The tertiary veins run out from the secondary veins at a right angle. The finest veins of all form a compact, uniform mesh of small polygons.[6]

Elm flowers[7] are always small. In contrast to those of some related trees, they are bisexual, with both stamens and ovary. The latter is surmounted by two conspicuous spreading styles. Flowers are not produced on young branches. Generally, a branch must be 7–8 years old before flowers develop. A heavy shower when elms are flowering occasionally results in 'blue rain', the drops being coloured by detached elm anthers.[8]

The elm pollen grain is highly distinctive and readily distinguishable from that of all other trees except *Zelkova*. It is roughly spherical with, on average, 5–6 unthickened pores arranged around an equatorial band. The surface is undulating.[9] What

Fig. 9. Pollen grain (× 2300).

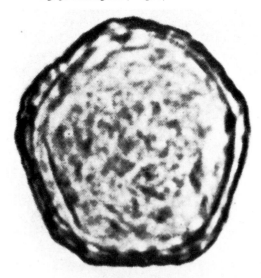

little variation there is in pollen structure appears not to correspond to the differences between elm species.[10] Nor, it appears, can elm pollen be reliably distinguished from that of *Zelkova*. This constancy in pollen characteristics extends to the remote geological past. The earliest pollen attributable to elm or *Zelkova* shows virtually no differences from that produced today. Once shed, elm pollen can travel great distances on air currents. It has been detected over Shetland, the nearest source being Sutherland.[11] It has been detected in arctic Canada and over the Greenland ice cap, the nearest sources being not less than 1500 km away.[12] The fact of long-distance pollen transport means that hybridization between elms far removed in location may occur. It has also to be borne in mind when interpreting fossil pollen profiles.

The principal microscopic character that distinguishes the elms is the anatomical structure of its wood. This determines the appearance of the grain and very largely determines the mechanical properties of the timber. Three characters, taken together, are diagnostic. The water-transporting vessels laid down in the spring wood are markedly larger than the vessels subsequently formed. A transversely cut end in any piece of elm furniture, examined under a hand lens, reveals these as a band of pin-prick holes. The smaller vessels laid down later in the season are in wavy bands. It is these that are responsible for the zigzag pattern between the main grain lines formed by the spring vessels. The zigzag pattern, especially conspicuous where the vessels are oblique to the surface of an elm artefact, provides instant identification of elm timber under most circumstances. The third character, visible under a hand lens, is the presence of relatively broad medullary rays. Elm wood, as a rule, is cross-grained. Its component cells do not run down exactly parallel to the axis of the trunk or branch. Consequently the timber cannot readily be split.

A quite different anatomical character is provided by the plastids in the sieve tubes that conduct organic material in the inner bark. In several plants related to elms, the plastids contain starch. In the elms they contain protein.[13]

Elms show wide variation, even within an individual plant, in the processes involved in the embryological development of the female reproductive cells. It seems to make no difference to subsequent development.[14]

The cell nucleus of the elms shows little visible

variation. There are 28 chromosomes in all species except *Ulmus americana*, in which the chromosome number has doubled.[15] There is not even much variation between species in chromosome size and shape. Of the 14 chromosomes of each set, one, with arms of equal length, is larger than the rest, and one is shorter. The remaining 12 chromosomes, with arms of unequal length, are almost indistinguishable from one another except for one which has a small satellite attached.[16]

Elms are not remarkable for a great array of chemical constituents peculiar to themselves. The inner bark is more or less mucilaginous and the mucilage, in the case of slippery elm, *Ulmus rubra*, has been shown to be composed of four units: galactose, 3-methylgalactose, rhamnose and galacturonic acid.[17] The second of these components has not been identified elsewhere in the plant kingdom. Another

peculiarity of the elms concerns the oil in their seeds. This has a high content of fatty acids with ten-carbon chains.[18] Temperate plants usually have longer carbon chains.

Elms are functionally self-sterile. The male and female organs mature at different times. Frequently, elms are also genetically self-sterile. No seed is set when a tree is pollinated by its own pollen or by pollen from a tree of genetically identical constitution. Crossing between different kinds of elm is consequently favoured. It seems, however, that most elm species will hybridize with all others, given suitable conditions for fertilization and seed development.[19] It is even possible, under laboratory conditions, to cross normally spring-flowering elms, such as occur in Europe, with the Asiatic autumn-flowering species *Ulmus parvifolia*.[20] The one elm which does not appear to hybridize at all easily with other species is the east European species

Fig. 10. Transverse section of the wood (× 20).

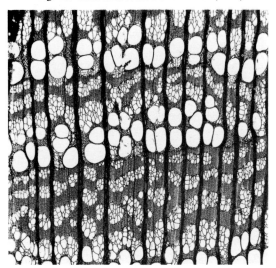

Fig. 11. Grain pattern.

Fig. 12. Chromosomes of Wych Elm (× 5000).
Re-drawn from Ehrenberg (1949).

U. laevis.[21] Elms produced by hybridization between different species appear usually to be fertile, except in crosses of other species with *U. americana*, with its doubled chromosome set.[22] The resulting hybrids have three chromosome sets, and like most such hybrids are highly sterile.

Recognition of hybrids is not always easy. The ideal method of matching against a controlled artificial cross is seldom practical. Generally speaking, a hybrid has to be recognized by a comparison between its probable parents. There must also be grounds for supposing that one could have pollinated the other, remembering that pollen clouds may travel great distances. One must in addition have reasonable confidence that the range of variation of the likely parents, which may be considerable, does not extend as far as the putative hybrid.

There is much evidence, assembled in chapters 12–14, that the various Field Elm introductions have hybridized amongst themselves and have also formed interspecific hybrids with Wych Elm (Hybrid Elms). But it appears, as would be expected, that the hybrids are greatly outnumbered by the parental forms and only occasionally dominate a landscape. It is also likely that more complex hybrids involving three or more kinds of elm have also been produced. In most crosses of suspected multiple origin, the chances of being able reliably to identify the parental kinds become vanishingly small.

Fertile elm seed does not retain its germination capacity for long under natural conditions. Germinability can, however, be retained for several years, by storing the fruits in sealed glass bottles at low temperature. The seeds germinate best in the dark.[23]

Variation

The most conspicuous variation, visible some kilometres away, is in habit. This blanked term covers silhouette, diameter and straightness of the trunk, and configuration of the subsidiary branches. A round-headed, mushroom-like silhouette is characteristic of the English Elm.[24] When the lower branches of this elm are well developed, the silhouette is often a figure-of-eight. The Guernsey Elm, in marked contrast, has a pyramidal silhouette, suggestive at first glance of a conifer. The trunk is stout and straight in the English Elm. In some of the Narrow-leaved Elms it is crooked.

100

6

2

The subsidiary branches may spread sideways or tend upwards. In the weeping elms they droop. *51*

Shoot growth in the elm depends greatly on the age of the tree and the position of the shoot. In young trees, in suckers and in new growth emerging from the trunk, the shoots are long, up to 1 m or more, and bear numerous leaves. In the main canopy of mature trees, shoot growth is much less. Typically, short shoots with *44, 45* no more than five leaves develop. A few kinds of elm *47* tend to form long shoots even in the mature canopy. *48* The principal English example is Lock's Elm.

Elm bark is smooth in shoots of 1–2 years old. In most elms it soon becomes rough. In the Wych Elm it remains smooth for up to 7–8 years. The bark of the trunk of mature trees differs considerably with the kind of elm. In the English Elm it is composed of small, *84, 13* plate-like pieces. In some of the Narrow-leaved Elms it is deeply grooved. Another variable feature of the bark is the production of corky flanges. This character is strongly affected by environmental conditions. Exposure to salt-laden winds, for example, as occurs in many coastal areas, promotes it. Some elms, however, irrespective of environmental conditions, are more liable to cork-flange production than others. The Dutch Elm shows the greatest tendency.[25] *49*

The most useful characters for discriminating between the different kinds of elms are provided by the leaf. The leaf is particularly useful to work with since it is available for a much longer period each year than the flower or fruit. It also provides a number of separate, precisely measurable characters that are largely independent of one another but relatively constant for a given tree or kind of elm. Great care has to be taken in choosing leaves for diagnostic purposes. The leaves on long shoots show far less sharp differentiation between different kinds of elms than those on short shoots. Comparative study is therefore based on the latter. Many elms, after the main emergence of leaves in the spring, show further growth in the summer. The latter growth, the so-called lammas shoots, is also discarded for diagnostic purposes, since the leaves on these are also less sharply differentiated than the leaves of the spring flush.

The five leaves of a typical short shoot also vary. The end leaf is likely to be considerably larger than the rest. The two leaves at the base of the shoot are normally rather small. The second and third leaves from the shoot apex are usually closely similar in all characteristics. The leaf chosen for comparative measurements is, whenever possible the second from the apex.

Seven leaf measurements have proved particularly useful for distinguishing between the different kinds of elms. The first is leaf length, measured from the apex to the point where the leaf blade margin runs into the leaf stalk. This measurement is taken along the longer side of the normally asymmetrical leaf.

The next two measurements are concerned with the size of tooth on the leaf margin. It has already been mentioned that elm leaves are doubly toothed, the principal teeth being themselves toothed. What is measured is the breadth and depth of the principal teeth along the shoulder of the leaf, usually about one-third along the leaf margin from the leaf tip. This is the part of the leaf where the teeth are most clearly developed.

The three measurements that follow are ratios. Relative breadth is leaf breadth divided by leaf length. Relative length of the leaf stalk is its length divided by leaf length. Leaf base asymmetry is determined by measuring the distance between the lowest points on each side of the leaf blade and dividing this by the leaf length.

The final measurement is tooth number along the entire leaf margin, each incision in the margin, whether a principal or a subsidiary tooth, being counted as one tooth.[26]

Two other variable leaf characters need a mention. The first is the nature of the upper leaf surface. This may be smooth, as in many Narrow-leaved Elms. It may be matt. It may be rough and resist a finger drawn lightly across it, as usually in Wych Elm. The roughness may be directional, being apparent on drawing the finger from leaf tip to leaf base, but not in the opposite direction. Roughness also depends on time of year, since, as the season advances, there is a tendency for the leaves in some kinds of elm to become smoother.

Fig. 13. Diagram of tooth measurements.

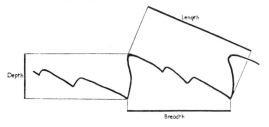

The second character is the nature of the leaf hairs. Four categories of hair need to be distinguished. First there are the relatively inconspicuous hairs of the upper leaf surface. These may have considerable amounts of silica in their cell walls, in which case they are rough to touch. Then, there are the usually softer hairs generally distributed over the lower surface. The last two categories are the more useful for discriminating between different kinds of elm. One is that of the relatively long hairs associated with the veins on the lower leaf surface. These occur in diagnostic patterns. They may be present as dense tufts or patches at the forks of veins. Sometimes they lie in bands along the veins. The other is that of the hairs on the leaf stalk. This may be devoid of hairs. It may be densely hairy. The hairs may be very short relative to the diameter of the leaf stalk. They may approach it in length.

Elm flowers vary considerably in the number and shape of the parts of the floral envelope. They vary also in the length of the flower stalk and in the number and arrangement of the flowers in the cluster.[27] All British elms have the flowers in dense clusters with a very short flower stalk. The eastern European species *Ulmus laevis* has a long flower stalk so that the flowers are in loose sheaves.

The elm fruit is the principal organ used in distinguishing between elm species. In the British elms the fruit is winged with a notch in the wing at the apex of the fruit. The wing varies in both size and shape. The position of the seed, whether central within the fruit or displaced upwards towards the apical notch, is a major point of discrimination. Wych Elm has a central seed. Field Elm has the seed displaced towards the apex. The presence and pattern of distribution of hairs on the fruit surface are very important. None of the British elms has hairs on the fruit surface, but some Wych Elms in the Caucasus and southern Russia have hairs over the seed. *Ulmus laevis*, mentioned above, has a fringe of hairs along the fruit wing.

The anatomy of the wood has already been described in general terms. There is some variation in detail but identification of kinds of elm from the appearance of the wood in microscopic section is difficult. Generally, it seems, the pores of the spring wood are somewhat smaller in Wych Elm than in Field Elm and the transition between spring and summer wood is more gradual.[28] There are differences between different kinds in degree of toughness and resistance to splitting.

Wheelwrights, as noted in chapter 8, would travel considerable distances to secure elm timber known from experience to be specially tough. It would also appear that the tendency of elm to drop large branches without warning is a varietal characteristic. English Elm is the worst in this respect. The heavy foliage of this kind is partly responsible. Some special property of the normally tough wood seems also to be involved.[29]

The chemistry of the elm has not been much studied. Two classes of compound have, however, proved of value in discriminating between different kinds of elm. The first is the assemblage of flavonoids of the leaf. It appears that a biochemical distinction based on flavonoids can be drawn between Wych Elm and Field Elm.[30] The latter contains the compound quercitin 3-glucuronide. The former does not. One should be cautious about this result since *Ulmus wallichiana*, a Himalayan elm related to Wych Elm, has the compound, while *U. wilsoniana*, a Chinese species related to Field Elm does not. Analysis of further material from Europe might upset the generalization drawn from the first results. European elms, in general, lack leucodelphinidin, which is a characteristic component of the flavonoid spectrum of some of the more primitive elms of North America and the Himalayas. It has turned up, however, without any corresponding visible feature, in some Field Elms in Corsica and Israel.[31]

The second group is that of the isoperoxidases of the leaf.[32] Peroxidase is a protein which, like many others, exists in a whole array of distinct but closely similar forms, the isoperoxidases. These are separable and identifiable by the technique of gel electrophoresis. Up to four diagnostic isoperoxidases may be present in the leaves of a single elm. In the discussion of elm reproduction below, it is pointed out that elms reproduce habitually by one of two methods: one sexually by means of seed, and the other vegetatively by root suckers. This difference in method of reproduction is reflected in the isoperoxidase complement. In elms reproducing by root suckers, like the English Elm, the isoperoxidase complement is usually constant. In elms reproducing sexually, as the Wych Elm, even outwardly similar adjacent trees may differ in their isoperoxidase complement. Wych Elm, consequently, exemplifies the phenomenon of protein polymorphism, that is, the coexistence of a range of proteins, in this case isoperoxidases, maintained in some sort of equilibrium in the population as a whole, and continually regrouped with every cycle of sexual reproduction. This pattern of conspicuous chemical variability appears to be totally independent of any variation in structural characters. While it is true that adjacent trees of a single sexually-reproducing stand may differ in their isoperoxidases, it appears that certain particular isoperoxidases occur in some regions and not in others. Differences of this sort could throw light on the history of the stands concerned.

It is possible to differentiate the European elms as a whole from the more primitive elms of southern USA by the composition of their heart wood extractives.[33] There appears however to be much variation within the various kinds of elms so that the diagnostic value of this character seems to be rather limited.

Elms also vary greatly in the organisms, fungi, insects and the like, which draw their nutriment from them. The basis of this variation is very largely chemical. This type of variation is considered in detail in chapter 6.

The principal structural and chemical characters of the elm having been considered, it is time to pass on to its physiological characters. Elms will only grow when certain basic requirements are satisfied. Temperature and water supply are the most salient of these. Each kind of elm has an optimum temperature for each phase in its annual developmental cycle and a corresponding minimum rainfall requirement. As the environment departs from the temperature optimum in either direction, so the elms become under increasing stress till finally they can no longer survive. Recession from the optimum rainfall requirement has a similar effect. Different kinds of elm differ markedly in these respects. Wych Elm, for example, has a much lower optimum July temperature than Field Elm. It mostly occurs where the maximum daily isotherm for August is below 69 °F (21 °C). It will not survive, however, if the mean July temperature is below 16 °C.[34] It has also a higher summer rainfall requirement.

Elms also have requirements as regards soil. They do not flourish on acidic soils low in calcium.

The distribution within England of Wych Elm can be largely explained in relation to its temperature, moisture and soil requirements. The same does not apply to the other elms which have been introduced by man. Environmental conditions are indeed important, but they affect the distribution of elms by having determined suitable sites for prehistoric settlement. In chapter 12, on the elms of the northern countries, physiographical factors directly determine the distribution of the principal elm, which is Wych Elm. In chapters 13 and 14, dealing respectively with the elms of southern and eastern England, the same factors are important, but only indirectly through their influence on the various peoples who have introduced and planted elms.

Elms are sufficiently sensitive to environmental stress to be usable in dendrochronology. The width of the annual rings of a tree reflects the incidence of environmental stress on the tree over the years. Since this stress often operates in the same way over wide areas, the sequences of ring widths can be used to identify a succession of seasonal conditions. These, in many cases, can be equated with a sequence of years AD or even BC. Oak is the deciduous tree that has been most used for this purpose. Some dendrochronological work has been done, however, with English Elm. In this elm, the width of the annual ring varies directly with rainfall in the season of growth of the ring, and varies directly with temperature and rainfall during the preceding autumn. The ring width varies inversely with the temperature during spring of the current growing season, and also with temperature during September of the previous year, and with rainfall during the previous summer.[35]

Two developmental characters that vary greatly between different kinds of elm are the date of leaf emergence and the date of leaf fall in the autumn. The English Elm flushes in spring earlier than other elms and holds its leaves several weeks longer in autumn. This is particularly noticeable in landscapes in which English and Narrow-leaved Elms occur together. The later autumnal leaf fall of the English Elm is also conspicuous in trees of this kind introduced into North America.[36] The Cornish Elm is amongst the latest to flush in the spring. The order in which the various elms flush and drop their leaves seldom varies. The actual dates are much influenced by seasonal conditions.[37]

There are two types of flowering behaviour amongst the elms. In some subtropical species, the flower clusters emerge in the autumn from buds formed in the current year and before the leaves are dropped. This mode of flowering does not occur in European elms. In all of these the flower clusters emerge in the spring from buds formed in the previous season. Flowering date is extremely sensitive to temperature. It may occur any time between December and May inclusive. Different kinds of elm react differently to seasonal conditions.[38] This differential behaviour is of special significance for hybridization. Only in some years will pollen of one kind be in the air while the ovaries of another are receptive.

The flowers remain open for 4–18 days.[39] They are mainly wind pollinated. Insects, especially honey bees, visit the flowers and presumably effect some pollination.

It was mentioned in the previous section that elms are normally self-sterile. Unless pollen from a different tree is available, no seed will set. But even if suitable pollen for fertilization is available, there is another potent cause of failure to set seed. This is low temperature during seed development. Over large areas of northern Europe, including the British Isles, the Field Elm only sets seed in exceptional seasons. The conditions most likely to promote elm seed setting are late flowering followed by a rapid rise in ambient temperature. Good seed setting years occurred in 1909 and 1942.[40] Field Elm sets seed readily in an adequately heated greenhouse. Wych Elm is not nearly so sensitive to low temperature and sets seed every year throughout the British Isles.

The failure of Field Elm to produce viable seed in most seasons is offset by its extreme efficiency in reproducing vegetatively by root suckers. It is probable that the greater part of the Field Elms now growing in England has resulted from continuous vegetative propagation extending back to the time of introduction of the various kinds, without any intervening sexual phase. If this is so, it has a very important consequence for historical studies. If a plant, say, had been introduced by the Romans and were propagated sexually from that day to this by seed, one could not infer a great deal about the precise characters of the original introduction. This is because, at each occasion of sexual reproduction, the characters under the control of the genes on the chromosomes reassort. Some characters will be favoured by the selective conditions then prevailing. Other characters will be selected against. The situation with a vegetatively propagated plant is quite different. There are no phases of character reassortment. New characters may arise as sports but many lines of the introduction will be likely to survive to the present day exactly in their original form. The Field Elm is one of the few introduced plants that comes into this category. Evidence will be provided in later chapters that it was introduced by man at various times back to the Bronze Age. It is the only plant introduced by man before the Middle Ages, some stocks of which appear to have survived to the present day in their original form.

It is not a unique example of such survival. Vine varieties are also propagated vegetatively. Some, such as Cabernet Sauvignon, are believed to have been preserved without essential change since Roman times.

Wych Elm, which reproduces by seed, does not sucker. For this reason, it is used by nurserymen as a stock upon which other kinds of elm are grafted. Elms are largely graft-compatible with each other, though some combinations are less easily effected than others.[41]

Very little indeed is known of elm genetics.[42]

3
Prehistory

Geological record

The remote origin and history of the elms have to be sought in the fossil record. Elm fossils are reasonably abundant. To obtain the fullest picture of the course of elm evolution, however, other lines of evidence have also to be considered. They include data on present-day geographical distribution and on the relationships

between elms and associated organisms, considered in detail in chapter 6.

The principal elm fossils are leaves, fruits, wood *14, 15* and pollen. These are usually found independently of one another. Connecting one with another is hazardous. Errors in associating particular organs with others have certainly been made in the past. In some cases, undoubted elm fossil parts have been associated with fossil organs of quite different plants.

Pollen grains from elms are frequent in the fossil record. Unfortunately, they cannot be distinguished from those of the related genus *Zelkova*. The pollen of both is covered by the blanket designation *Ulmipollenites undulosus*.[1] Even less can fossil elm species be distinguished by means of their pollen.

Fossil woods can sometimes be confidentially identified as from elm. Again, a closer identification with particular species is seldom possible. The genus of fossil wood designated *Ulminium* has nothing to do with elm. It applies to the wood of various members of the unrelated family Lauraceae.

It is fossil leaves and fruits of elm that provide the essential evidence for reconstructing its evolutionary history. The major difficulty here is not the incidence of gaps in the fossil record. It is the difficulty, mentioned in general above, of associating particular fossil leaves with particular fossil fruits. They are not shed at the same time. The diagnosis of elm species today usually depends on an association of characters in these two organs.

Many so-called elm species have been erected on the basis of fossil elm leaves or leaf fragments. Many of these are near worthless as biological entities.

Fig. 14. Elm leaf of Miocene age (× ⅘).

Fig. 15. Elm fruit of Miocene age (× ⅘).

Several fossil *Ulmus* species have turned out not to be elms at all. In fact, it is usually possible to be quite confident whether a fossil leaf is or is not an elm leaf. The combination of doubly toothed leaf margin, conspicuous leaf asymmetry, and the fine veination pattern characteristic of elms is found only in the genus *Ulmus*. The principal confusion is with the related genus *Zelkova*, which has leaves with mainly singly toothed margins, the teeth being coarser than in *Ulmus*.[2]

Fossil elm fruits have survived more rarely than leaves. They can be identified to species to some extent by their size and shape. The important diagnostic character, distribution of fruit hairs, is not usually available, since the hairs only survive fossilization under very exceptional conditions.

It can be inferred on distributional grounds that the genus *Ulmus* had become differentiated from its nearest relative *Zelkova*, and that some at least of the sections[3] of the genus had evolved before continental drift had led to the separation of America from Eurasia. Section *Ulmus* of the genus *Ulmus*, occurs today in Eurasia, for example, Wych Elm, and in North America, where the corresponding elm is *U. rubra*, the slippery elm. Section *Blepharocarpus* is represented by *U. laevis* in Europe and *U. americana* in North America. *U. villosa* in northern India is most closely related to elms such as *U. mexicana* in tropical and subtropical America. In several cases, as noted in chapter 6, related elms on either side of the Atlantic Ocean have associated organisms that are also related and have presumably evolved with the elms on which they occur since before the continents separated.

Fossil pollen of the elm–*Zelkova* type is known from upper Cretaceous times (*c.* 80 million years ago), principally from what is now eastern Asia or Alaska. Fossil leaves that are undoubted elms become abundant in Europe in Oligocene times (*c.* 40 million years ago) onwards. They show a rather wider range of variation than exists in Europe now. Much of what has been designated *Ulmus carpinoides*[4] appears to be ancestral to and some of it within the present range of variation of present-day *U. laevis*. *U. carpinoides* is recorded from the Oligocene onwards. *U. braunii*[5] covers material, much of which is practically indistinguishable from present-day Field Elm. It is recorded from Miocene times (*c.* 25 million years ago) onwards.

That relatively little evolution has occurred in the elms since Miocene times is confirmed by the chemical analyses that have been possible of the remarkable Miocene Creek fossil leaves in the USA. These have retained their green colour and are largely chemically unaltered since their deposition. They show little chemical difference from corresponding elm species growing in North America today.[6]

The principal difference between the European elm flora in Miocene and earlier times and that of today, is the former occurrence of elms with long, narrow leaves usually designated *Ulmus longifolia*.[7] These did not persist in Europe after the end of the Pliocene period (*c.* 1 million years ago). Their relationships have been very diversely interpreted.[8]

In the Pleistocene period following the Pliocene, the European tree flora suffered continued depletion, presumably owing to climatic deterioration. With each glaciation, the elms and other temperate trees disappeared from most of Europe. Some temperate trees became extinct as far as Europe was concerned, but the elms survived each glaciation in a number of southern refugia. From these they emigrated northward in each of the interglacial periods. In the last glaciation, the Weichselian, the fossil pollen record indicates that elm had disappeared once again from almost the whole of Europe.

Natural distribution

Since elms, along with other trees, were eliminated over most of Europe by the last glaciation, it follows that the subsequent pattern of natural distribution was determined by the number and location of their refugia. It is not yet possible to make assured statements about these. However, it appears difficult to account for the present pattern of distribution and variation, and for the numerous fossil pollen profiles now available, without postulating a minimum of four refugia for the Wych Elm, one refugium for *Ulmus laevis* and nine refugia for the Field Elm.

The Wych Elm is adapted to cool summer temperatures. In western Europe, the first trees to appear in fossil pollen profiles as temperatures rose after the last glaciation were birch and pine. With further rise in temperature, trees with a higher heat requirement appeared, including elm. This can hardly have been other than Wych Elm. This made its first post-glacial appearance in Preboreal times (*c.* 8000 to 7000 BC) in southern Europe and in Boreal times (*c.*

7000 to 5000 BC) further north. The northern
16 European limit of this species today runs from
northern Scotland, through southern Norway,
Sweden and Finland, to north central Russia.

The Wych Elm in northern Spain differs from
that elsewhere in western Europe in comprising many
trees having relatively long leaf stalks. This could most
easily be explained if Spanish Wych Elm originated
from a separate west European refugium. But there are
difficulties. Spanish fossil pollen profiles[9] reveal very
little elm, either in the post-glacial period or in the
preceding glacial period, or even in the preceding
interglacial period. It seems rather unlikely therefore
that Spanish Wych Elm can have originated from a
refugium to the south of Spain. A more likely
refugium would be somewhere farther east, perhaps
on the longitude of Corsica or Italy.

In France, as the temperatures rose after the last

glaciation, Wych Elm appears to have penetrated up
the Rhone valley. From here there is a likely route to
the Rhine via the Saône and the Belfort Gap. Again,
a Wych Elm refugium on the longitude of Italy, but
distinct from that supplying the Spanish Wych Elm
population, seems probable.

There are two possible immigration routes along
which Wych Elm may have travelled to reach
northern Germany. One is via the Moravian Gate in
the Carpathians from the Danube. The other is from
the east, from the region along the northern flank of
the Carpathians. The distribution of the bug *Asciodema
fieberi*, mentioned further in chapter 6, which feeds
exclusively on Wych Elm, suggests immigration via
the Moravian Gate since its occurrence in the
European mainland is mainly in Austria, Switzerland
and northern Germany. However, another insect
feeding exclusively on Wych Elm in Europe, the moth
Discoloxia blomeri, also mentioned in chapter 6, 34
principally occurs on the European mainland in
Germany and further to the north-east. This distribu-

Fig. 16. Northern and southern limits of Wych Elm
and Field Elm. Toothed line: Wych Elm; barred line:
Field Elm.

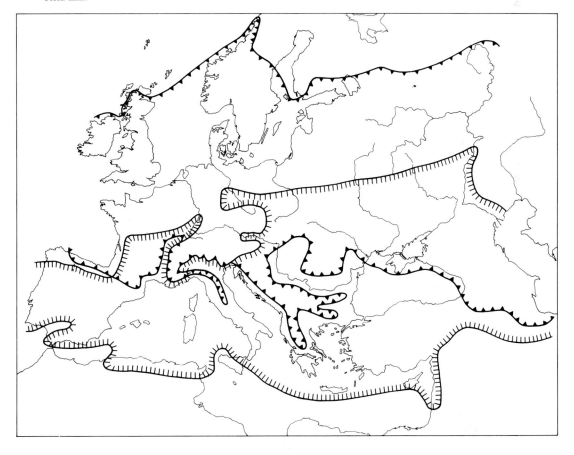

tion suggests that the Wych Elm host plant had entered from the east. The two distributions taken together point to two sources for the north German Wych Elm population. The refugium that supplied the putative Danube source of north German Wych Elm is likely to have been on the longitude of the Balkans or west Asia Minor. Possible immigration routes are up the Vardar River from Macedonia, thence to the Danube via its tributary the Morava, or along the Black Sea coast to the mouth of the Danube. *Discoloxia blomeri* is mainly an eastern species. Its distribution extends to Japan. It seems rather improbable that it should have survived the last glaciation in the putative Balkan–Asia Minor refugium of Wych Elm. It is more likely that it survived in a Wych Elm refugium farther east, perhaps in the Transcaucasus, to be considered shortly. It could then, assuming an immigration route into northern Germany from the east, have transferred from Wych Elm populations coming from some such source to those coming from the Balkan–Asia Minor refugium.

Wych Elm with hairs over the seed, occurs only in eastern Europe, mainly in the Transcaucasus and southern Russia. But they have also been found on the Swedish mainland and on the island of Gotland. On Öland, Hybrid Elms have been found with the hairy-fruited character, presumably derived from a hairy-fruited Wych Elm parent. The obvious conclusion is that the Transcaucasus was a Wych Elm refugium and the sole source of the hairy-fruited character.[10]

It is significant that both *Asciodema fieberi* and *Discoloxia blomeri* occur on Wych Elm in England, though not, or only very rarely, in France, Belgium or Holland. This suggests that the English Wych Elm population, or at least some of it, emigrated across the North Sea bed while it was still dry, from northern Germany.[11]

Ulmus laevis does not occur in England. It is an elm of markedly east European distribution, only just reaching eastern France. It reaches its northern limits in southern Finland and north-central Russia. Since this elm is absent from the Transcaucasus, the only feasible location for its glacial refugium would seem to be somewhere on the longitude of the Balkans.

Wych Elm and *Ulmus laevis* are high-latitude elms in comparison with the Field Elm. This last species, when truly wild, occurs in three rather different situations. In the Mediterranean region, it occurs along the banks of relatively small streams. In

central Europe it is a prominent component of the woods flanking the Danube and of similar woods along the major north-flowing rivers, namely the Rhine, the Elbe and its larger tributaries, and the Oder and Vistula. It also occurs in the wooded steppe of southern Russia and the north-east Balkans. There is conspicuous longitudinal variation.

In western Spain and northern Portugal, the local Field Elm population closely resembles the English Elm. Apart from England, where it is not native, and other places where it is known to have been introduced, elm of this type is only known along the Mediterranean coast of southern France. It has already been pointed out that a glacial refugium of elm in the south of Spain is unlikely. It seems necessary therefore to suppose, as with Spanish Wych Elm, that the west Spanish Field Elm immigrated from the east from a refugium also on the longitude of Corsica or Italy.

The Field Elms of southern France, Italy and the Balkans and central Europe are highly variable and still insufficiently known to permit a confident reconstruction of their post-glacial history. It seems doubtful if the present range of variation could be explained without postulating a minimum of three further refugia on the longitude of Italy, one for the small-leaved Field Elm of France and Spain, one for the narrow-leaved Field Elm of north and central Italy, and one for the densely hairy Field Elm of southern Italy. Three separate refugia are envisaged since it would be expected that in a single refugium hybridization would result in a single, undifferentiated, even if highly variable population. At least two Balkan refugia seem to be required, one for the Field Elm of Yugoslavia with small-toothed leaves, and one for the Danube Field Elm with large-toothed leaves. For the small-leaved Field Elm of southern Russia, a Transcaucasian refugium is likely. In addition to all these, there are distinctive Field Elms in Crete and in Cyprus. These are likely to have survived the last glaciation close to where they now grow. So, a total of nine post-glacial refugia for the Field Elm seems to be the minimum.

Several of the organisms associated with Field Elm, described in chapter 6, are strongly localized. It is likely that they survived in only one or a few of the postulated Field Elm refugia.

The Field Elm of the upper Rhine is likely to have come from France via the Belfort Gap, as with the

Rhine Wych Elm. For the Danube Field Elm, as again for Wych Elm, immigration could have been along the Vadar–Morava route or from the Black Sea coast. The Field Elm in the southern courses of the large north-flowing rivers of Central Europe could have arrived via the Moravian Gate or from the east along the northern flank of the Carpathians. The northern limit of the Field Elm at the time of the post-glacial climatic optimum is likely to have been on the latitude of central France and southern Germany. If it had emigrated further north, extensive hybridization with Wych Elm would have been expected. The hybridization that did occur appears to have taken place much later.

All over north-western Europe, Wych Elm was a conspicuous component of the deciduous forest flora, in some cases the most conspicuous, until the Atlantic Period (*c.* 5000 to 3000 BC). Then, practically all the fossil pollen profiles reveal a sharp decline in the relative frequency of elm pollen. Thereafter elm pollen values continue at a markedly lower level or become discontinuous. This drop in elm pollen frequency is referred to as the 'Elm Decline'. The inference that elm suddenly became much less frequent, or at least that elm pollen production suddenly dropped, is inescapable. There has been much speculation why.[12]

The first point that needs to be made is that the Elm Decline is a phenomenon that affected Wych Elm. This was the species that first immigrated from southern glacial refugia as temperatures rose and this was the species that was undoubtedly native in the areas where the Elm Decline was marked.

The Elm Decline is approximately but not precisely synchronous over the whole area in which it occurs. One explanation is that a climatic change took place. This is likely to be part of the explanation. Wych Elm today is clearly limited in England and in Europe generally by conditions bringing about desiccation when it is in leaf. Heat, drought and high winds, alone or in combination, can all act as limiting factors. Wych Elm has the largest individual leaves of any British tree and these have a drip-tip such as characterizes plants adapted to humid growing conditions. It is generally supposed that the climate became more continental at the time of the Elm Decline. The recession of other species at this time, such as holly and ivy, is usually explained as a consequence of lower winter temperatures. It is far more likely that any effect of climatic change on Wych Elm would have been via diminution of humidity in spring and summer.

Approximately, but again not exactly, coincident with the Elm Decline was the beginning of Neolithic farming, for which the presence of cereal and weed pollens in the fossil pollen profiles provides evidence. Quite likely, there was much Wych Elm on rich calcareous soils that held promise for prehistoric farmers. It might be supposed therefore that it was selectively felled when they appeared. Elm foliage in addition, has been used throughout the area of distribution of elm as animal feed as described in chapter 8. It is a reasonable inference that this use went back to prehistoric times. Taking these two operations together or separately, it can be argued that the inroad into the wild Wych Elm in northern Europe may have been on such a scale that pollen production was drastically reduced. It is doubtful whether these operations could have had more than a subsidiary effect. Much of the Wych Elm growing in prehistoric times must have been inaccessible or remote from farming sites and can hardly have been affected by man. Nor, assuming some climatic change, for which there is independent evidence, would it be easy to separate its effect from that of human activity. Perhaps the weightiest argument in favour of climatic change as the principal cause of the Elm Decline is the situation today. No one today selectively fells Wych Elm to exploit the soils on which it grows, nor does anyone lop it for animal feed. None the less, its current natural distribution is narrowly restricted within the British Isles and the causes are mainly climatic. Soil type has a subsidiary influence. It has also been suggested that an early epidemic of Dutch elm disease might have occurred at the Elm Decline. This would be difficult to disprove. Since however the explanations already considered seem sufficient to account for what happened, a third explanation involving an event for which there is no independent evidence would multiply hypotheses unnecessarily.

The effect of the Elm Decline was to restrict Wych Elm, which had previously been widespread over north-western Europe, to the northern part of its former range and to higher ground further south. So, today it is a frequent tree on low ground in southern Norway, south and central Sweden, Denmark, the extreme north of Germany, southern Finland, the Baltic States and north-central Russia. South of this, the main areas of Wych Elm are in the Ardennes, the

Vosges, the lower Pyrenees, the Cantabrian mountains and Basque valleys of northern Spain, the Eifel, Hunsrück, Harz, Hohe Rhön, Thuringian Forest and Erzgebirge and Black Forest uplands in Germany, and sheltered sites at moderate elevations in the Alps, the Carpathians and the northern Balkan mountains.

Before the Elm Decline, Wych Elm was very generally distributed in England. Thereafter it largely disappeared from the south and east. It is still widespread in the north, especially in the Pennine dales and the Eden valley. It is also frequent on the Lincolnshire Wolds and along the Welsh border. *17* Further south, Wych Elm is mainly found on higher ground, as on the Cotswolds and Mendips.

In Wales, Wych Elm is most frequent in the east, towards the English border. Its principal occurrence further west is along the Tywi valley. It appears that Wych Elm never grew in the far north-east of

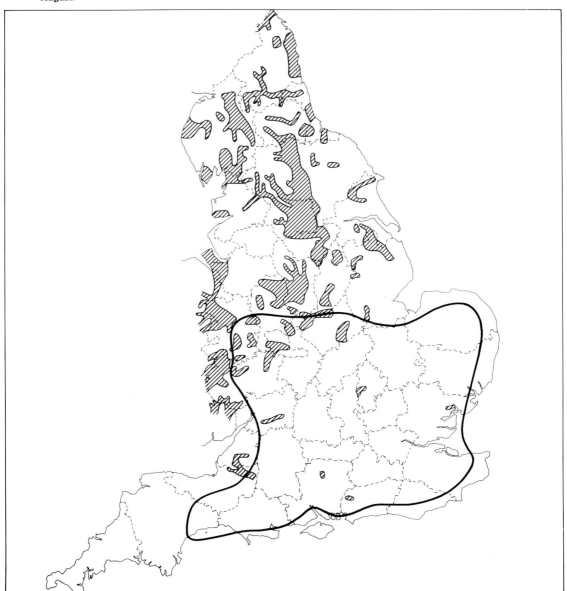

Fig. 17. High-frequency distribution of Wych Elm in England. Diagonal shading: Wych Elm; Heavy line: 69 °F (21 °C) average daily minimum isotherm for August.

Scotland. It was present elsewhere in Scotland till the Elm Decline. This appeared to have had most effect in south-west Scotland where Wych Elm, apart from recent plantings, is now infrequent. In the rest of Scotland the effect of the Elm Decline seems to have been to restrict Wych Elm to the more sheltered river valleys.

What happened in Ireland is obscure. Wych Elm was present in great abundance till the Elm Decline. It then diminished in frequency, but was still sufficiently common to receive several mentions in early Irish literature.[13] A further diminution then appears to have taken place and elm became extremely rare till the landscape planting of the eighteenth century. In the sixteenth century wars, the English army had to import elm from England for its gunstocks.[14] Such Wych Elm as survived was mainly in the Antrim glens.

It is not possible, as already stated, to distinguish with certainty between Wych Elm and Field Elm pollen. Consequently, it is not possible to tell from the fossil pollen profiles whether the reduction in frequency of Wych Elm at the Elm Decline was accompanied by any change in the area of distribution of Field Elm. A climatic change in the direction of significantly hotter springs and summers would, presumably, induce a northward expansion of Field Elm, but there is little evidence that this occurred. At the present day, conditions over most of the south of England are favourable for the growth of Field Elm but not for its reproduction by seed. The limiting factor, mentioned in chapter 2, is spring frost. If such has been the case since the climatic changes at the Elm Decline, no natural expansion northward of Field Elm would be expected. The fact that the Wych Elm and Field Elm population appear not to have been sufficiently close till late prehistoric times for extensive hybridization to occur supports the conclusion that the Elm Decline was not followed by any significant movement northward of Field Elm.

The large north-flowing rivers of central Europe present a problem. The Field Elm present in the riverain woods in the southern courses of the rivers seems to be natural. What is uncertain is the status of Field Elm along the northern courses of these rivers where seed is not normally set. The possibility that man was responsible for taking it northward will have to be considered shortly. It is also possible that branches may have broken off trees in the south and

that these were carried seaward by the current to become stranded and root further north. Should this have happened, it would be expected that Field Elm would become an important member of the riverain woodland, reproducing itself vegetatively though seldom sexually. The status of Field Elm on the Baltic islands Gotland and Öland and on some of the Danish islands is also problematic. There are strong grounds for supposing that man is responsible for its presence in some of these locations. However, it sometimes occurs as a strand plant growing just above high water mark. It is not inconceivable that branches broken off in the southern section of the large north-flowing rivers of central Europe were carried out to sea. If these were then thrown up on a beach, say after a storm, and if the interval of time since they were broken off were not too great, they could root and establish themselves.

There is no convincing evidence that Field Elm ever entered the British Isles naturally at any time since the last glaciation.[15] It does not occur in natural woodland. It fails to reproduce by seed except on rare occasions. Although some of the Field Elm in the British Isles is akin to that in northern France, most is not. The Continental populations most closely related to the rest are as far away as Spain and central Europe. The distribution of the different Field Elm populations in the counties bordering the English Channel corresponds very little with that in the French departments south of the Channel. Finally, the overall distribution of the different Field Elm populations in England bears relatively little relationship to the climatic and soil factors determining plant distributions in general. It has, however, a very close relationship to human settlement patterns in the last two-and-a-half millennia.

What has been said above on the status of the Field Elm in England applies with even greater force to Wales, Scotland and Ireland. Field Elm is native in none of them.

Prehistoric utilization

The uses to which elm has been put are considered in detail in chapter 8. What needs to be considered here is whether the uses to which elm was put by prehistoric man in Europe were such that its distribution was radically affected.

There are two lines of evidence that can be pursued. One is the survival in archeological sites of elm artefacts or artefacts associated with elm utilization. The other is utilization of elm in later times for which prehistoric antecedents are probable. It is not necessary to go into all of the wide range of applications to which elm was put in earlier times. Attention will be confined to four: elm bows, elm wheel hubs, elm leaf fodder, and elm as a cult object.

Elm bows are the earliest elm artefacts to have survived. They have been found in Mesolithic deposits in Denmark and in Neolithic deposits in Denmark and Schleswig-Holstein.[16] The use of the bow seems however to have declined greatly in the Bronze and Iron Ages in Europe. It regained importance around the third century AD. It is not therefore likely that elm distribution in the first millennium BC would have been affected by requirement for bow wood. As will appear shortly, this is the period during which elm planting by prehistoric man seems to have taken place.

The earliest wheels to be made were solid wheels such as are still in use in north-west Spain. These are of no present concern. The spoked wheel is a much more sophisticated object. The remarkable uniformity in its construction over the centuries and throughout its range suggests strongly that it originated in single centre from which it spread by cultural diffusion. The centre of origin appears to have been the Middle East and the date of invention the second millennium BC.[17]

Of its three components, hub, spokes and felloes, the hub has been preferentially of elm, wherever available, in historic times. In prehistoric times other timbers such as oak, ash and birch were used. It is interesting that the first reference to elm in writing is in the inventories of military stores at Knossus, Crete, in Mycenean times.[18] Some of the chariots are specified as of elm. More particularly, there are two mentions of elm wheels.

Elm hubs have seldom been found in archeological excavations. In Scotland an elm hub of the Roman period was found at Newstead.[19] A hub of similar age, said to be probably of elm, was excavated at Bar Hill.[20] To what extent the infrequency of prehistoric elm hubs is a matter of chance discovery cannot be assessed. It is unlikely that the merits of elm for hubs were not widely known in earlier times. More likely there was not enough elm locally available in prehistoric times for all the hubs that were required. Elm was sometimes used for other wheel parts in prehistoric times. At the Iron Age site at La Tène in Switzerland, a wheel was found in which the rim was made of a single piece of elm bent into a circle.[21] Such wheels would not be expected to stand up to much wear.

The relative infrequency of prehistoric elm hubs rather suggests that prehistoric wheelwrights were not responsible for any distribution of elm by prehistoric man. The relatively late arrival and comparative rarity of spoked-wheel vehicles in northern Europe point in the same direction.

In the nature of the case, elm leaves collected for fodder would not be expected to survive very often in archeological sites. Heaps of leaves, presumably intended for fodder, were found in one archeological site in Switzerland. Elm was one of six trees species involved.[22]

Implements believed to be prehistoric leaf knives are known,[23] but it cannot be argued that they were used, in particular, for elm leaves. A rather tenuous argument for the utilization of elm foliage in Denmark

Fig. 18. Mesolithic elm bow from Denmark.

has been built around a Neolithic elm bowl found at Christiansholms Mose.[16] The grain of this showed a bird's eye pattern, indicating that it was formed from burr wood. Burrs are formed when elms are pollarded, and it could be that the bowl was made from a pollard and that the pollard was used for fodder production. Burrs however are formed under other circumstances so this argument is far from coercive.

Utilization of elm foliage as fodder is associated in some areas with special agrarian systems which are possibly of great antiquity. In the Swedish islands of Gotland and Öland, in particular, the *löv äng*, literally leaf pasture, consisted of a rather open stand of trees, often elms, in grassland. The leaves from the trees were harvested.[24] Possibly the Cambridgeshire village elm groves mentioned in chapter 5 had comparable antecedents.

Further, as mentioned above, the Elm Decline may have been caused, in part, by foliage gathering by prehistoric man. If this were so, it would provide further grounds for regarding leaf fodder production as a major use of elm by prehistoric man.

Fig. 19. Romano–British wheel with elm hub from Newstead, Scotland.

How important elm was as a cult object in early times is impossible to assess. It is possibly significant that amongst the very rare occurrences of elm timber in prehistoric sites in England are two Bronze Age burials in Salisbury Plain. In one there is a burial in an elm trunk, and in the other a burial on an elm plank.[25] References by Homer and Virgil to elm as a feature in the cult of the dead are quoted in chapter 10. Pliny mentioned elm in a grove dedicated to Juno at Nocera.[26]

The widespread distribution of elm in churchyards is mentioned in chapter 9. Its role seems to be very similar to that of yew. Both have a particular association with the dead. Both, also are bowyer's trees and a further connection is mentioned shortly. The significant ethnological difference between the two trees is that elm foliage is palatable while yew foliage is poisonous. That the cult role of the elm in recent times had prehistoric antecedents is quite feasible.

Distinguishing between the practical and cult use of a tree is, however, projecting modern ways of looking at things into a milieu where they did not apply. It is likely that cultic significance in prehistoric times usually attached to an object that was also utilized in some practical way. Conversely, a cultic aura could quite easily become attached to objects that were of particular practical value.

That elm was important for some prehistoric Celtic communities is shown by the tribal name Lemovices borne by the former inhabitants in the Limoges (Haute-Vienne) area. Limoges is in fact derived from Lemovices. The designation Lemovices, as mentioned in the next chapter, means elm warriors. It is exactly analagous to another Celtic tribal name, Eburovices, yew warriors, who gave their name to the present Évreux (Eure). Once again, elm and yew are seen in parallel. Perhaps they were totems. Perhaps the name indicated the bow wood formerly preferred, but caution is necessary here since when these tribes flourished the bow seems to have been largely disused. There might have been some other cultic significance. Whatever the role of the two trees, their importance in this Celtic area seems clear.

The conclusion that emerges is that elm was important to prehistoric man in Europe, most probably as a source of animal feed. It might also have had cultic significance.

Distribution by prehistoric man

It has already been concluded that the Field Elm populations of northern Europe are outside their area of natural distribution and that prehistoric man utilized elm. The way is now open to enquire which Field Elm populations, if any, were established by man in prehistoric times in what is now England, and from what sources.[27]

Direct evidence on elm planting by prehistoric man is not to hand. It is therefore necessary, if the events responsible for the present English elmscape are to be uncovered, to devise a set of reasonable postulates as to how early and later man would treat the elm, granted that this tree were of some importance to him. Then, having accepted the postulates as a working model, it is possible, in many cases, to derive a simplest solution to explain all the information available, systematic, distributional, and historic, on a particular elm population. This is the procedure followed in the rest of this chapter, in chapter 5 and in chapters 12–14 which cover the populations county by county.

It is always possible· that a simplest solution, while correctly derived, does not correspond with what really happened, since the same result may have been achieved by a more complex set of events, for which however there is no evidence. Confronted with this situation, most people reach for Occam's razor and cut away all but the simplest solution. There is no coercion, however, and if anyone prefers to leave the vast array of facts now assembled on the elm, unexplained and uninterpreted, they are at liberty to do so.

Having said this, the postulates that are to provide the working model in this book can be unfolded.

It is possible, for each Field Elm population, to establish the earliest date for its present siting as attested by historic and linguistic evidence. Going back in time from this base line, search can then be made for the latest time, before the base line, when the area of distribution of the elm population shows a meaningful relationship with a historical or prehistoric distribution. Historical distributions could be diocesan or archidiaconal boundaries, which may indicate earlier tribal arrangements, linguistic regions, forest extents, and so on. Prehistoric distributions would be mainly maps of discoveries of particular artefacts.

Economy of hypothesis requires that the date of establishment of the elm population should be no later than the latest historical or prehistoric distribution to which it is meaningfully related.

The earliest possible date of establishment of the elm population, granted the hypothesis that man has introduced it, corresponds to the date of human settlement within the area of its distribution. Hopefully, the archeological record will provide a reasonable estimate of this.

Having obtained estimates of the latest and earliest possible dates of introduction of a Field Elm population, the range of possibilities can be shortened by the application of an age-and-area hypothesis. Other things being equal, it would be expected that the earlier an elm population was introduced, the larger would be its present area of distribution.

The source populations of some introduced Field Elm populations have been identified by the sort of detailed biometrical studies described in chapter 2. When this is the case, it is reasonable to assume that the date of introduction must fall within a period when there were human connections, such as immigration routes or trading contracts, between the source area and the site of introduction.

If the overall distribution of elm populations is known or can be reasonably conjectured at a particular time, it is possible to apply a further working hypothesis. This is that, other things being equal, it would be expected that an introduced population will have been obtained from the nearest known source within convenient access.

The simplest hypothesis as to where, within the total area of distribution of an introduced Field Elm population, was the site of first introduction is that this should be at a location where human settlement is evidenced and that this is more likely to be central to the total area than peripheral. It also seems reasonable to postulate that, within a continuous area of distribution of an elm, spread has, in general, been outward from the centre of introduction towards the periphery of the total area occupied.

Occasional use can be made of negative evidence. The hazards of so doing need no elaboration. It can only be used with confidence with data believed to offer complete coverage or an adequate sample of information on a particular matter.

Sufficient distributional data are not at present available to analyse the Field Elm populations of the European mainland in detail along the lines just described. These will have to be dealt with rather sketchily. Field Elm is clearly outside its area of natural distribution in northern France, Holland, northern Germany and Scandinavia. It is however widespread in northern France today, and historical evidence attests its presence there in the twelfth century. The famous elm at Gisors (Eure) between which the kings of France and England had their acrimonious parleys in that century is mentioned in chapter 9. North French place names[28] based on Latin *ulmus*, such as Ormes (Marne), advance the data to the ninth century. Place names supposedly based on the Celtic root for elm, *lem-, are, as explained in the next chapter, of rather poor evidential value. The only such northern French names which can be confidently used as an indicator of elm in Romano–Gallic times are Limeil-Brévannes (Seine-et-Oise) and Lumeau (Eure-et-Loir).

Some of the northern French Field Elm populations are markedly coastal in distribution and appear to have been established by small coastal communities. Such a pattern of human settlement would not be expected later than the Iron Age. It has been suggested that Field Elm reached northern France by two routes, one by sea and the other inland.[29]

The very different kinds and distributional pattern of Field Elm on the two sides of the English Channel are strong evidence that some at least of the Field Elm population of England were not derived from northern France. The postulate that Field Elm is likely to have been introduced from the nearest convenient source, when applied to England and France, leads to the conclusion that there was no Field Elm in northern France when it was first imported into England. Consequently it seems unlikely that Field Elm had reached northern France in the Bronze Age, when, as will appear shortly, it seems that it was first introduced into England.

At the present day, apart from recent amenity planting, Field Elm is of quite restricted occurrence in Belgium. It is mainly to be found in Brabant and Hainaut and to the west of Ostend. In the sixteenth century, the distribution seems to have been much the same.[30] Of early place names,[31] Elmt in Brabant, recorded in the thirteenth century, is probably based on Germanic *elm-. Wiekevorst, in Antwerpen, recorded in the same century, has been thought to be based on the other Germanic root for elm, *wik-, but it is outside the present area of high frequency of Field Elm and there are several objections to the proposed etymology.[32]

Apart again from relatively recent amenity planting, Field Elm is very rare in the Netherlands. The only early place name possibly based on *elm- is Elmt in Noord Brabant, first recorded in the twelfth century.[31]

In both Dutch and Flemish, as described more fully in the next chapter, there are no words for elm based on the Germanic roots *elm- and *wik-. They were replaced, in the Middle Ages, by the word *olm*, borrowed from French, and this, in its turn, has been replaced by *iep*, first recorded in the sixteenth century. The other principal trees retain their Germanic names in Dutch and Flemish and the most likely explanation for the name shift for elm would seem to be the infrequency of elm over the major part of these two countries. The occurrence of Wych Elm in the Ardennes has already been noted. *Ulmus laevis* also occurs there. It is highly doubtful whether Field Elm reached what is now Belgium or Holland in the prehistoric period.

The principal occurrence of Field Elm in Germany is along the major rivers. It is virtually absent from the north-west corner of the country, including the basins of the Ems and Weser rivers. The early place names supposedly based on the two Germanic elm roots *elm- and *wik- are mainly in the area of distribution of Wych Elm and do not help much to define the early distribution of Field Elm. Along the Rhine, where Field Elm now occurs, the old Germanic word *elm* survives in dialectal use as far north as Dusseldorf. North of here, the usual dialectal word is *ip* or *iper*, corresponding to the Dutch *iep*. South of Dusseldorf, the most usual dialectal word is not *elm* but *effe*, cognate with *ip*. A possible explanation for the new terms *ip* or *effe* is that elm may have been previously used for Wych Elm on the high ground on either side of the middle Rhine, Field Elm being absent. When Field Elm came northward, it may have been recognized as a distinct kind and the borrowed words *ip* or *effe* coined for it. In the nature of the case, it is very difficult to be sure whether the present stands of Field Elm along the lower Rhine are of human establishment or not. A combination of natural and artificial establishment is also quite feasible. As regards the Field Elm of England, the precise history of the Rhineland Field Elm may not be relevant, as this does not seem to be a source of any Field Elm population now growing in England.

It has been mentioned above that the various Field Elm stands growing behind beaches on the

Danish and Baltic islands might be natural. The extensive inland stands on Gotland and Öland are a different matter. For these, human introduction is perhaps more likely.

With this brief summary of the possible role of man in distributing Field Elm on the Continent in early times concluded, the course of events in England can be tackled.

The historical evidence assembled in chapters 12–14 provides good grounds for supposing that the kinds of Field Elm extant today and their areas of distribution were established by the Norman Conquest. Field Elm of some sort must also have been present in southern England at the time of the Anglo-Saxon settlement, otherwise the Germanic elm names that gave Anglo-Saxon *elm* and *wice* would not have been transmitted. It can be argued further that Field Elm was present both in the east of England where *elm* was the current name, and in the east Midlands and the south of England where *wice* was the name used, and in the West Midlands where, again, *elm* was current. For the larger populations of Field Elm, then, the dates of introduction can hardly be later than the end of the Roman occupation.

It will be convenient to take the main Field Elm populations in chronological order of their probable date of introduction. The first is the English Elm, the kind of Field Elm prevalent in the south of England and much of the Midlands. It is presumably this kind that is mentioned in an Anglo-Saxon charter relating to Weston super Mare (Av).[33] It must also be the elm that grew at Tyburn, outside London, in the twelfth century.[34] Place names such as Elmley Castle (HW) take the evidence back to the eighth century. The main area of distribution is markedly western though excluding Cornwall. It is roughly centred on Wiltshire.

The absence of English Elm from most of Cornwall suggests that its boundary in the south-west must have been determined before Cornwall was incorporated into England. Its absence from East Anglia and eastern Kent reflects a time before London had become a main centre of communication. It is not likely, on distributional grounds alone, that the English Elm was of Roman introduction.

English Elm has the largest area of distribution of all kinds of Field Elm and age-and-area considerations would suggest that it was the earliest Field Elm to arrive. The source population is known. It is in the north-west of the Iberian peninsula where elms identical with English Elm are frequent. As indicated in chapter 6, the Spanish elms carry many of the associated organisms common on English Elm in England. Some of these could have been introduced with the elm. There have been trading contacts between England and Galicia at various times. The period most relevant in the present connection is the Later Bronze Age, the date of the manufacture in Galicia of double-looped palstaves. These were exported in quantity to the British Isles,[35] the main site of importation apparently being the south coast of Dorset and Hampshire. The majority of the palstaves of this sort found in England were within the main area of distribution of English Elm.

The correspondence between the distribution of double-looped palstave finds and the distribution of English Elm in the Iberian Peninsula and England is very striking. The conclusion that English Elm was introduced from as distant a source as Galicia suggests that Field Elm had not, in the Later Bronze Age, arrived in northern France. It may be significant in this connection, as will be made evident shortly, that the north of France appears to have received Celtic-speaking peoples relatively late.[36]

Wiltshire is roughly the centre of the area of distribution of the English Elm. Salisbury Plain, which lies therein, is one of the longest settled areas of the British Isles and would receive overseas goods via ports on the coasts of Dorset and Hampshire. Bronze Age burials in elm trees or on elm planks in Salisbury Plain, already mentioned, may also indicate some particular interest in elm. The use of elm for the Bronze Age burials just mentioned is an important exception to the general rule that elm does not feature in prehistoric timber identifications.[37] The exceptions are so few that any introduction of elm before the Bronze Age seems most unlikely. That English Elm, in particular, is unlikely to be older than the Later Bronze Age is suggested by two pieces of negative evidence. The extensive assortment of wooden artefacts from the Somerset Iron Age lake settlements include no objects of elm.[38] Neither does elm appear in the extensive remains of Roman pilework excavated along the London waterfront.[39]

The situation in the north-west Spanish peninsula at the end of the Bronze Age requires further consideration. Galicia was inhabited by Celtic speakers when the Romans arrived. The archeological

22

3, 63, 108

24

20

evidence suggests that the infiltration into Spain of peoples of central European origin began in the Later Bronze Age. There is even a suggestion that the Galician bronze industry, which traded with the British Isles, owed its development to immigrants from over the Pyrenees.[40] It is also the case that one of the Celtic tribes in Galicia in Roman times was the Lemavii and it has been suggested that this tribe was of the same stock as the Lemovices of southern France mentioned in the previous section.[41] Both names appear to be based on the Celtic word for elm, *lem-*, indicating some particular interest in this tree.

The conclusion that is emerging is that interest in elm for animal feed, discussed in the previous section, was particularly strong amongst the Celtic-speaking peoples of central Europe during the Later Bronze Age and Iron Age, and that this interest persisted during their emigration to France and the Spanish peninsula. Finding a good elm in Galicia, they traded it to Britain along with the products of their bronze industry.

The next largest English elm population and the one that one would anticipate on age-and-area considerations to be nearest in age of introduction to the English Elm is the Narrow-leaved Elm population 2, 152 of East Anglia, eastern Cambridgeshire and some parts of northern Essex. It is not continuously distributed but is found in a number of discrete areas. The centres of the largest areas are by the eastern tributaries of the Great Ouse, and in eastern Suffolk, an area settled continuously since before Bronze Age times.

Historic evidence attests the presence of elm in East Anglia in the seventh century, when what was later known as the Bishopric of Elmham was founded. Evidence is assembled in chapter 14 that this name

Fig. 20. Continental distribution of English Elm and of Later Bronze Age double-looped palstaves. Heavy line: boundary of main distribution of English Elm; dots: finds of double-looped palstaves.

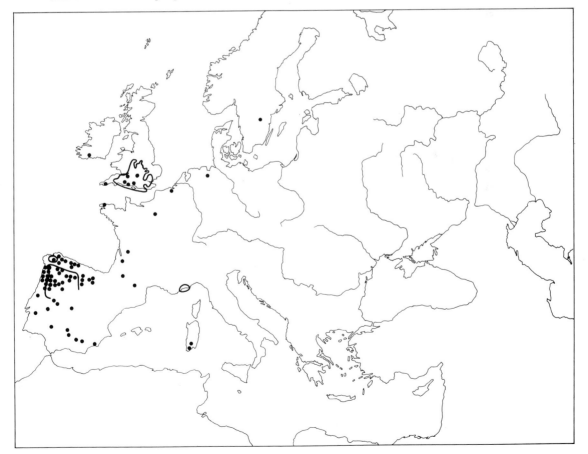

originally applied to the South Elmhams (Sf) rather than to North Elmham (Nf) to which it is believed it was transferred. For present purposes it does not matter which. The fact that two Anglo-Saxon words for elm were used in the area under consideration, *elm* in the east and *wice* in the west, suggests that elm was encountered on arrival both by the people who became the Middle Angles who used *wice*, and the precursors of the East Angles who used *elm*. There is then every reason to suppose that the present population was established by the end of the Roman occupation. Again, as with English Elm, the area of distribution of the present population bears no apparent relation to settlement patterns in Roman Britain.

The probable source population is central Europe. No elm of the sort under consideration has been located in either France, Spain, or the Rhine Valley. Closely similar elms are to be found, however, in the basin of the upper Elbe and its tributary the Saale, in the Danube valley and in the upper parts of some of the Polish rivers. The nearest source to England would be the upper Elbe.

Trading contacts between England and the Elbe valley existed in Later Bronze Age times, as evidenced in particular by the distribution of finds of Taunton– Hademarschen socketed axes.[42] In Great Britain they are scattered, but with principal concentration in East Anglia. It is significant that after this period, during the Iron Age and the Roman occupation, there is no evidence of trading links between England and the Elbe.[43] Contact was only resumed in Anglo-Saxon times. As with double-looped palstaves and English

Fig. 21. Continental distribution of the East Anglian form of Narrow-leaved Elm and of Later Bronze Age Taunton–Hademarschen socketed axes. Heavy line: boundary of main distribution of the East Anglian form of Narrow-leaved Elm; dots: finds of the socketed axes.

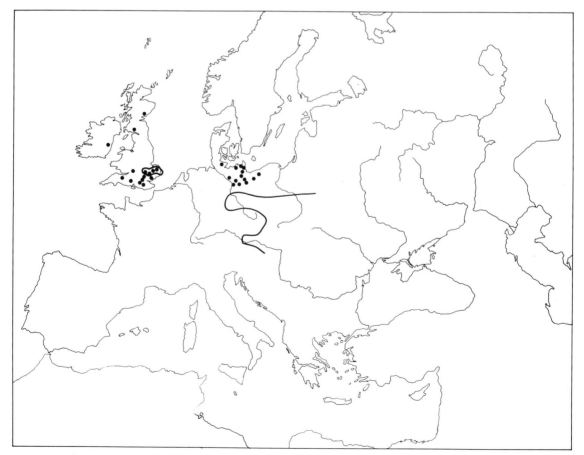

Elm, an equally striking similarity of distribution patterns links the Taunton–Hademarschen axes with East Anglian Narrow-leaved Elm.

If the Narrow-leaved Elm were introduced via the Wash, as seems most likely, the question again arises why no nearer source than the Elbe was tapped. The answer again would be that Field Elms may not have been introduced to Northern France by the Later Bronze Age. The dates of introduction of the English Elm and the East Anglian Narrow-leaved Elms would seem to be close. It is not surprising that elms were not exchanged across England. Contact between East Anglia and Salisbury Plain, say via the Icknield Way, is likely to have been tenuous. In early days the sea was an easier route than land, and East Anglia, right till the end of the Middle Ages, was in more intimate contact by sea with the Continent opposite than with the Midlands behind.

It was mentioned when discussing the origin of the English Elm that, by a coincidence too striking to be ignored, one of the Celtic tribes in its source area bore the name Lemavii, quite likely based on Celtic *lem-*, meaning elm. In Roman times, there was a tribe called Lemovii living near the mouth of the Oder. The suggestion has been made[41] that there was a Celtic tribe, from which both the Lemavii and the Lemovii originated, which lived somewhere around the lower Elbe in the Later Bronze Age. It is further supposed that, as a result of pressure from Germanic-speaking peoples from the north, this tribe was displaced. One body moved to central France where it remained as the Lemovices, whose capital was Limoges (Haute-Vienne). Another went to Galicia and became the Lemavii of Roman times. A third moved to the Oder valley and became Germanized. There are a number of queries about this theory. In particular it assumes the existence of Celtic speakers some distance to the north of the boundary of Celtic speech in the late Iron Age. However, some intrusion of Celtic influence even as far as Scandinavia is known to have occurred, perhaps in connection with the amber trade.

What is remarkable is the double coincidence of two likely source areas for British elms, far apart from one another and from the British Isles, each associated with a Celtic tribe named apparently after elm, and all this at the time when the importation of the elms is thought to have occurred.

A likely centre of introduction of the East Anglian Narrow-leaved Elm would be the valleys of the Lark and Cam where the finds of the socketed axes are concentrated.

There is another population of Narrow-leaved Elm whose centre of distribution lies in the valley of the Great Ouse in western Cambridgeshire. This was certainly established by the seventeenth century. It can be inferred that it is much older since it extends into the west part of the Isle of Ely, the east part of which has an elmscape of the East Anglian form of the Narrow-leaved Elm considered above. The place names of the Isle of Ely discussed in chapter 14 include several based on *wice*. The earliest record is of the tenth century. Since the contact zone of the two types of Narrow-leaved Elm is presumably of this age, it is reasonable to argue the elm stands at the centres of introduction of the two types would be considerably older than this. The Great Ouse valley between St Ives and Huntingdon, which is the centre of distribution of the second type, has been settled continuously since the Bronze Age. The source population on the Continent has not yet been located. It seems not to be in France, Spain or the Rhineland. Central Europe, again, seems more likely. In support of this is the absence of the gall mite *Eriophyes campestricola* on both the Narrow-leaved Elms just considered. The mite occurs on all the Narrow-leaved Elms discussed below, for which an origin in northern France is probable.

The existence of two quite distinct forms of Narrow-leaved Elm with centres of distribution as close as the Great Ouse, on the one hand, and the Cam and the Lark on the other, suggests a date of introduction for each when communications even over relatively short overland distances was poor, that is before the Roman period. Communication, as already noted, between England and central Europe lapsed during the Iron Age, so one is taken back again to the Later Bronze Age as the most likely time of introduction. As early a date as this would then explain why the East Anglian form of Narrow-leaved Elm was not taken to the west of Cambridgeshire in Roman times. Another elm would have been there already. The considerably smaller area of distribution of the western Cambridgeshire population is understandable since it would not have been taken further east where the East Anglian population is believed to have been already established. Its spread further west would have been impeded by the unsettled forest terrain of Northamptonshire.

The elm population of eastern Kent demands

attention next. The place name evidence is good. Elmstead is recorded in the ninth century. Lemanae, now Lympne, is recorded in the fifth century, and if the usual derivation of this from the Celtic *lem-*, meaning elm, is correct, then elm, probably Narrow-leaved Elm, must have been growing in eastern Kent during the Roman occupation. Elm charcoal, very rare in Roman archeological sites, has been found in Roman Canterbury.

The east Kent Narrow-leaved Elm can be matched by elms now growing in northern France, most particularly in the Cotentin peninsula and in northern Brittany. The Kent elm is heavily galled by the mite *Eriophyes campestricola*, also abundant in northern France.

The small East Sussex population of Narrow-leaved Elm is very similar to the east Kentish population, but not so similar as to suggest that the one was derived from the other. An origin for both in northern France seems more probable.

The area of distribution of the Kentish and East Sussex populations suggests an association with coastal settlement. It corresponds far more closely with the settlement pattern in the Iron Age than in the Roman period. Population M11, defined in chapter 14, does not include Canterbury though it comes close to it on its northern and eastern sides. This distributional feature reinforces the conclusion that M11 was established before Canterbury was the focal point of human settlement in eastern Kent, that is, well before the Roman period.

Evidence for contact between eastern Kent and northern France in the Iron Age is far too voluminous to require setting out. The fact that the Kentish Elm resembles populations in northern France, whereas the Narrow-leaved Elms considered earlier do not, would be explained if Narrow-leaved Elm had been taken from southern to northern France during the Iron Age and was then available for import to England by communities still lacking elm. It would be natural for Kentish settlers to look to France rather than elsewhere in England for their elms. Communication by sea would be easier than overland in England.

Three different Narrow-leaved Elm populations occur along the Essex coast: in the Tendring peninsula; around Colchester; and in the Dengie peninsula and Crouch estuary. The northern two are well recorded in earlier times. In the Tendring peninsula the surname atte Elme is on record in the

fourteenth century at Great Clacton. Near to Colchester, the place name Tey ad Ulmos, the present Marks Tey, is recorded from the thirteenth century. The fact that the east Saxon settlers on the coastal strip used the word *elm* while the Anglo-Saxons further inland used *wice* strengthens the supposition that the former found elm already along the coast when they arrived.

The elms of the Tendring peninsula can be matched with elm populations in central Normandy and Brittany.[44] The elms around Colchester may be of hybrid origin but one component is likely to be a Field Elm from somewhere in northern France. All these Essex populations are galled by *Eriophyes campestricola*.

Again, the pattern is of a number of small coastal settlements, with apparently poor communication between them, otherwise different types of Narrow-leaved Elm in each would not be expected. The pattern seems to reflect the conditions of the Iron Age, not the more centralized circumstances of the Roman period. Contact between Essex and northern France is well attested. There is, in addition, specific evidence of trading contacts between Brittany and Essex, revealed by the discovery of coins of a Breton tribe, the Coriosolites, at Chigwell, Colchester and Great Bardfield.[45]

The Essex coast elm populations lie behind the site of Iron Age salt workings, whose remains are the 'red hills' of the Essex marshes. The putative source area of the Tendring peninsula elm population in southern Brittany is similar behind salt workings of comparable antiquity. One cannot help wondering whether one is dealing with movements and trading contacts of Celtic-speaking peoples who, in addition to planting elm for animal feed, were also involved in salt production. The coincidence is heightened by recalling how important salt working was in the Celtic homeland in central Europe in prehistoric times, as at Hallstadt and along the Saale valley, where the place name Halle likewise refers to salt working.[46]

There are also a number of different populations of Narrow-leaved Elm, both along the north bank of the lower Thames, behind Dagenham, and along its northern tributaries, the Roding and Lea, with its branches the Stort, Ash, Rib and Maran.

The lower Thames population was certainly there in the early seventeenth century. It is the *Ulmus folio glabro* of the second edition of Gerard's herbal,[47] stated to occur in North Ockendon (GL). The surname de Ulmis or atte Elme in Havering atte Bower (GL) is first encountered in the twelfth century. It probably refers to the same Narrow-leaved Elm population.

The elm populations along the Lea and its branches are recorded from the sixteenth century onwards. Elm in the Roding Valley is cited in a fifteenth century document. The evidence is set out in detail in chapter 14.

The area north of the Thames in the modern counties of Greater London, Essex and Hertfordshire was settled late by Anglo-Saxon immigrants. There are no reasons for attributing to them the various Narrow-leaved Elm populations now in this area. These almost certainly were in place by the end of the Roman occupation. Again, the distribution pattern of small isolated populations, centred on river sites known to have been occupied in the Iron Age, suggests that this was the date of introduction. Elm populations in northern France provide close or fairly close matches for the Narrow-leaved Elm populations north of the Thames. Like in others supposedly of French origin, galling by *Eriophyes campestricola* is frequent. There is much evidence of contact between these Iron Age sites and northern France.

Last of the possible pre-Roman elm introductions is the Narrow-leaved Elm population of southern Hampshire. This is also mentioned in the second edition of Gerard's herbal.[47] It is *Ulmus minor folio angusto scabro* from Lymington. Should the usual etymology of this name be accepted, from Celtic *lem-, meaning elm, then its presence here much earlier would be attested. However, the etymology is not secure. The coastal distribution of this small population recalls that of the various other coastal sites already described. It suggests an introduction by an early, relatively isolated settlement of Iron Age date. Domesday Book records salt workings at Lymington. It has been suggested but not demonstrated that these were in existence in the Iron Age.[48] The source population has been identified in the Cotentin peninsula on the opposite side of the English Channel.[44] A cross-channel connection here has been postulated as one of the possible Iron Age tin trade routes.[49]

Contrary to what has become a traditional view, namely that the Field Elm was of Roman introduction, no evidence has been obtained that suggests a Roman origin for any kind of Field Elm. One further Field Elm population, that of Cornwall, is shown in chapter 5 to have been probably introduced in the Anglo-Saxon period. It is surmised, however, that the areas of distribution of the various elm populations introduced in the Bronze and Iron Ages expanded with the agricultural development and good interior communications due to the Romans.

Much timber work of Roman age has survived, notably the wharfage of the Roman London waterfront. Elm appears not to be represented. This is in marked contrast with medieval pilework, which is mainly of elm. It seems reasonable to conclude that there was not much elm surrounding Roman London. An elm pile of Roman age, most likely Wych Elm, is noted in chapter 13 as underpinning the wall at Gloucester. The Roman age charcoal at Canterbury (Ke) is mentioned above.

An expansion of elm in the Roman period would bring with it the probability of some hybridization. The different introduced Field Elm populations might be expected to have hybridized along their contact zones. It is also likely that they would hybridize with native Wych Elm when brought into its vicinity.

Some hybridization did probably occur in Roman times. However, the main hybridization zones, those involving crossing between Wych Elm and Field Elm, in Essex and along the Bedfordshire–Northamptonshire border, are in areas sparsely settled in Roman times. It is more likely therefore that most of the hybridization in these zones took place later. Further consideration of this point is therefore deferred till chapter 5.

4
Vernacular names

There is general agreement today as to what trees are designated by the word *elm* or its equivalents in other languages. It would be unjustified to assume, without further investigation, that this was the situation in earlier times. There are two not mutually exclusive possibilities to consider. The first is that in former times a more general concept was current, elms being included in some more comprehensive category. The second is that a narrower concept was adopted, some kinds of elm being recognized as one entity, and others as another.

The first possibility will be taken first. It appears that in many areas elm and lime (*Tilia* spp.) were regarded as a single entity and given a single designation. In the original Irish version of the ninth–tenth century life of St Finian of Clonard, a tree is mentioned for which a form of the word *lem* is used. This word is normally used to designate elm. Cognate terms of similar meaning occur in other Celtic languages. Yet in the Latin version of the same life, *lem* is translated *tylia* (*tilia*), the usual Latin word for lime.[1]

Medieval Breton provides a parallel case. The main dialects of Breton each have and had a different name for elm. In the dialect of Tréguier, the word used was *tilh*, derived ultimately from the Latin *tilia*. This word had entered old Breton as early as the eighth century, and a medieval Breton–French dictionary translates it as *orme*, the standard French word for elm.[2]

Bartholomew the Englishman, the medieval encyclopedist, was confused too. He contrasts elm with lime but then goes on to state that the former has whitish flowers frequented by bees, which could apply only to lime.[3]

In Pont Canavese in northern Italy in the sixteenth century, there was a tree indifferently referred to as *olmo* or *tiglio*. Possibly, however, in this case, *tiglio* did not have its usual meaning of lime but denoted, rather, a marked tree, deriving from medieval Latin *teclatus*, not from *tilia*. Reference is, in fact, made to an *ulmum teglum* at Carpice, in the same region, in the thirteenth century.[4]

In northern France in the eighteenth century, some kinds of large-leaved Hybrid Elm went under the name *orme-teille* or *orme-tilleul*, literally, elm-lime.[5] This name also appears in Flemish as *lindolm*. Such names, others of which will come up shortly, are relatively late, compared with single names. They would appear to reflect some uncertainty about the distinction between the trees so coupled.

The sixteenth century botanists, who might have been expected to know better, aggravated this particular uncertainty. Their predecessors in ancient Greece had elaborated the notion that some trees exist in male and female forms, not indeed as sexual partners, but, respectively, as tougher and softer variants of the same tree. One tree so treated was φιλύρα [filura].[6] The names of the two forms were respectively rendered into Latin as *Tilia mas* and *Tilia foemina*. In de Lobel's illustration of these,[7] the former is undoubtedly an elm and the leaves carry galls as could only be produced by the aphid *Tetraneura ulmi*. The latter is just as certainly a lime with its characteristic inflorescence.

In Welsh, till the sixteenth century, the same word, *llwyf*, had to do both for elm and lime, and when, later, a special designation, *palalwyf*, was coined for lime, it was constructed from *llwyf*. Further afield still, amongst the Osage Indians, the word *hindse* covered both lime and *Ulmus rubra*, the slippery elm.[8]

The tree whose identity has caused most perplexity is the hornbeam, for which cognate names are often absent, even with closely related pairs of languages. The Renaissance botanists, who might have been expected to clear up the confusion, were completely at sea. The Flemish botanist R. Dodoens, whose writings were to be utilized in the herbals of both H. Lyte and J. Gerard, flounders about with what seems to be hornbeam. He thinks it might be the *ulmus sylvestris* of Pliny and the same as what in Flanders is called *herseleer*, usually taken to be a kind of elm.[9]

Lyte takes all this over into his English herbal[10] and compounds the problem by suggesting that the tree is what in England is called *wych* or *wych hazel*, both names for elms. Gerard treats hornbeam separately from elm[11] but remarks: 'my selfe better like that it should be one of the Elmes'. He states that it grows at Gravesend, where it is taken for an elm. He follows Lyte in asserting that in some places it is called *wych hazel*. In the second edition of Gerard's herbal, put together by the generally percipient T. Johnson,[12] he repeats what Gerard had to say about hornbeam and adds that it is in Essex that it is called *wych hazel*.

Elm and poplar have also been confused. The large-leaved Hybrid Elm of Picardy, later introduced into England and called the Dutch Elm, was originally known in its place of origin as *ypereau*. Later, in the nineteenth century, in northern France, this name was transferred to poplar. In the Iberian peninsula, the Castilian and Galician word *negrillo* [black tree], and its Portuguese equivalent *negrilho*, are all applied indifferently to elm and to black poplar. It will be seen in due course that the qualification black is applied to elm in other, quite unrelated languages. Another Spanish word for poplar is *álamo*. In various parts of Spain, it means elm.

Even such superficially dissimilar trees as elm and yew have been confused. In the seventeenth century, the inhabitants of Amsterdam used the word *ipen(boom)*, corresponding to the modern *iep*, to denote elm. But in Leiden it meant yew. The point of resemblance between the two trees is the dark red colour of the freshly sawn heartwood.[13]

Wych Elm has been taken for a kind of hazel. In Belgium, for instance, it has been called *noisetier sauvage*, literally wild hazel.[14] In many parts of England, its oldest designation was *wych hazel*. As is made clear below, *wych* was originally a name for any elm. The coupling of the names of the two trees suggests, as with elm and lime above, some uncertainty as to their distinctiveness.

What these various examples suggest is that, in earlier times, ideas as to which trees were distinct may have been quite different from what is now accepted. In particular, trees like elm, which had undivided deciduous leaves and did not produce a distinctive utilizable fruit, were likely to be confused with each other and receive a common name. This might have been the case in the early history of the speakers of the Indoeuropean language family. There are two linguistic roots *lm- and *wig-,[15] derivatives of which occur in several different Indoeuropean language groups, and in all of which they denote elm. It is reasonable to infer that corresponding ancestral terms, also meaning elm, were current among Indoeuropean speakers before the language groups differentiated. It is further inferred, on a variety of grounds, that the home territory of the first Indoeuropean speakers was in the deciduous tree zone of eastern Europe.[16] There are, however, no Indoeuropean root terms for either lime or hornbeam, both of which would have occurred in the putative ancestral area. This is understandable if the linguistic roots for elm were originally umbrella terms covering lime and hornbeam also. Subsequently, lime was recognized as distinct in several language groups, but the hornbeam had to wait till recent times for adequate recognition.

The alternative possibility mentioned above was that a narrower concept of tree categories was held, different kinds of elms having quite different names. This is the case in Finnish and the related languages Estonian, Livonian and Vot, which belong to the Finno-Ugrian language family. These have different words for Wych Elm and the east European elm species *Ulmus laevis*, in Finnish *jalava* and *kynneppää* respectively. To each of these words there corresponds a related word of similar significance in each of the other languages. It may be that these words and their ancestral forms always applied to different kinds of elm. This would be a different situation from that in which separate names, originally applying to elm in general, were later applied to particular kinds. In England, the words *elm* and *wych* were both originally applied to elm in general, though they appear to have been current in different areas. Later they came together, and then *elm* was used for English Elm and *wych* either for Wych Elm or Narrow-leaved Elm.[17] A comparable situation is found in the Slavonic language groups where up to three words for elm occur in particular Slav languages. Again it seems that originally all applied to elm in general. Later they are found attached to particular kinds of elm. The comparison, however, is not exact. The English situation arose in the ordinary course of language evolution. The Slav applications appear to be artefacts of dictionary compilers. It can be taken as a general rule that a lexicographer, given two plant names for the same plant, will endeavour to find a difference of meaning. If he cannot, he will create it.

The changes in a tree name that occur in the course of time under the operation of laws of phonetic shift, throw no light on the tree denoted. Other changes may be significant. Sometimes, a name for elm in a particular language disappears and is not replaced. It can be inferred in some cases that the tree had gone, perhaps through climatic change or through human eradication. In other cases, the speakers of the language may have migrated from a region where elm grows to one where it does not. The original name for elm might also disappear but be replaced by another. This is likely to be because the older name has ceased to be regarded as appropriate for any of a variety of reasons. Finally, an elm name may persist in use while another is introduced alongside. Again, a range of explanations for acceptance of a new name are to hand. All these types of name change will feature later in this chapter.

If a tree word persists with its original meaning, it is fair to infer that the tree it denotes has remained continuously known, even if, sometimes, it may no longer grow where the users of the word live. An important corollary is that if an immigrant group gives its own word to a tree in a new environment, then the tree concerned must have been present when the immigrants arrived.[18]

It is necessary to keep in mind the degree of certainty that attaches to etymological relationships. The one situation which inspires full confidence is where a word is represented in several languages within a language family by variants with the same meaning and where the differences between the variants conform to the rules of phonetic shift inferred to have taken place in the language family. If the meaning is not the same, or if the phonetic shifts are irregular, then confidence in any etymological statements shrinks greatly.

Minimum confidence attaches to situations where there is doubt as to what language family a word pertains. The principal cases where this occurs in the present context are where Indoeuropean speakers have entered territory in the Mediterranean region or in the Caucasus. A skilled Indoeuropean etymologist can construct a phonetically acceptable etymology for many words which perhaps were taken over from non-Indoeuropean languages. The skill is self-defeating. It undermines the possibility of identifying the sources of such words. Plant names are

particularly vulnerable to uncertainties of this sort since they are very liable to be taken over by an immigrant people from the inhabitants of the territory they have invaded.

What have been discussed till now are words denoting the elm tree. When these enter into the formation of place names, they are particularly useful since they provide more precise information on localization. A disadvantage is the uncertainty about the etymology of many place names that perhaps contain a tree element. This uncertainty is reduced by ascertaining for each language group the grammatical forms shown by place names derived regularly from trees, and the particular elements with which tree words may combine. It is also necessary to identify what other words could be confused with a particular tree name.

With very few exceptions, English place names are either Celtic or Germanic. The Celtic names will be considered first. These changed in form early in the first millennium AD so it will be necessary to deal with the earlier and later place names separately. No examples of tree names used as place names without modification seem to have survived from the earlier Celtic period. On the Continental mainland, however, earlier Celtic tree names with various suffixes are not infrequent. In France, Lemoialo, now Limeuil (Dordogne) represents *lem-+-ialo*, the former element probably designating elm and the suffix perhaps meaning a clearing. Lemausus, now Limeux (Cher) represents *lem-+-ausus*. The suffix is of uncertain meaning, and probably originally pre-Celtic, but is found in combination with other elements designating trees. The commonest category of earlier Celtic place names consists of two elements, the first qualifying the second. Examples involving tree names are, however, rare. Cassinomagus, now Chassenon (Charente), represents *cassinos+magos*. The first perhaps means oak and the latter a clearing. It is also possible, however, that *cassinos* is a totemic personal name. All these examples are French. Earlier Celtic names in England believed to be based on *lem-*, meaning elm, are nearly all irregular or with some complication or uncertainty in meaning that seriously reduces whatever evidential value that they might have had.

Later Celtic names, Welsh, Cornish and Gaelic in Britain, and Breton in Brittany, differ from the foregoing in that the commonest form of place name again consists of two elements but with the qualifying term following, not preceding, the main term. Uncompounded tree names may still form place names, either singular or plural, such as Derwen [oak] in Welsh, or they may carry suffixes. In Welsh, a preposition may precede, such as Tandderwen [tan + derwen], below the oak. Most usually, the place names found are compounds of the type indicated above, such as Eglosallow, perhaps church of the elms, in Cornish, or Auchleven [achadh + leamhain], field of the elm, in Gaelic. In Breton, trees perhaps masquerade as saints. It has been suggested, for instance, that the place name Saint-Onen has nothing to do with a saint so called, but was derived from *san onenn* [valley of the ash].[19]

Germanic place names vastly outnumber all others in England. A tree name may be used alone, such as Elm, now Great and Little Elm (So). Suffixed derivatives of tree names are very rare in England. They occur more frequently on the Continent, as German Elmpt [elm + ith], the suffix having a collective force. The most frequent form is the compound with the qualifier first, as in older Celtic names, as Elmley (Ke) [elm + leah], elm clearing, with the tree name first, or Radnage (Bu) [readan + aec], red oak, with the tree name second.

Very rarely, inverted compounds of the later Celtic pattern are found, usually in names attributable to Scandinavian settlers from Ireland, as Aspatria (Cu) [askr + patric], Patrick's ash.

Place names in Romance languages are rare in England. Overseas, they are of great importance when attempting to discover Continental sources of the English elm populations. Frequently, the tree name is used alone as Ormes (Aube) in France or Olmos in Spain. Also frequent is the Latin collective *ulmetum* which becomes Ormoy (Eure-et-Loir) in northern France, Olmet (Puy-de-Dôme) in southern France and Olmedo in Spain. Compounds such as Longus Ulmus, now Lancôme (Loir-et-Cher), are rare. Rather more frequent are such forms as St-Pierre-des-Ormes (Sarthe).

Slavonic place names also need a mention since they cover one of the probable source areas of English elm material. Those based on trees are typically composed of the tree name alone, as Czech Jilem, from the Slavonic root *ilm-, and Sorb Briest, from another Slavonic root for elm, *berst-. Suffixed tree names are

very common, as Polish Brzostów and Germanized Sorb Bristow, both from *berst-. Doubly suffixed names may be encountered as Russian Il'movka and Il'movitsiï, both from *ilm-.

Surnames provide useful ancillary information to place names for the relatively short period before they became hereditary and their bearers mobile, that is for the fourteenth–fifteenth centuries. Those relevant to present purposes are based on one of the two English words for elm, *elm* and *wych*. In the south of England the forms commonly found are atte Elme, or atte Elmes, or rarely atte Nelme, and atte Wyche. In the north of England, surnames based on trees were in Anglo-French form as del Asshes. In the case of surnames based on *wych* there is often the possibility that *wych* is being used in some other sense. This is particularly the case in the north-west Midlands where *wych* is more likely to mean a salt working. Surnames such as de Elme are not likely to refer to the tree but to one of the places called Elm. Surnames of the various types described are often recorded in Latinized form. This is normally *de Ulmo* or *ad Ulmum*.

Indoeuropean *lm-

Comparison linguistic investigation leads to the conclusion that there were two words for elm among the first Indoeuropean speakers, say about the fourth millennium BC.[20] These will be represented as *lm- and *wig-. The first question is why there should be two names, in particular since it has already been suggested that the original Indoeuropean elm concept probably extended over other trees as well. The subsequent history of the derivations of these two words does not reveal any general tendency for one or the other to be associated with particular elm species. It seems, on the contrary, that *lm- was originally a western Indoeuropean term, descending mainly into the western language groups, and *wig- an eastern term, descending mainly into the eastern languages.

The root *lm- is certainly ancestral to words for elm in the Italic and Germanic language groups and probably ancestral to elm words in the Celtic and Slavonic language groups. It can be inferred from this that elm had been known continuously by speakers of the former two groups since before these groups differentiated.

Among the early Italic languages, the word for

elm is only known in Latin where it is *ulmus*. It entered into several place names either as such or as the collective Ulmetum during the period of the Roman Empire. From *ulmus* descend the words for elm in the Romance languages. In Italian it became *olmo* but in Sardinian *ulmu*. The former is widespread in place names.

In southern France *ulmus* became *olme* or *oume*. In central France it became *orme*, eventually the form in the standard language, and in Picardy *omme*. All forms occur in place names. Among French speakers in medieval England, elm was designated by such forms as *olmeu*, *(h)oulmeu*, *ulmeu*, *(h)omeu* and *humeau*, corresponding to the derivative standard French form *ormeau*.

In Spanish and Portuguese *ulmus* became *olmo* and in Catalan *om*. Both are widespread in place names. The Romansch derivative is *ulm*. In the Balkan peninsula *ulmus* became *ulm* in Romanian and *ulmu* in Vlach. The former is frequent in place names outside the mountain zone.

Derivatives of *ulmus* also passed into other languages. The older word for elm in Dutch is *olm* and this occurs occasionally in place names. Its adoption by the Dutch suggests that the elm to be found in Holland in the early Middle Ages had been introduced from the French-speaking area to the south-west. The German place name Ollmuth is based on the collective Latin noun *ulmetum*. Further afield, *ulmus* provided the Maltese with *ulmu* and the Berbers with ولم [ulm]. The Vannes dialect of Breton borrowed *oulm* from French.

The prestige of Latin as a language of culture led to *ulmus* being borrowed as a learned term in some non-Romance languages. Such borrowings do not provide place names. In England, the monastic lexicographer Aelfric proposed that *ulmtreow* be used for elm.[21] Wycliffe used *ulm* in his translation of the Vulgate Bible into English.[22] However, *ulm* did not take on. It was different in Germany. The *ulme* found in the medieval literary language eventually became the standard term. In Denmark, *ulm* was tried but dropped.

Place names in Romance languages derived ultimately from *ulmus* are usually readily recognizable. A Norman name such as Le Houlme (Seine-Maritime) could, however, derive either from *ulmus* or from Scandinavian *holmr*, flat land by a river.

Indoeuropean *lm-* gave rise to a root in the Germanic language group that can be represented as *elm-*. It is usually supposed that the first Germanic speakers were located along the Baltic coast of Germany, and in Jutland and perhaps also in southern Scandinavia. The only elm likely to have been encountered there would have been Wych Elm. The early German speakers also inherited a word for elm derived from *wig-*, so presumably both words at this period denoted the same elm species. Most of the Germanic speaking peoples moved from their original homeland and the history of their elm nomenclature has to be unravelled in the light of their movements. There were extensive movements southward. In the late Iron Age German speakers got as far as the Erzgebirge, south of which Celtic languages were spoken.[23] Later they took over much of the land formerly occupied by Celtic speakers in central Europe and crossed the Rhine into north-west Europe. At the end of the Roman empire, linguistic differentiation appears to have taken place between the coastal Frisians and the Germanic speakers inland. When the Roman Empire collapsed, and to some extent before, Frisian and Germanic speakers passed to England, and Germanic speakers entered France.

The Germanic speakers outside the Scandinavian zone retained their two words for elm. Derived from *elm-* are such place names as Elmenhorst in Germany, Elmt in Holland and Elmt in Belgium. In central Europe, Germanic speakers would have encountered *Ulmus laevis* and the Field Elm. They appear to have applied their two elm words indifferently to whatever elm they met. In the later Middle Ages, *elm*, through sound shift, became *ilm* in southern Germany. In what is now Holland and Belgium, not much elm would have been encountered and what there was would most likely have been Wych Elm. At all events, by the time that Dutch together with Flemish had differentiated as a separate language from German, it had lost both the original Germanic elm words, suggesting that there was not much elm around to keep them in currency. In Germany proper, both the original Germanic elm words remained, but they disappeared from the standard language and sank to dialectical status. The introduction of the term *ulme* from prestigious Latin might partly account for their downgrading. However, other elm words, to be considered later, were also introduced, which might imply that different forms of elm were being recognized and separate names for these were being sought.

Neither of the two Germanic words for elm passed into Frisian. This suggests that there was no elm in early times along the North Sea coastline occupied by Frisian speakers. The English language is most closely related to Frisian. It is consequently supposed that the Frisians supplied the major contingent of Germanic-speaking settlers in England towards and at the end of the Roman occupation. Since they would have been unfamiliar with elm, they would of necessity have had to borrow a word for it from such other immigrants as knew it. In fact they borrowed the two available words, which became Anglo-Saxon *elm* and *wice* respectively.

There is nothing to suggest that either term, in Anglo-Saxon times, denoted any particular kind of elm. The map of the distribution of the two terms in 22

Fig. 22. Distribution of *elm* versus *wice*. Unshaded: *elm*; horizontal shading: *wice*; vertical shading: Celtic speech. The heavy lines enclose the medieval dioceses of Lincoln (Middle Angles and Lindissi), Winchester, Salisbury and Exeter (West Saxons) and Chichester (South Saxons).

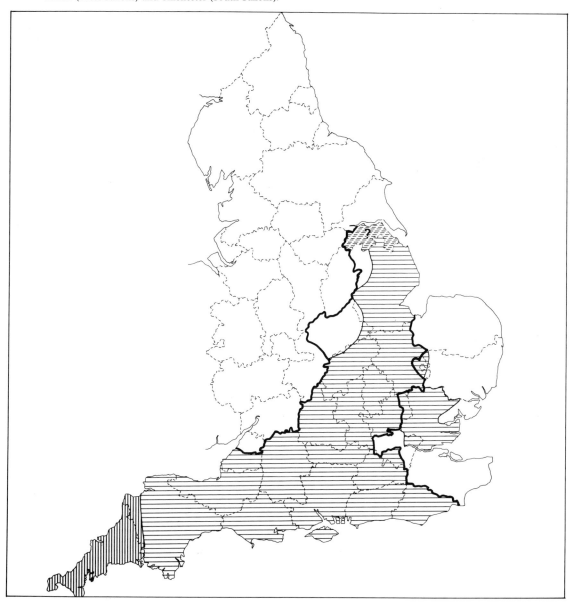

England, which together cover the entire country, suggests that each centre of settlement borrowed its own term.

It is probably not relevant to attempt to relate the distribution of the words to the areas traditionally allocated to Angles, Saxons and Jutes. Quite likely, the Germanic terms had already been introduced with the auxiliary forces introduced by the Romans themselves towards the end of their occupation. The word *elm*, at all events, was in general use by Germanic speakers outside the Frisian area. In the Jutland peninsula, the homeland traditionally attributed to the Angles and Jutes, *elm* was the only word that survived in later use. The distribution of the derivatives of **wig-* will be taken up later.

The two words *elm* and *wice* persisted in England till the spread of standard English from London. The first Germanic settlers must have found elm already present, otherwise the Germanic words for it would not have remained in currency. Since *elm* had been the term chosen presumably by the Anglo-Saxons who took over London, it eventually displaced *wice* (*wych*), which then passed into dialectical obscurity.

The word *elvin* was used for elm in parts of Kent, Sussex, and Hereford and Worcester. The fact that it appears to have interchanged with elm in at least two Kentish place names, as shown in chapter 14, suggests that it may be an aberrant derivative of *elm*.

In Scandinavia, Germanic **elm-* became *almr*, whence Norwegian and Swedish *alm*. The Danes have *elm*, probably from early German, since *elm* is the form used in place names in Jutland, originally German, and only occupied by Scandinavian speakers after the post-Roman migration period. The term *alm* was introduced into England with the Viking settlers and displaced *elm* where Scandinavian settlement was thickest, namely in parts of Cumbria and Yorkshire. It remained in dialectal use till the present century. It referred mainly, perhaps exclusively, to Wych Elm since the Scandinavian settlement areas where *alm* became current were all in the north of England.

In southern Wales, *elm* had been borrowed into Welsh by the sixteenth century and was then inflected as if a native Welsh word.[24] A little later it was further acclimatized by entering into the compound words *elmwydd* or *elmenwydd*, also denoting elm. It is possible that when introduced into southern Welsh, *elm* was reserved for the English Elm, only frequent in Wales

in the extreme south, and associated with areas of English influence.

Though English place names based on elm are usually instantly recognizable, there are snags. Anglo-Saxon *aewielm*, a spring, is rendered as *elm* in some later texts of Anglo-Saxon charters.[25] The West Yorkshire district name Elmet, preserved in Berwick in Elmet and Sherburn in Elmet, is probably not analogous to the Dutch Elmt, a collective name based on *elm*. It is the name of a small Celtic kingdom mentioned by Bede, identical in form with Elfet in Wales. Both are of uncertain meaning. Personal names entering into place names lay some traps, in particular personal names compounded from *aeðel-* and a second component beginning with *m*. Thus Elmstone (Ke) is Aeðelmaer's village; Elmstree (Gl) is Aeðelmund's tree. Initial *h-* is notoriously labile in the north of England. A word *helm*, meaning shed, is current in parts of northern England. It may be the source of Elmswell (Hu). The personal name Helm appears to be the donor in Helmsley (NY), but in Upper Helmsley, earlier Hemelseya, in the same county, the meaning appears to be Hemele's island. If Elmhurst in Brenchley (Ke) is not based on *elm*, it may be Helma's grove, Helma being a short form of names beginning or ending in *helm*.

The Celtic words for elm present difficulties, philologists disagreeing greatly as to their correct derivations.[26] The two key words in any discussion are Old Irish *lem* and Welsh *llwyf*. There are five views about their etymologies. The first is that each is derived from Indoeuropean **lm-*. The second is that *lem* came from **lm-* but that *llwyf* originated from an independent Indoeuropean root. In the third view, *lem* is again derived from **lm-* but *llwyf* is regarded as cognate with the derivatives of the Baltic-Slav linguistic root **leip-*, meaning lime. The fourth view amalgamates views two and three and interprets *llwyf* as a conflation of a derivative of **lm-* with one from **leip-*. The fifth view is the most radical. It supposes that the Celtic elm words are derived from a pre-Indoeuropean or at least pre-Celtic Mediterranean language. In favour of this last view is the occurrence of place names in which *lem-* is combined with pre-Celtic suffixes such as *-annus* or *-uris*. Thus in the Ligurian cultural area, place names such as Lemannus, now Léman, the French name of Lake Geneva, and the river anciently known as Lemuris, occurred. Against views three and four is the fact that no word for lime cognate with *leip-* occurs in either the Germanic or Italic language groups. Against the last view is the improbability that Celtic speakers, who must have been in constant contact with elm, would have discarded the Indoeuropean word they had inherited in favour of a very similar word from some pre-existing language. If one were to be guided by simplicity of hypothesis, the first view, deriving the Irish and Welsh elm words from Indoeuropean *lm-* via a Celtic root *lem-* would come out on top.

When Celtic speakers can be first located they are in central Europe and southern Germany. They would probably have been familiar with both Wych Elm and Field Elm and perhaps also with *Ulmus laevis*. There is no evidence that they had more than one word for elm. Subsequent movements of Celtic speakers in the first millennium BC were extensive. They entered France. Some passed on to Spain and some to the British Isles. Others entered northern Italy. Their migration to Asia Minor is not of concern here. The fact that the Old Irish term for elm is *lem* implies that the Celtic speakers who emigrated there had retained acquaintance with the tree and that a similar term had been used by their presumed ancestors in France. The Asturian dialectal name for elm, *llamera*, has been supposed to derive also from Celtic *lem-*.[27] A corresponding implication could be drawn.

The Celtic tribal name Lemovices, generally interpreted as elm warriors, analogous with the Eburovices, yew warriors, is further evidence that a derivative of *lem-* was current in Celtic France. Yew and elm, as is shown in chapter 8, are the principal bowyers' woods, and this may account for the tribal designations. French place names believed to be based on *lem-* are mentioned in the previous section. The Celtic tribes Lemavii in Galicia and Lemovii in northern Germany are discussed in the preceding chapter. In Strathclyde and Wales the word for elm was *llwyf*, just mentioned above.

No derivative of the Celtic elm word survives in Cornish or Breton, which both, otherwise, retain common Celtic forms to name the principal trees. It is a reasonable inference that there was little elm in Cornwall in the early post-Roman period, so the Celtic word for it was lost. When the Cornish later came in contact with elm, they had to find a new word and so also did the Bretons who emigrated from the Cornish

peninsula. They in fact coined four words, differing with dialect, and none corresponding with that adopted in Cornwall.

Unlike the situation with Romance and Germanic place names, the presence of *lem-* meaning elm is not often demonstrable with any degree of confidence. There is no cogent evidence that the root *lem-* appearing in conjunction with pre-Celtic suffixes means elm. It has, in fact, been suggested that there was a pre-Celtic Indoeuropean root *lim-*, denoting river or marsh, which could be behind some of the later *lem-* place names. Some writers suppose that such a root was present in the Celtic language group too.[28] One reason for this conclusion is the occurrence of *lem-* place names where elm is absent or infrequent today and for which an earlier presence cannot be presumed. The upshot of all this uncertainty is that *lem-* place names can only be used very tentatively in attempting to unravel the history of elm.

Names usually thought to contain earlier Celtic *lem-* are widespread in England, often, as indicated above, not in regions where elm is present today nor in regions where place names based on the Germanic words for elm occur. Welsh place names containing *llwyf* are very rare, even in Wales, and though Welsh names occur along the English border, none involving *llwyf* have been found. Gaelic names supposedly containing *leamhain* are more frequent but do not occur amongst the few Gaelic names south of the Scottish border.

The difficulties of interpreting names of assured Celtic origin containing *lem-* have been sufficiently emphasized. In some cases it is not even certain whether a place name is of Celtic or Germanic origin and *lem-* has to compete with such Germanic roots as *hlim-*, to roar, or even with personal names such as Leofwine.

The fourth language group to which *lm-* contributes an elm word is the Slavonic. While not immediately relevant to English place names, it is apposite when considering the Continental origins of some forms of Narrow-leaved Elm. There is an initial difficulty here in that it is not certain whether the Slavonic root *ilm-* was derived directly from Indoeuropean *lm-* or was borrowed at a later date from High German *ilm*. The latter was the older view, based on the assumption that Indoeuropean *lm-* would not give a form with initial *i-* in Slavonic. Against this view are three facts. The first is that shift from *elm* to *ilm* in High German took place after the Slavonic form had become established. The second is that place names, such as Il'movo and Il'movka based on **ilm-* are widespread in Russia.[29] The third is that derivatives of **ilm-* passed from Russian into other, non-Indoeuropean, languages.

If it is accepted that the homeland of Slavonic speakers was to the north-east of the Carpathians,[30] they would have been in contact with all three of the European elm species. It so happens that there are three sets of elm names in the Slavonic languages, but there is nothing in their subsequent history to suggest that either was attached to any particular kind of elm. The great expansion of the Slavs occurred in the sixth–seventh centuries. In general it was within terrain where all the three European elm species were to be found, though there is likely to have been little Field Elm in north-west Germany, northern Poland or central and northern Russia.

In the later differentiated Slavonic languages, derivatives of **ilm-* are only found in the eastern and western Slavonic languages. Of the former, Russian has ильм [*il'm*] and Ukrainian ільм [*il'm*]. Of the latter, Czech has *jilm*, Polish *ilm*, Sorb *lom* and Polabian *yelm*. There is no derivative of **ilm-* in the southern Slavonic languages. Further, its presence in Polish, Sorb and Polabian is problematic. It does not occur in Slavonic place names in most of the area where these languages were spoken. It may in fact be only a word of literary origin borrowed from Czech or Slovak.

In view of the many uncertainties about the history of **ilm-*, it is probably unwise to speculate as to why its principal occurrence is in Czechoslovakia and Russia.

In Russia, ильм passed into three non-Indoeuropean languages. The Finno–Ugrian language family has several roots for elm, but one language, Zyrian, has supplemented its own Permian word, сир(пу) [*sir(pu)*], by ильм from Russian. Over a wide area of central Asia the various Turki languages have equivalents of the same term for elm represented by Turkish karaağaç, literally black tree. However, in Russia, Turki speakers had a different word, карама [karama], among the Kazan Tatars and хурама [khurama] among the Chuvash Tatars. Both groups, in addition borrowed ильм from Russian. It became элмэ [elme] and йёлме [ielme] respectively. It seems likely, in all these cases, that the original elm words lost out to the superior prestige of Russian.

Indoeuropean **wig-*

The second of the original Indoeuropean words for elm, represented by **wig-*, has entered into the vocabularies of four language groups, Germanic, Slavonic, Baltic and Albanian, and possibly a fifth, Iranian. Its spatial distribution therefore is more eastern than that of **lm-*. Two of these groups, Germanic and Slavonic, have a word based on **lm-* in addition to their derivative from **wig-*. There is no evidence that **wig-* applied to any particular elm.

As already stated, it has to be assumed that the Germanic derivative, which can be represented as **wik-*, was current alongside **elm-* when Germanic speakers were settled in the Jutland peninsula and around the Baltic, where the only elm species was Wych Elm. Their subsequent movements are described in the preceding section. On the European mainland, derivatives of **wik-* seem to have died out everywhere except in north-west Germany where it is apparently represented by such place names as Wickede. As mentioned in chapter 3, a derivation of the Belgian place name Wiekevorst from **wik-* is improbable. In north-west Germany **wik-* ultimately became the Low German dialectal term *wieke*. No equivalent word exists in modern Dutch or Frisian. Nor are equivalents known in any of the Scandinavian languages. It would seem to follow from this that the immigrants who carried the word to England are most likely to have come from that part of the restricted Low German area where *wieke* was current.

The distribution of Anglo-Saxon *wice*, later *wych*, and Latinized as *wycheus*, as a word for elm in England bore no relation originally to the kind of elm present, but only to the Anglo-Saxon settlement [22] pattern. It was the usual term in Wessex, Sussex and Middle Anglia. There is no evidence of its currency amongst the Northern or East Angles, or the Kentish Jutes. However, a word of warning has already been sounded against interpreting the distribution too closely in terms of the traditional terrains of the Angles, Saxons and Jutes, as the word may have already been current in late Roman times. There is no difficulty in supposing that the West and South Saxons may have brought the word with them, but the Middle Angles are unlikely to have brought it from Angeln, if that is where they came from. They could however have got it from the stray Saxons that found their way to the Middle Anglian area or indeed from Saxon auxiliaries already in the country before the Romans left.

In its original area of distribution, *wice*, later *wych*, long remained in use, but was replaced gradually in most places by *elm*, the word in the standard English of the London area. It persisted longest in vernacular use in Cambridgeshire, where it was still current in the first decades of the present century.

There appears to have been a variant of *wych*, *wick*, with a further derivation *wicken*. This perhaps resulted from Scandinavian influence, for it is recorded in Lincolnshire and probably occurred also in northern Cambridgeshire and Bedfordshire.[31] There is independent evidence, mentioned in chapter 14, that elm was known as *wych* at Elton (Ca) in the fourteenth century. An individual known as Richard atte Wych in 1327, was Richard atte Wick five years later.[32]

Place names and surnames incorporating *wice* or *wych* are frequent in the area in which the terms were current, but their identification is quite specially difficult because of the various other place name elements with which confusion is possible. The rowan, *Sorbus aucuparia*, seems to have been called *cwic(beam)* in Anglo-Saxon. In southern England this became *quick(beam)*, but in the north *wicken* or *wiggen*. Confusion with *wych* or *wicken* meaning elm is clearly easy.

The Latin word *vicus* was borrowed by the Germanic languages before the migrations at the break up of the Roman Empire. In Anglo-Saxon it appeared as *wic*. In the Middle Ages it became *wick*. It acquired a wide range of meanings, of which dairy farm seems to have been the most widespread. In compound place names it may be either the first element or the second. As first element it is difficult to distinguish it from *wice*, meaning elm. It has recently been argued that in the compound Wickham, a Romano-British settlement is referred to.[33] As second element, it often becomes *-wich* as in Norwich. In the form *wych* it had become the standard term for a salt working in Cheshire and in Hereford and Worcester. As second element too, confusion with *wych* meaning elm is easy.

Confusion of *wych*, meaning elm, with *wicce*, later *witch*, a practitioner of witchcraft, may also have occurred, particularly in the fairly frequent place name Witch Pit. On the analogy of the frequent Withy Pit, this could mean a water-filled pit fringed with elm. Perhaps, however, in most cases, it was a pit into which a suspected witch was thrown to sink or swim.

Other possible confusions are with Scandinavian words. Lowick (Cu) and Wigtoft, earlier Wiketoft (Li), may be based on Scandinavian *vik*, a bay. Wickerby (Li) appears to incorporate the Scandinavian personal name Viking.

It is known that the *k* in *wice* or in *wic* may become voiced in its compounds. Former Wikeholt (WS), perhaps based on *wice*, is now Wiggonholt. Other confusions are then possible, notably with *wicga*, later *wig*, insect, the personal names Wicga and *Wicg and Scandinavian *veggr*, a wall. Wigley (Db), for example, is thought to be derived from *wicga*, Wigton (Cu) from *Wicga*, Wigsley (Ng) from *Wicg and Stanwick (NY) from *veggr*.

A final confusion results from the Anglo-Saxon tribal name Hwicce. The aspiration is liable to disappear in place names. The Hwicce wudu of the Anglo-Saxons is now Wychwood (Ox).

The Slavonic root derived from Indoeuropean *wig-* can be represented as *vyaz-*. Its derivatives occur in the East, West and South Slavonic language groups. In Russian it takes the form вяз [*vyaz*] with the corresponding Ukrainian в'яз [*vyaz*]. Czech has *vaz*. Polish has *wiąz* and Sorb *wjaz*. In Serbocroat and Slovene *vez* is the form. It is not represented in Bulgarian or Macedonian. The status of these words is not clear however. Place names based on *vyaz-* are exemplified by Vazovnice in Czechoslovakia, Wiązownice in southern Poland, and Vezseny in Hungary, in an area where a Slavonic language is no longer spoken.[34] In Slav areas outside these countries, derivatives of *vyaz-* appeared not to maintain themselves. They were probably reintroduced as literary borrowings. As pointed out under *ilm-*, no ancient connections between any of the Slav words for elm and any particular kind have been discovered.

Both extant Baltic languages have a word for elm derived from *wig-*, *vinkšna* in Lithuanian and *vīksna* in Latvian. Other elm words in the Baltic languages are mentioned in the next section. There is, as usual, no satisfactory evidence that particular words were originally applied restrictively to either of the two elms, Wych Elm and *Ulmus laevis*, that occur in the area where Baltic languages are spoken.

Albanian belongs to a group of its own amongst Indoeuropean languages. Its principal word for elm, *vidh* is derived from *wig-*. Owing to the great uncertainty about the history of Albanian-speaking peoples, it would be fruitless to consider its significance further in the present context.

The last language group to require attention in this section is the Iranian. The principal Kurdish word for elm is وزم [*wuzm*]. Closely similar words for elm occur in other Iranian languages: وزم [*vezm*] in the Persian dialect of Kazvin, вызм [*vyzm*] in Talyshi, *guzm* in the Persian dialect of Seistan, and гучум [*gujum*] in Tajik. There is also a dialectal word *wiz* meaning elm in Kurdish, and it has been suggested that this is a derivative of **wig-*. From *wiz*, other Iranian elm words, such as the *wuzm* group, *buz*, another Kurdish dialectal word, and *visk*, a word for elm in the Persian dialect of Esfahan, can be supposed to have originated.[35]

A derivation of *wiz* from **wig-* would be phonetically regular. The rest of the scenario inspires no confidence. If *wiz* is the Iranian base word, it would not be expected that the most widespread elm words would be the *wuzm* group. Further, there is no phonetic progression in the Iranian languages that involves the addition of a final -*m*.

Botanical considerations also tell against a derivation from **wig-*. An original Indoeuropean word would only be expected to retain its original meaning when its speakers remained continuously acquainted with the object to which it refers. Emigrant Indoeuropean speakers passing from their presumed homeland in eastern Europe to Iran would most likely go through Asia Minor where elm is largely absent from feasible immigration routes. If they took this route they would not be expected to have retained a knowledge of elm. It is also the case that no other tree name in any modern language in Iran corresponds to an original Indoeuropean term. It seems unlikely therefore that any of the Iranian elm words were derived from **wig-*.

Other Indoeuropean names

It has already been mentioned that both Germanic words for elm disappeared from Frisian and Dutch, presumably because of the rarity of elm along the northern seaboard. A new elm word had meanwhile appeared in north-west Germany. In the Rhine valley this appeared as *effe* south of Cologne, and as *ip* to the north of it. Along the German coast, between the Dutch border and the Jutland peninsula, it took the form *iper*. It entered Dutch as *iep* in the sixteenth century. In Frisian it became *iperen*(*beam*). What appears to be the same word also entered

north-west France, presumably from Flemish, as *ypereau* or *ypreau*. It was applied here, in the sixteenth–eighteenth centuries, to a local Hybrid Elm, later introduced to England as the Dutch Elm. These names may have derived ultimately from a German dialectal word for yew. The likely point of resemblance linking yew and elm is probably the dark red colour of the freshly sawn heartwood as already mentioned. The elm word is not directly connected with the place name Ieper (Ypres).

Along the Rhine valley, these words seem to have been used for Field Elm, and the older term *elm* may have been restricted here to Wych Elm. The Dutch had borrowed the word *olm* from north-west France where many of the elms were probably Hybrid Elms. At any rate, in the seventeeth century, in the Netherlands, *olm* seems to have been mainly used for Hybrid Elms and *iep* for Field Elm. Subsequently *olm* died out of general use, and *iep* is now used for elm in general.

Two other elm words have been current in the area of the Germanic languages. In the High German area, the word *rüster* appeared in the later Middle Ages. It is not found in older place names and the reason why it should have been found necessary to add this to the other elm names in circulation has not been discovered. In Flanders, the term *herselaar* was current in the sixteenth–eighteenth centuries.

The German word *baum* is the general term for tree. It was borrowed into the Finno-Ugrian language Livonian as *puom* and used for Wych Elm. Livonians were under German rule in the Middle Ages and a German term could have appeared prestigious. They retained a Finno-Ugrian word, *kundaba*, for *Ulmus laevis*.

The Baltic language group, in addition to the Indoeuropean words derived from **wig-*, has another set of elm words represented by *guoba* in Lithuanian and *guôba* in Latvian. Both Wych Elm and *U. laevis* are present in the Baltic States, but elm names appear not to attach to particular kinds. The Finno-Ugrian Livonians just mentioned as having *baum* unloaded on them from German, also took the present term in the form *guobu*. It even entered Polish, with its three Slavonic elm words, in the form *gab*.

The multiplicity of words for elm in the Baltic language group does not stop here. Lithuanian has a third word for elm *skirpstas* and there was a cognate word *skerptus* in Old Prussian, now extinct.[36] These words seem to be derived from a Baltic root whose original meaning was hornbeam. A distant relationship with Latin *carpinus* has been proposed.[37] The puzzle as to why so many names should have been current for only two elm species is unresolved.

Historically, the principal words overall for elm in the Slavonic languages are derived from a root confined to this language group. It can be represented as **berst-*. It has been derived from the Slavonic root **berz-* for birch, though the nature of the supposed link between birch and elm is totally obscure. In Russian and Ukrainian it is берест [*berest*]. In the Western Slavonic languages it is *brěst* in Czech, *brest* in Slovak, *brzost* in Polish, and place names show that a similar word was formerly current in Sorb, though it has not survived in the language now spoken. Amongst the southern Slavonic languages, Serbocroat and Slovene have *brest*, and Bulgarian and Macedonian брест [*brest*]. It is clear that in the period of Slav expansion, this was the principal term used everywhere, except Russia. It occurs in such place names as Břešt'any in Czechoslovakia, Brzostówka in Poland, and Briješći in Yugoslavia. It has even survived the local disappearance of a Slavonic language, as in Bristow in northern Germany,[38] and Breasta in Romania.[39] This term seems not to be restricted to any particular kind of elm, and its wide distribution indicates that speakers of Slavonic languages have been continuously in contact with elm over their extensive movements within Europe.

The oldest word for elm on record is Greek πτελέα [*ptelea*]. This appears in Mycenean inventories of military stores as *pte-re-wa*. No elm word has been the subject of more conflicting etymologies than this.[40] One view is that it is derived from an Indoeuropean root that gave Armenian [θ]եղի [teghi], which also means elm. Another view is that the Armenians borrowed the word from Greek. It has been suggested that the Latin *tilia*, meaning lime, was also a borrowing from Greek. It is generally supposed that however the Armenians got [θ]եղի, they passed it over to the Georgians whose word for elm is თელა [tegha].[41] Other suggested relations for πτελέα are with Latin *populus*, meaning poplar, with Old High German *felawa*, denoting willow, and with Ossetian фæрв [færv] which stands for alder.

Other etymologies link πτελέα, not with tree names, but with other categories of object or even with verbal roots. It has been suggested for instance that πτελέα might be related to πτέλας [ptelas], wild boar, the thought being that wild boars might fancy elms. Another suggestion derives πτελέα from an Indoeuropean root **pet-* meaning to spread, and related in consequence to Greek πέταλος [petalos], spread out, and to Latin *patulus* meaning much the same. Many of the connections suggested are strained. None carries conviction. It is also the case that very few Indoeuropean tree names passed in Greek with their original meaning. If they have been taken over, they have usually been attached to different trees or to artefacts. The upshot of all this indecisiveness is that little use can be made of πτελέα for elucidating the history of elm in eastern Europe.

Cornish speakers, as noted in the next chapter, seem to have lost the Celtic term for elm. It was inferred from this that elm was relatively unknown in Cornwall and adjacent parts at the end of the Roman occupation of Britain. It is also noted in that chapter that the immigrants from Cornwall to Brittany coined four new words for the elm whose acquaintance they now made. One term in the Vannes dialect, *oulm*, is mentioned above. It has also been pointed out that *tilh*, the word in the Tréguier dialect, was borrowed ultimately from the Latin *tilia*, which originally meant lime. If it were true that Latin *tilia* was derived from the Greek πτελέα [ptelea] for elm, it would have to be concluded that the Bretons had restored it to its primeval significance. This is thought unlikely. The Breton dialect of Vannes had a second word for elm. This was *onn*, a derivative of Cornish *on*, meaning ash. The word for elm in the remaining Breton dialects, those of Cornouailles and Léon, was *evlac'h*. The origin of this word is obscure. The only suggestion with any plausibility is that it is a variant of *elv*, meaning poplar. Instances of confusion between elm and poplar have already been cited.

Meanwhile in Cornwall, elm had been introduced, it seems, from Brittany. What name it was given is uncertain. Elm is not mentioned in medieval Cornish texts. For late Cornish, two somewhat divergent forms, *elaw* and *ula*, are cited.[42] These may indicate a Cornish root **el-* for elm. Some place names appear at first sight to include a component **elwyth*, meaning elm, but the earliest form of each seems to suggest that a personal name akin to Welsh Eluoid is more likely to be involved. The upshot is that putative Cornish elm names have to be treated with extreme reserve.

Non-Indoeuropean names

The major non-Indoeuropean language family surviving in Europe is the Finno-Ugrian. There are two linguistic roots for elm going back to the time before the various language groups of this family had become differentiated, that is to around the fourth millennium BC.[43] The first is represented by the word for elm in both languages of the Volga group: шоло [sholo] in Cheremis and селей [seleĭ] in Mordvin. The speakers of Cheremis and Mordvin have remained within the area long inhabited by Finno-Ugrian speakers. One group of speakers of the Ugrian language group, however, emigrated into central Europe in the ninth century. These were the Hungarians whose corresponding term for elm is *szil*, normally compounded as *szilfa*. The persistence of derivatives from this linguistic root, all retaining the original meaning, indicates the occurrence of elm in the Finno-Ugrian homeland and along the migration path of the Hungarians. All three elm species occur in the Volga basin and in Hungary. The elm words derived from this linguistic root appear to apply to any of them.

The other linguistic root is represented by нюло [*nyulo*] in Voytak, in the Permian language group, and by нолго [*nolgo*] in Cheremis, in the Volga language group. The Baltic Finnish language group developed an elm nomenclature of its own, differentiating sharply between the two Baltic elms, Wych Elm and *Ulmus laevis*. The former is *jalava* in Finnish and *jalaka* in Estonian. The latter is *kynnepää* in Finnish and *künnapuu* in Estonian. Analogous terms for each occur in most of the Finnish Baltic languages.

What may be a pre-Indoeuropean elm name, *atinia*, is mentioned by Columella and following him, Pliny. The basis for this inference is the occurrence of a number of Latin words concerned with viticulture with no obviously related forms in other Indoeuropean languages.[44] It is assumed therefore that the immigrant Indoeuropeans into Italy took over the viticultural vocabulary of the people already in Italy, having no such words in their own language. Elm, being used as a support for vines, comes into this category. It has been suggested that *atinia* may enter into the Italian place name Tignasco, with the Ligurian suffix *-ask-*.[45] It would follow that elm, presumably Field Elm, was

already in use as a vine stock at the time when the Indoeuropeans descended into Italy.

In the early part of the first millennium BC, western Europe was still largely occupied by non-Indoeuropean speakers. Today only the Basques retain their non-Indoeuropean language. Their usual word for elm is *zumar*, with its variants *zugar*, *zuhar* and *zunhar*. It is found in such Spanish place names as Zumárraga. The history of the elm in Spain is obscure. This country is, however, almost certainly the source of the English Elm, so the history of its elms is relevant to unravelling the history of this particular elm. The Basque name unfortunately helps very little. The elm of the Basque valleys is mainly Wych Elm, much, at least, of a kind different from the Wych Elm in the British Isles, northern France and central Europe. The elm population related to the English Elm occurs in the north-west of Spain. If it were there in pre-Celtic times and were named by a people speaking a non-Indoeuropean language, it is not known what name was applied.

It has already been noted that both Celtic and Latin speakers in Spain were probably acquainted with elm, but the earliest contemporary allusions to it that have come down are not in a Celtic language, Latin or Spanish, but in Arabic. The name given by Arabian writers in the Iberian peninsula to elm was نشم [*nasham*] or more precisely نشم أسود *nasham aswad*, black *nasham*. The term نشم was also applied to various other trees, in particular to *Celtis australis*. Used alone, however, or qualified as above, it appears always to have denoted elm. It is as نشم that elm is described by Spanish Arabic poets, as mentioned in chapter 10. It cannot be doubted that elm, mainly Field Elm, was well known to the Arabs in Spain in the period of their ascendancy.

Kinds of elm

It has already been noted that when two or more names for elm have become current, for whatever reason, there is a tendency, irrespective of original meaning, to apply the names to different kinds, if such are recognized.

The notion of kind of elm did not earlier have the same significance as today. The designation *rock elm* is a case in point. In North America it applies to a particular species, *Ulmus thomasii*. In the north of England, however, and in the Channel Islands,[46] it applies to elm, Wych Elm in the first case and Field Elm in the second, which has grown in infertile soil and has, in consequence, developed timber of hard rock-like quality.

In England the two main words for elm, *elm* and *wych*, both originally meant elm in general. The desire to discriminate naturally arose when two or more easily differentiated kinds occurred together. In Bedfordshire and Cambridgeshire, *wych* and *elm* were latterly treated as different entities.[47] Apparently *elm* meant English Elm. In Cambridgeshire, *wych* came to mean Narrow-leaved Elm. In Bedfordshire it could have denoted either Narrow-leaved Elm or Wych Elm. Elsewhere, where Wych Elm and English Elm coexisted, the term *wych hazel*, noted earlier, was applied to the former and *elm* reserved for the latter. This is perhaps not really a case of distinguishing what were thought of as two kinds of elm, since *wych hazel* seems to have been considered as something intermediate between elm and hazel.

Sometimes a distinction was drawn between *elm* and *wych elm*. Again, *elm* meant English Elm. In some places, such as Essex and Hertfordshire, *wych elm* meant Narrow-leaved Elm. In others it meant Wych Elm, the meaning current today and used in this book with capitalized spelling.

In northern France, the term *orme*, or its variant *omme*, and *ypereau*, discussed above, came together. The term *orme* became the name for the local Field Elm while *ypereau* was used for the local Hybrid Elm, later introduced into England as the Dutch Elm. In Holland *olm* and *iep* came together. Then, as mentioned in the preceding section, *iep* was reserved for the Field Elm and *olm* for the Hybrid Elms. Eventually *iep* displaced *olm* entirely.

Some particularly interesting distinctions were worked out in north-east France and Belgium, with parallel terms in French and Flemish. Two categories of elm were popularly recognized. One corresponds to Field Elm, which is slower growing and more irregular in branching pattern, and has a darker heartwood. The other corresponds to the Hybrid Elms. These are faster growing and have a more regular branching pattern and a lighter heartwood. Accordingly, the Field Elm was called *orme maigre* [lean elm], in French and *magere olm* in Flemish. The corresponding terms for Hybrid Elms were *orme gras* [fat elm] and *vetten olm*.

On heartwood colour, the same two kinds of elm were distinguished as *orme rouge* [red elm] and *orme blanc* [white elm]. On heartwood texture, they were differentiated as *orme dur* [hard elm] and *steenolm* as against *orme doux* [soft elm] and *zoeten olm*. Based on heartwood texture and habit, the distinction was between *orme mâle* [male elm] and *orme femelle* [female elm]. Occasionally, the terms got reversed, and in Jersey, for instance, *orme rouge* has become *orme femelle* and *orme blanc* has become *orme mâle*.[48]

5
History

It is convenient to take the withdrawal of the Romans as the dividing line between the prehistoric and historical eras. Written records hardly bear on the former at all. For the latter they become progressively more and more important.

Anglo-Saxon period

Elm artefacts, should they be preserved, would be of the same evidential value as in the prehistoric era. In fact, very few have survived. Documentary evidence however now becomes important. It falls into three categories. The most useful is that of land charters. In was common practice in these to define the boundaries of estates in Anglo-Saxon vernacular by citing the landmarks that stood along them. Very frequently these were trees. It is remarkable how very seldom elm occurs in this context. One of the few exceptions is the tenth century charter relating to Weston super Mare (Av), referred to in chapter 13. Elm is mentioned under the name *wice*.

The Latin–Anglo-Saxon glossaries are the second category.[1] The earliest goes back to the seventh century and provides the Anglo-Saxon equivalent *elm* for Latin *ulmus*. The later glossaries repeat this entry with the exception of Aelfric's glossary which, as mentioned in the previous chapter, renders *ulmus* as *ulm(treow)*. This is almost certainly a lexicographer's word with no existence outside the world of glossaries. For some reason Aelfric decided to transliterate rather than translate. He did the same for Latin *pinus*, which appears as *pin(treow)*. The other Anglo-Saxon word for elm is *wice*. This appears in some of the later glossaries

as a rendering of Latin *cariscus*, which is usually translated *cwic(beam)*, which became dialectical *quick-beam*, that is rowan (*Sorbus aucuparia*). Clearly the confusion discussed in chapter 4 between *cwic(beam)* meaning rowan, and *wice*, denoting *elm*, goes back well before the Norman Conquest.

The third category is that of the medical texts. These are referred to in chapter 8. They mention elm both as *elm* and *wice*.

Far more copious than the direct evidence of these documents is the indirect evidence of place names for which an Anglo-Saxon origin is certain or probable. These are discussed from the linguistic angle in the last chapter. The detailed evidence that they provide is assembled in chapters 12–14. The significance of the historical and place name evidence cannot however be evaluated without some understanding of the course of events involved in the Anglo-Saxon settlement.[2]

This had begun during the latter part of the Roman occupation. Auxiliary troops of Germanic origin appear to have been settled by the Roman authorities in or near a number of towns in eastern Britain during the fourth century. The massive immigration of Germanic speaking peoples principally took place in the fifth century, after the departure of the Roman armies. The sources of the immigrants extended from Jutland, along the coastal districts of what are now Germany and the Netherlands, to the mouth of the Rhine. Smaller contingents came from as far afield as Sweden and southern Germany. The invaders would mainly have spoken either Old Low German or Old Frisian. What has been described as 'the first certain fact in English history'[3] is that the language of the Anglo-Saxons, when it is first recorded, is related far more closely to Frisian than to German. The inference is that the bulk of the settlers were from the Frisian coastal area. This has an immediate significance in the present context. There appears to have been no elm along the Frisian coast at the time of the invasion. It is certainly the case that neither of the words used for elm in Anglo-Saxon and later times in England had any relation to the word eventually used by Frisian speakers.

The Anglo-Saxon words for elm, *elm* and *wice*, are related respectively to German *elm* and *wieke*. However, the two German words appear not to have similar distributions. In the Jutland peninsula, where Wych Elm occurs, the only term used for elm appears to have been *elm*. At the south of the North German plain, where Wych Elm also occurs, both *elm* and *wieke* were current.

Anglo-Saxon tradition derived the East and Middle Angles, Mercians and Northumbrians from Angeln in the Jutland peninsula.[4] The source of the East, South and West Saxons was placed in Saxony. The Jutes of Kent and the Isle of Wight were attributed with a homeland in northern Jutland. Some archeological backing for this schema can be found from pottery and brooches which can be identified as Anglian, Saxon or Jutish. In a few cases, objects associated with one or other of these groups can be traced back to the postulated Continental homeland.

It has long been apparent, however, that the schema, if true in outline, needs extensive modification in detail. The various Continental groups had become much intermixed before the invasion of Britain and the important role of the Frisians is not even mentioned. The Jutes of Kent, judging from their field systems and laws, had been in, or had had contact with peoples dwelling in, the Rhineland.

The evidence of the earliest Anglo-Saxon cemeteries makes clear that the earliest settlement was in the east of England. The Wash appears to have been a major point of entry. Penetration into the heart of England appears to have been along the Icknield Way to the upper Thames valley, though a separate penetration up the Thames itself is also likely. Much of the cultural differentiation between Anglo-Saxon kingdoms, as for example dialectal differences, appear to have arisen subsequent to settlement, and not before.

This being said, it is possible to assess the significance of the two Anglo-Saxon words for elm. The boundaries of the early Anglo-Saxon kingdoms can, with reasonable assurance, be reconstructed from the boundaries of the medieval dioceses. When the word distributions are mapped against these it is apparent that speakers using *wice* were mainly in the dioceses of Lincoln (Middle Angles and Lindissi), Winchester, Salisbury and Exeter (West Saxons) and Chichester (South Saxons). Speakers using *elm* were in York, Durham and Carlisle (Northumbria), Lichfield (Mercia), Worcester (Hwicce), Hereford (Magonsaete), Norwich (East Anglia) and Rochester and Canterbury (Jutes). The association between the elm terms and medieval dioceses breaks down, however, for the diocese of London (East and Middle Saxons). In London

itself and north-east Essex, *elm* was used. In the rest of Essex *wice* appears to have been current. In the diocese of Lincoln, though *wice* was the usual term, *elm* was current in Leicestershire and north-west Lincolnshire.

The overall pattern is of a great wedge of *wice* speakers separating three regions of *elm* speakers. If it is true that the bulk of the immigrants were Frisian speakers, unfamiliar with elm, then the adoption of the elm words must have been from the German-speaking minority. The word *elm* presents no problem. It was used on the Continent both in Jutland and Saxony. It would be appropriate alike for Angles, Saxons and Jutes. A different situation arises about *wice*. Its use by the West and South Saxons is consonant with its occurrence in Saxony. The Middle Angles and Lindissi of Lincolnshire constitute the anomaly. If the main part of these came from Angeln, they would not have been expected to produce the word *wice*. It seems likely therefore that there were sufficient immigrants from Saxony in the Middle Anglian and Lincolnshire areas to determine the choice of word for elm. In fact, there is independent evidence that the population of Middle Angles was mixed. Whether this happened before or after immigration, or perhaps both, would be difficult to determine.

The implication that can be drawn from the word distribution is that elm was on the ground in each of the four areas that the linguistic map reveals. In the northern area, this indicates the existence of native Wych Elm. In the other three, it provides evidence for the existence of various forms of Field Elm at the end of the Roman occupation. In East Anglia, one can be confident that it was the Narrow-leaved Elm that was encountered. In the other two it would have been either this or the English Elm. The fact that the English Elm has the larger total area of distribution and that this bears no relation, except in Devon, to Anglo-Saxon movements or groupings, suggests strongly that the relative distribution of the two elms in settled areas had become established before the Anglo-Saxon invasion.

The medieval diocesan boundaries need to be considered in a further connection. It seems probable that the earliest stage of Anglo-Saxon settlement was characterized by the occupation of a number of centres or cells, between which the terrain was largely unoccupied. Many of them would have had elm growing in them, derived from before the invasion.

With the expansion into unoccupied land that occurred between the invasion and the Norman Conquest, the expectation would be that, as each immigration cell expanded, it took with it the elm that it first found. Since the medieval diocesan boundaries appear to represent the contact zone between Anglo-Saxon groups from different settlement cells, it could be expected that, when these had different elms, the boundaries between these might coincide with those of the medieval dioceses. A typical example is where the boundary between the medieval dioceses of Lincoln and London runs down eastern Hertfordshire, roughly on the line of the Roman Ermine Street. This represents the boundary between the Middle Angles and the East and Middle Saxons. It should be noticed that this lies some way west of the boundary between the present and previous counties of Hertfordshire and Essex. The elm in the north Hertfordshire part of Middle Anglia was English Elm. This now extends roughly to Ermine Street. The elm found by the East or Middle Saxon settlers in what is now north-west Essex would have been a Narrow-leaved Elm. This now extends westward to Ermine Street, which forms the contact zone between it and the English Elm of the Middle Angles.

Sometimes the distribution of elms seems to reflect a stage in Anglo-Saxon settlement prior to the formation of the various kingdoms. Essex again provides a case in point. Although the east of England in general was the part first settled by Anglo-Saxon immigrants, there are very few sites of the Pagan period in Essex and the London area. These appear to have been avoided by the first settlers and the course of events in these areas in the fourth century is largely unknown. When information comes to hand in the wake of Augustine's mission, what is now Essex is the domain, as its name indicates, of the East Saxons. They are also in possession of London. However, what archeological evidence there is suggests cultural affinities between the Anglo-Saxon settlers in the Thames plains of Essex and those in northern Kent. Also, the dispersed settlement pattern of southern Essex, as compared with the nucleate settlements in northern Essex, has its counterpart in the dispersed settlements in northern Kent. The elm pattern of Essex reveals a strong contrast between the Thames plain east of the Ockendens, with English Elm, and northern Essex with Narrow-leaved Elm and Hybrid Elms. This seems to reflect a stage in Anglo-Saxon settlement prior to the establishment of the East Saxon kingdom.

The medieval diocesan boundaries were far more stable than any civil boundaries, hence their particular value in the present context. They may in fact represent not only the boundaries of Anglo-Saxon units but of still earlier divisions which the Anglo-Saxons inherited. Thus East Anglia appears to correspond fairly closely with the territory of the Iron Age Iceni. Sussex corresponds just as closely with the domain of the Regni.

The medieval dioceses were later divided into archdeaconries whose boundaries seem usually not to be of high antiquity. A possible exception is the Archdeaconry of Middlesex whose strange configuration, discussed in chapter 14, may throw some light on the elusive history of the Middle Saxons.

There is a great difference between the extent of terrain occupied by the earlier Anglo-Saxon immigrants, as defined by the distribution of Pagan cemeteries, and the settled pattern at the time of the Norman Conquest, as revealed by Domesday Book. Assuming that the various forms of Field Elm can only have been taken to their present sites by human agency, they cannot have reached areas first opened up by Anglo-Saxon expansion till this had occurred. The historical evidence surveyed in the next section and presented in detail in chapters 12–14, provides strong evidence that, in the south of England at least, the present pattern of elm distribution had been reached by the Norman Conquest.

The expansion during the Anglo-Saxon period was very largely into fertile but heavy clay lowlands. It is probable that the high frequency of elm in so many of the clay lowlands only goes back to the Anglo-Saxon colonization of these areas.

The principal areas of unoccupied land in the southern half of England before the Norman Conquest were on terrain that presented difficulties to cultivation. Much of this lay in what became the royal forests after the Conquest. These are discussed in the next section.

Some indication of the woodland cover of the earlier stages of Anglo-Saxon settlement can be obtained from the distribution of certain place name elements, in particular initial *wald-* and *wood-*, and final *-field*, *-hanger*, *-holt*, *-hurst*, *-ley*, *-wald* and *-wood*.

The extent of woodland at the time of the

Conquest can be computed from figures given in Domesday Book on the number of pigs that the various woodland areas could feed. As would be expected, the woodland cover had decreased significantly during the Anglo-Saxon era.

A different category of terrain that was little settled in the Anglo-Saxon period was marsh. Its extent can be inferred from the present distribution of land with artificial drainage channels or liable to flood. The detailed evidence provided in chapters 12–14 indicates how very frequently royal forest, heavily wooded terrain and marshland correspond with regions of low frequency of elm in southern England. This is so striking that it is reasonable to suspect that when Field Elm is found today in former royal forest, wooded terrain or marshland areas, it is of recent introduction.

Elm hybridization is considered in chapters 2 and 7. The principal areas where Hybrid Elms abound, central Essex and northern Bedfordshire, are regions settled relatively late in the pre-Conquest period. Consequently, it seems unlikely that the hybridization which produced these elms would have occurred before then. What appears to have happened was that the Anglo-Saxon settlers moving in to these previously unoccupied areas took with them the Narrow-leaved Elms they had encountered on previously farmed land. Wych Elm was presumably present in the uncleared woodland, and massive hybridization between this and the introduced Narrow-leaved Elm seems to have taken place.

One new elm population, that of Cornish Elm, appears to have been introduced in the Anglo-Saxon period. Cornish Elm occurs in sheltered lowland areas in Cornwall and adjoining parts of Devon. It was certainly present there in the sixteenth century. Cornish place names possibly attest its occurrence in Cornwall centuries before this.

4, 127

The overall area of occurrence corresponds very clearly with the county of Cornwall and so with the kingdom of Cornwall which was incorporated into Anglo-Saxon England in the ninth century. Celtic church dedications are frequent in the part of Devon into which Cornish Elm extends, suggesting that this fringe area was in some cultural relationship with Cornwall proper. The working hypothesis was elaborated in the previous chapter that the date of introduction of an elm is likely to be before the latest

time at which its pattern of distribution is significantly related to a human distribution. If this principle be applied here, it indicates that Cornish Elm had been introduced before Cornwall was annexed to Wessex.

The reputed Cornish word for elm, *elaw* or *ula*, differs from all four of the Breton names. For all other of the main trees, the Breton term is clearly derived from the Celtic term used in Cornwall. The simplest explanation is that there was no or little elm in south-west Britain at the time of the immigration into Brittany, which was roughly contemporary with the Anglo-Saxon invasion of eastern England. If no elm were present, the emigrants would most likely have had no word for it. On meeting elm in Brittany, they would have had to coin or borrow a new word for it. So the date of introduction should be after the fifth and before the ninth century.

The source population has been identified. It occupies a relatively small area in western Brittany.[5] There is no doubt as to the close connections between Cornwall and Brittany at the period of putative

Fig. 23. High-frequency distribution of Cornish Elm and of churches dedicated to St Winwaloe and St Budoc. Heavy line: boundary of the main distribution of Cornish Elm; squares: dedications to St Winwaloe; circles: dedications to St Budoc.

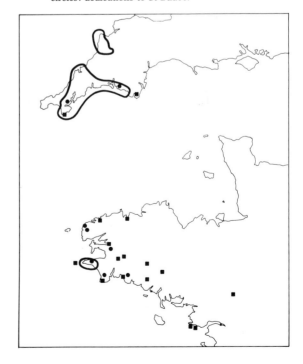

introduction. It is exemplified by the distribution of church dedications to such Breton saints as Budoc and Winwaloe which occur both in Cornwall and the source area in Brittany. The lives of several Celtic saints attribute them with tree-planting activities. Like other elm populations presumed to be derived from France, the Cornish are galled by the mite *Eriophyes campestricola*.

Medieval period

The number of lines of evidence concerning elm multiplies after the Norman Conquest. It is possible that a few trees planted or originating in the Middle Ages survived into the present century. Elms however do not usually live for more than three hundred years. Girth measurements of large trees and ring counts of felled trees made subsequent to the sixteenth century provide some direct evidence of the occurrence of elms in medieval times.

The elm groves in villages in south-west Cambridgeshire can be traced back to the Middle Ages.[6] How much further back they go is unknown. They could even be the remnants of something corresponding to the Scandinavian *löv äng*, described in chapter 3. These elm stands are usually called groves, sometimes spinneys. They are composed almost entirely of Narrow-leaved Elm, often pollarded. The groves are usually at or near the compact nuclear settlements that characterize the parishes in this part of England. They contrast strikingly with the residual native woods, largely of oak and ash. These are usually on the periphery of the parishes, away from habitation. The term grove, in this region, is not applied to these.

The groves survived into the present century, though much diminished. They encountered the severe disapproval of the eighteenth century improvers.[7]

Medieval elm artefacts survive in some quantity. Those that remain in use include floor boards in houses and churches, elm timber framing where this was in vogue, headstocks to bells, and coffins. Other artefacts survive in archeological contexts such as the medieval elm pilework along the London waterfront mentioned in chapter 14.

The medieval place name evidence attesting the presence of elm is copious. It extends to the very important category of field names. Medieval surnames such as atte Elme are particularly useful for the

thirteenth–fourteenth centuries. As surnames become hereditary and families mobile, they obviously attest much less. Surnames appear on a great many categories of medieval document. The most useful, for present purposes, are the hundred rolls and the various lay subsidy rolls.

One of the most valuable types of document for locating medieval elms is the manorial court roll. It is explained in chapter 9 that elm is one of the three trees, oak and ash being the other two, that is legally recognized everywhere as timber. The proprietor of timber is the landlord, and his licence is normally required before a tenant may cut it. The principal tenure in rural areas till recently was copyhold, the tenants holding their land from the lord of the manor by virtue of an entry in the manorial court roll. Except when there was an overriding custom to the contrary, the copyholder could not fell timber, including elm, either on his own land or elsewhere. Unlicensed felling was one of the commonest infringements of the rights of the lord of the manor and is recorded whenever reported in the manorial courts. Since the kind of tree felled is normally specified, a most useful source for the kinds of tree growing in each manor is to hand. It is normally entered in some such form as:

> Henricus Cage per servientes suos loppavit ij arbores vocatas wyches in communa juxta portam suam sine licencia.[8]
>
> *Henry Cage's servants, at his behest, lopped two trees called wyches on the common by his gate without licence.*

Records of the various borough courts are a parallel source of information for the countryside around towns. They vary greatly in the amount of relevant detail they record.

Churchwardens' accounts are another major source of information on local trees though those that survive are mainly sixteenth century and later. Purchase of timber for repairs, and planting, felling or trimming churchyard trees are the sorts of circumstances likely to be recorded. Some of the early churchwardens' accounts are in Latin but they soon went over to the vernacular. The spelling characteristically makes no concessions to whatever orthographic conventions might exist outside the parish. Consequently, they are an invaluable source for local tree names as locally pronounced.

Sometimes bound up with churchwardens'

accounts but more often on separate sheets of paper are the descriptions of perambulation routes taken when the parish boundaries were walked. Other geographical units, such as manors and forests might also be perambulated. Such perambulations as have survived are more likely to be postmedieval than medieval, but a number of medieval ones are extant. These records specify the various landmarks along the boundary, especially where it changes direction. In the north of England, the landmarks tend to be inanimate, streams, hill tops, rocks, cairns and the like. In the south, trees are frequently cited, sometimes several hundred for a single parish. If elm is present along a boundary, it is likely to be picked up. However, the observation made in respect of the boundaries in the Anglo-Saxon land charters is relevant too for the later perambulation routes. The boundaries tend to cover terrain at the furthest distance from the parish centre. They will quite probably run across wooded terrain or across country reclaimed relatively recently from the wild. Their trees, therefore, are more likely to be those of the original flora than the introduced elms that occur near the village centres of parishes in southern England.

Monastic and lay chartularies, estate rentals, estate surveys and leases all occasionally yield an elm. In general, however, they are unrewarding to peruse. Care has also to be taken lest legal forms, commonly drawn up from a precedent, be interpreted as necessarily biologically significant. An injunction in a lease that the lessee plant oak, ash and elm may not mean that oak, ash and elm grew locally. It may indicate nothing more than the existence of these trees in the precedent which the estate lawyers made do for all circumstances.

Occasionally surveys devoted specifically to trees are encountered, though not so frequently in medieval times as later. If these distinguish between tree species, as some do, they are of great use. Building accounts also survive from the Middle Ages onwards, especially for royal or monastic buildings. Sometimes they particularize the species of timber used either for construction or as scaffolding. These are often of local provenance.

No medieval records exceed the various bridge-wardens' accounts in usefulness for present purposes. Such accounts are very few. Since, however, in many areas, bridge piles had come to be made exclusively of elm in the medieval period, the value of the accounts is very great. They supply detailed information on the sources of elm in the catchment area of the bridgewardens. The accounts for London and Rochester bridges are utilized in chapter 14.

Elm features seldom in the records of the High Courts. The *de banco* rolls occasionally specify elm in actions for waste of timber.[9] The *coram rege* rolls also have some elms. Sometimes they particularize the timber of offensive weapons.[10] Coroners' rolls may also produce information on the timber of an offensive weapon or the precise kind of tree whose fall occasions a death.

Extensive records remain of the proceedings of the courts of the royal forests. These tracts were often woodland, though not necessarily. On infertile soils, tree cover could be light, and on high ground, there might be no trees at all. The royal forests in the Middle Ages were managed with hunting as the prime consideration. They were under forest law and had their own courts. Information about trees is mainly to be found in the presentments for offences against *vert*, the general designation of trees and shrubs growing within the forest. The kind of tree is normally specified in cases of unlawful felling or lopping. There is, however, as pointed out in the previous section, a strong inverse correlation between elm and royal forest. So, the records of the forest courts are amongst the least likely places to come across mentions of elm.

The boundaries of the royal forests fluctuated during the medieval period. They were greatly enlarged under the early Plantagenet kings but after the Forest Charter of 1217 were cut back to what they were supposed to have been before the reign of Henry II. Thereafter, changes were relatively minor. The Duchy of Lancaster became a largely self-governing separate entity in the Middle Ages. Its forests had much the same status as royal forest. Sometimes a royal forest would be granted to a subject and then, technically, ceased to be forest. Some tracts were referred to as forests although they never had the status of royal forest. Such were Charnwood Forest (Le) and the Forest of Arden (Wa).

Modern period

The opening of the sixteenth century is a convenient starting point for the modern era. Many elms standing in the present century would have been planted over 200 years earlier. The number of surviving elm artefacts is naturally much larger than for the Middle Ages.

One new artefact, the elm water main, survives underground in many towns from seventeenth–eighteenth century installations. The towns that laid elm water mains were mainly in the south of England

Fig. 24. High-frequency distribution of English Elm in England, and the locations of towns known to have had elm water mains. Horizontal shading: English Elm; dots: towns with elm water mains.

and within the main area of distribution of English Elm which supplied the timber utilized. Some northern towns, such as Hull and York, also had elm pipes, and it is significant that these two, and perhaps others also, had detached populations of English Elm nearby which probably provided the timber. It is not at all likely that northern waterwork companies deferred main laying till English Elm had been planted and had matured sufficiently to provide boles of sufficient diameter for boring. More likely, mains were first laid with locally

24

available inferior timbers such as oak, alder or even imported deal. Meanwhile, the planting of English Elm was probably arranged, and as it matured it would be used to replace such earlier pipes as would have rotted. If this is what happened, the distribution of English Elm in the north of England might well have been largely determined by the distribution of towns that wanted elm mains. Certainly there is a correlation between the two distributions.

Most of the types of document which featured elm in the Middle Ages continued to do so in the Modern period, or at least the earlier centuries of it. The parliamentary surveys of royal and ecclesiastical property made during the Commonwealth are specially useful. They often particularize the kinds of trees on these properties. This was to expedite the quick realization of the exploitable value of confiscated estates by felling and selling as much standing timber as possible.

Later estate records and surveys survive in great quantity. Their value for present purposes is diverse. When an owner or his agent took an interest in timber, their value can be considerable. Wood books and records of timber sales can be useful but are liable not to specify the kind of tree. Those that survive tend to be rather late. Records of quarter sessions are a new occasional source of information. Elms may appear for such offences as obstructing highways.

Many commonplace books and diaries have survived. When their compilers had an eye for trees, which was not often, they contribute considerably to knowledge of elm distribution. A type of record that seldom survives is the nurseryman's order book. An example that has survived provides useful information on the sources of elms in southern Cambridgeshire and Bedfordshire.[11]

There is virtually no type of document that may not throw up an elm from time to time. Even marginal notes in parish registers and jottings in printed books sometimes lead to an elm.[12]

Records of remarkable storms, whether in manuscript or printed, are always worth perusing. Trees that have been blown over are nearly always subject for remark and the kind of tree is frequently recorded.

Maps are often useful as a source of information. Best of all, but very rare, are large-scale estate plans in which each kind of tree is marked by an appropriate symbol. The 6 inch to 1 mile Ordnance Survey maps, especially the first edition, often particularize what kind of tree a free-standing specimen is. This is most often done for trees on parish boundaries.

Paintings and drawings might be thought to be a major source of information on trees in landscape. This is not so. One of the findings of chapter 11 is that it is exceptional for an artist to represent a tree in such a way that it can be identified. There are exceptions and these are valuable. A few of the early topographical artists represent trees adequately and so do some of the least esteemed Victorian landscape painters and water colourists. Sometimes, an elm can be identified by the close trimming, mentioned in chapter 8, that was used to produce boles suitable for water mains. This feature was often grasped by artists for whom the niceties of foliage texture were an unplumbed mystery.

Printed books are a new source of information. However, botanical books such as floras and the herbals that preceded them, are not usually the best sources. Botanists tend to be 'intent on herbs'[13] and trees are often underlooked. Also, botanists over the years have been liable to be more interested in quoting their predecessors and in recondite nomenclatural exercises than in the trees around them. Books on forestry and agriculture can be more informative. The *General views of agriculture* of the eighteenth century are of particular use. Often a general writer can be more informative than a specialist. More can be gleaned about elms from Francis Kilvert's diaries[14] than from many floras by his contemporaries.

Three kinds of elm definitely appeared in the English landscape in the Modern period. One other, Lock's Elm probably did so.

The earliest of the three was the Dutch Elm, not to be confused with the *hollandse iep* of the Netherlands. It is well documented. A consistent tradition associates its introduction with the accession of the Dutch monarch William III to the English throne. Only one year afterwards, 12 Dutch Elms were planted at the Meads in Winchester (Ha).[15] The first scientific writer to mention the Dutch Elm was Queen Mary II's botanist, L. Plukenet, in 1696.[16] He merely cites it by name as *Ulmus major Hollandica angustis magis acuminatis samarris folio latissimo scabro*. Three years later, Henry Wise, the Brompton nurseryman, had sufficient stock to be able to send a thousand

7

Dutch Elms to Melbourne (Db).[17] Alexander Pope discussed its merits in 1719.[18] Writing in 1748, P. Miller says explicitly that the Dutch Elm was introduced at the same time as the Revolution which put the Dutch king William III on the English throne.[19] It had reached Scotland from London by 1723.[20]

The Dutch Elm is principally to be found in estate plantations of the eighteenth century. It also occurs, mixed with sycamore, in exposed places near the sea in the west of England. Its source area has been located.[21] It is in Picardy where the elm was known in earlier times as *ypereau*. Strangely, the first scientific report of its occurrence there under this name was in Lyte's sixteenth century English herbal, mentioned further in chapter 7. Cotgrave's French–English dictionary, published in 1611,[22] states that *ypereau* is a large-leaved elm found in northern France, in contrast, presumably, to the small-leaved Field Elm that also occurs here. The word is first recorded in 1432 from Béthune (Pas-de-Calais).[23] It remained in general use in the departments of Pas-de-Calais and Somme, with the same meaning, till the nineteenth century.

It is a rare tree in the Netherlands today, where the elm most commonly planted is a quite different Hybrid Elm, the *hollandse iep*. The Dutch Elm, as understood in England, was present at Het Loo in 1699 and was noticed by an English visitor.[24]

5   A note of uncertainty attaches to Lock's Elm. It lacks any documentation before the present century. Its pattern of distribution is strange. It is present in quantity in a few villages along the upper course of the River Witham (Li). Elsewhere its distribution is very scattered over the whole of the north Midlands. Very rarely it occurs outside these limits as in Humberside and Cambridgeshire. When it does occur, it is often as a single tree, or perhaps a single hedgerow or around a single field. Many of the locations are alongside a road or in a hedge running into a road.

Lock's Elm is not known on the Continent and one is driven to conclude that it must have arisen as a sport from a local Narrow-leaved Elm. This could have occurred almost any time between prehistoric times and the beginning of the eighteenth century.

Age-and-area considerations suggest that, outside the Witham valley, this elm is of relatively late appearance, otherwise it would have been expected to spread further and more evenly. How it has spread is not known. One line of enquiry, however, seems worth pursuing. Scattered elm trees of types not known elsewhere in the locality occur in Wales, sometimes along cattle drovers' routes. It is also stated that farmers planted pines to indicate overnight pasture.[25] It seems worth investigating whether the distribution of Lock's Elm is significantly associated with drovers' routes. If so, its odd distribution could be accounted for. Drovers' routes have not been mapped comprehensively since they are largely known from such few drovers' account books as have chanced to survive. The north Midlands is not well covered.

The origin of the Huntingdon Elm is the best   8 documented of all. It was introduced by the firm of Wood and Ingram (now Laxton Bros) of Brampton (Ca). J. Ingram, of this firm, writing in 1847, says that it had been raised over one hundred years earlier from seed from trees of 'true English or Field Elm (*Ulmus campestris*)' in Hinchingbrook Park.[26] The trees were still extant when he wrote. They would probably be the Narrow-leaved Elm of south-western Cambridgeshire. The pollen parent must have been Wych Elm, but the pollen cloud could have originated from trees a considerable distance away. Bedfordshire would be a likely source.

The distribution of the elm has been almost entirely through nurserymen. The Huntingdon Elm is especially characteristic of Cambridge college gardens. It was used to replace English Elm in the Broad Walk in Christ Church Meadow, Oxford.

Last of the recently introduced elms considered here is the Guernsey Elm. Occasionally it occurs in   6 hedgerows. Usually it has been planted as a roadside or park tree, situations for which its tidy, compact habit recommends it. It is first on record as an elm of peculiar habit confined to Guernsey.[27] Exactly who brought it to England and when are uncertain. It is listed without description in the 1836 catalogue of Loddiges of Hackney[28] and described inadequately and inaccurately, by J. C. Loudon in 1838, from Loddiges material.[29]

The ultimate source of the Guernsey Elm is probably the Cotentin peninsula of Normandy or northern Brittany, where elms of similar leaf form occur. Whether the distinctive habit first developed on the mainland or in Guernsey is uncertain.

The agents chiefly responsible for distributing it in the earlier years of the nineteenth century were two Hampshire nurseries: Rogers of Red Lodge Nursery, Eastleigh, and Hillier of Winchester. In neither case have early business records survived so that tracing the history of this elm in detail is difficult. W. H. Rogers, in 1869,[30] was one of the first people to give an informative horticultural description of this elm.

The Guernsey Elm is extensively planted around Osborne (IW). Possibly it took the fancy of the Prince Consort. He was actively concerned in tree planting on the royal estates and is known to have planted trees with his own hand at Osborne.[31] The prim behaviour of this elm would presumably have recommended it in the royal circle of the time.

Though often miscalled the Jersey elm, the Guernsey Elm is rare in Jersey. When it does occur there it can usually be traced to relatively late introduction from Guernsey. The Guernsey Elm is also known as the Wheatley elm, supposedly after one of the various English villages so called. The nature of the connection is not now known, nor indeed which Wheatley is supposed to have been involved. There appears to have been much confusion in the minds of nurserymen between the Guernsey Elm and the Cornish Elm. The latter is rarely planted outside Cornwall and a number of nineteenth century references to this elm probably relate to the Guernsey Elm.

The history of elm so far uncovered has been largely one of introduction by man from Continental Europe to England, followed by farther spread by man in England itself. It remains briefly to describe the reverse process.

One kind of elm, the English Elm, its origin long forgotten, came to be regarded as the quintessence of the English landscape. It was thus admired by foreign visitors. It was taken abroad by English emigrants to other temperate climes to remind them of home.

The first foreign visitor to fall under its spell was Philip II of Spain. An early tradition, attested in the seventeenth and eighteenth centuries by several writers, asserts that the much admired elms in the royal garden at Aranjuez, south of Madrid, were introduced from England at the behest of Philip II.[32] The fact that similar elms were to hand in Spain itself does not upset this story. Spain is a large country and communications were indifferent.

The overseas landscape that accepted English Elm most gratefully was that of New England, around Boston.[33] The English Elm growing there is of frequent note by nineteenth century writers. It is carefully distinguished from and contrasted with native *Ulmus americana*, as shown in chapter 10. It was even planted in the heart of New York in 5th Avenue.[34] English Elm was introduced similarly into Australia and New Zealand.

6
Associated organisms

A large number of organisms, plant and animal, are associated with elm. The nature of the association is diverse. Most often the elm is the source of nutrients. It may perform other roles, for instance, providing a site for nesting for birds.

The complete range of organisms associated in any way with elm is too large to be covered here. Much association is only casual. What will be attempted is to describe the principal organisms whose association with elm is obligatory, that is, those which cannot live or complete a life cycle unless they have access to elm. In this book, the terms associated organism, or associate, are restricted to cases where the association is of this obligatory nature.

These organisms are of great value for unravelling the evolution and later history of the elm. Since they have evolved on and with the elms that have supported them, it may prove possible to demonstrate that two elms are related because they carry the same or two different but closely related associated organisms. Some associates have limited power of distributing themselves. If an elm population is found on which such organisms occur, the associates are likely to be derived from some other elm population on which they also occur.[1]

There are degrees of obligatory association. Some obligatory associates can live on any elm. Others are more exacting and can live only on certain kinds of elm. Some obligatory associates show the phenomenon of physiological specialization. They exist in a number of different strains, identical in appearance, but differing in the kind of elm they require to complete their life cycle.

Some organisms, obligatorily associated with elm at one stage in their life cycle, move over to some other plant, with which they may also be obligatorily associated, at another. The latter is the so-called alternate host to elm. Clearly the survival of such organisms depends on the co-existence of both host plants.

Two particular kinds of associate require special attention: leaf miners and gall formers. Leaf miners are insects whose larvae live within the tissues of the leaf. They do not stimulate the leaf to grow in any abnormal way. Their presence is indicated by the trace of their excavation or mine, apparent when the leaf is held up against the light, and often by some surface discoloration. The configurations of mines are highly distinctive and the organism responsible can usually be identified to its species. Since mines remain after the miner has departed, they provide useful distributional data on organisms which are themselves inconspicuous.

Three types of mine are distinguishable. Linear mines are long narrow tunnels in the leaf tissues. 37 Blotch mines are excavations which spread out in all 33 directions. Tent mines are those in which the leaf 39 surface becomes gathered into a number of parallel pleats. A key to the elm leaf mines occurring in England is given in Appendix 2.

Galls are abnormal plant growths brought about by an associated organism. The abnormality may be very slight, such as a dimple on a leaf or a slight 32 rolling of the leaf margin. It may be a very striking departure from normality, as in the large pouches of plant tissue caused by some of the elm aphids. A wide range of organisms, bacteria, fungi, nematodes and insects, may cause gall formation. Like the leaf mines, the galls remain after the organism responsible has departed. They also provide valuable distributional data on organisms not otherwise likely to be detected. A key to galls on elms in England is given in Appendix 3.

It has been assumed till now that the associated organism depends directly on elm, that is, it is a primary associate. This need not be the case. The associated organism may be directly dependent on some other organism which is directly associated with elm. For instance, there are nematode parasites of elm beetles boring bark. These are secondary associates. There may even be tertiary or high-order degrees of association. Secondary and higher-order associates may throw light on the evolution and later history of the elms on which they ultimately depend in the same way as primary associates.

The general supposition behind the detailed accounts of particular groups of elm associates that are given in the following sections is that, if two kinds of elm carry the same or closely related associates, they are likely to be themselves closely related.[1] There are some exceptions to this generalization, as for example, when an elm is introduced by man to a new environment and the elm associates, if any, in the new environment, transfer to the introduced elm.

Viruses, mycoplasmas and bacteria

Viruses are infective entities small enough to pass through bacterial filters. They only multiply within a host cell. Several infect elms, but only one, the elm zonate canker virus of America, has not been shown to infect other plants too.

Mycoplasmas are the smallest known cellular organisms. They differ from bacteria in having no cell wall, only a bounding membrane. No mycoplasma confined to elm is known in Europe. In North America, a mycoplasma is responsible for the serious elm disease phloem necrosis. The bark of trees so affected loosens, and the trees quickly die. The mycoplasma is transmitted from tree to tree by the leaf hopper *Scaphoideus luteolus*, mentioned further below.

Bacterial associates of elm might be expected to occur either as primary associates or as pathogens of primary associates. In fact, very few bacteria directly affecting elm have been recorded. Various trees, but most particularly elm, are liable to the condition known as wetwood. Excessive sap accumulates in the trunk and oozes out under pressure from fissures in the bark. The external trickle is called a slime flux and normally ferments under the action of colonizing yeasts. The stinking brew, known poetically as *marmelade ulcéreuse* by the French, attracts various flies and is further discussed in the section of this chapter concerned with these. The organism chiefly responsible for wetwood is the bacterium *Erwinia nimipressuralis*.[2] It was first described for elm but has a wide host range.

A little-known bacterium, *Pseudomonas lignicola*, causes black stripes in elm wood.[3] This was first described in the Netherlands but is believed to occur elsewhere in north-western Europe, including Britain.

Two bacteria have been isolated from elm bark beetles in France: *Aerobacter scolyti* and *Escherichia klebsiellaeformis*. These can cause the death of the beetles through septicemia but they may be also present as part of the normal bacterial flora of the alimentary system of the beetles. Next to nothing is known about the distribution of these bacteria nor whether they have a wider host range.[4]

Fungi

A large number of fungi occur on elm. Many are imperfectly known and there is often inadequate information on host ranges. Attention is confined here to fungi that are relatively well known and whose host ranges can be stated with some degree of reliability.

Fungal associates of trees can be conveniently divided into three classes: those principally affecting the roots, those living on the wood or bark, and those occurring on the leaves.

The first category, the root fungi, includes the mycorrhizal fungi. These are organisms, often toadstools, whose mycelial threads enter the superficial cells of tree roots or form a network around them. The fungi are parasitic on the roots from which they obtain nutrients but there is benefit to the tree also since phosphorus absorption from the soil is enhanced. Some species of mycorrhizal toadstool are exclusive to particular trees. A number of toadstools forming mycorrhiza with elm are known,[5] but no toadstool appears to be confined to elm in the British Isles.

The fungi living on wood or bark can in turn be subdivided into three groups. There are those that grow on the living tree, their mycelia either penetrating deeply into the wood and causing a rot, often killing the tree, or growing only in the bark and relatively harmless. There are those that cause wilts by blocking the passage of water through the vessels of the wood. Finally there are the fungi which cause decay of wood in dead trunks or branches.

The principal fungi in the first group are inconspicuous till they form fructifications. These are toadstools or bracket-like structures growing out from the trunk or major branches and may reach considerable size.

Fomes ulmarius causes a butt rot which may kill

Fig. 25. *Fomes ulmarius* (× ⅔)

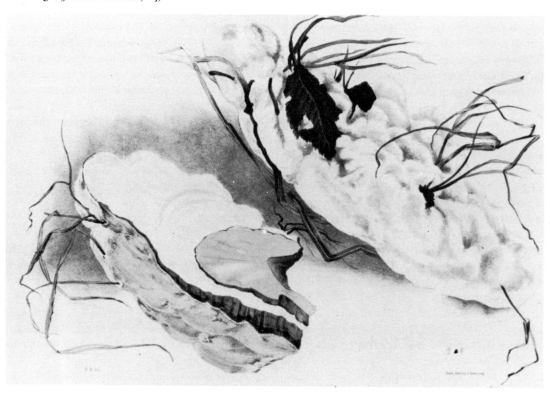

Fig. 26. *Lyophyllum ulmarium* (× ⅓).

the entire tree. The fructifications are hard, whitish, corky, somewhat hoof-shaped brackets with the spores produced on the underside. This fungus is of world wide distribution. In England it is confined to elm. Abroad it may be found on other trees. 25

Three gill fungi are mainly found on elm trunks or branches in the British Isles. The first is *Lyophyllum ulmarium*. In this the fructification is a stalked toadstool up to 20 cm across. The cap is grey. The stalk is markedly out of centre. The next is *Pleurotus cornucopiae*. The distinguishing character of this species is the branched stem, each branch carrying a funnel-shaped cap. The fungus is white when fresh, the whole fructification rather resembling a candelabrum. This species is closely related to *P. ostreatus* which is a common fungus on tree trunks, especially beech. Last, in *Rhodotus palmatus*, the fructification is in the form of an aggregation of overlapping caps, up to 12 cm across. The upper surfaces are orange brown. 26 27 28

The principal fungus causing wilting is *Cerato-cystis ulmi*, the cause of Dutch elm disease.[6] This name does not indicate the place of origin of the disease, but merely where it was first thoroughly investigated.[7]

Since this fungus can be so lethal to mature trees, it has been exhaustively studied.

The fungus is inconspicuous growing in the water-conducting vessels of the wood. It produces a toxin, ceratoulmin, that acts on the parenchyma cells next to the vessels. These cells, in turn cause blockage of the vessels, partly by sending cellular outgrowths, tyloses, into their cavities and partly by secreting gum. The blocked vessels appear as a ring of black dots in a cross section of the wood. Since the passage of water from the roots to the leafy shoots is blocked, the latter wilt, usually in late summer. A tree may survive a first attack or the overground parts only may be killed. If the fungus goes down into the roots, which only happens in some cases, the whole tree may be exterminated.

The spores of the fungus are distributed by bark-boring beetle. In the British Isles, the beetles responsible are *Scolytus scolytus* and *S. multistriatus*.

Fig. 27. *Pleurotus cornucopiae* (× ⅔).

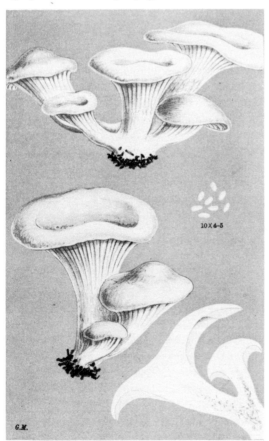

Other *Scolytus* species, in addition, spread the fungus on the European mainland. In North America, the fungus is spread by the native bark beetle *Hylurgopinus opaculus* and also by *S. multistriatus* which has been introduced from Europe. These beetles are discussed in more detail below.

There are numerous uncertainties about the fungus. It was unknown till 1918. Where it came from is a mystery. It has been supposed that it might be East Asia since *Ulmus pumila*, which is resistant to it, is native there. It occurs now throughout Europe. The first European epidemic of Dutch elm disease reached a peak in the 1930s. It then declined in severity and ceased to be regarded as a serious problem. Meanwhile, around 1930, the fungus had been introduced into North America where it started a serious epidemic which is still in progress. The second European epidemic began in the early 1970s, following the importation of diseased elm logs from Canada. It has proved far more serious than the first. In England, mature elms have been eliminated from many parts of the country.

It is now known that there are three races of the fungus. One is a non-aggressive strain that only causes serious damage occasionally. The other two are aggressive. One is the strain responsible for the second

Fig. 28. *Rhodotus palmatus* (× ½).

epidemic in England. The other, which appears to have originated either in Asia or eastern Europe, has appeared, by means unexplained, in south-west Ireland.

Both the non-aggressive strain and the first mentioned of the aggressive strains appear to have been present in the first European epidemic. Both were carried to North America. In Europe, the aggressive strain seems to have died out. The reason is not known. Possibly it is associated with the optimum temperature for growth, which is some 10 °C higher for the aggressive strain. In North America both strains maintained themselves, and around 1970 the aggressive strain was brought back to Europe with catastrophic results. Why it has proved more virulent than in the first epidemic is again unknown. It remains to be seen whether it will disappear again or if its virulence will attenuate.

No British elm is immune against Dutch elm disease, but there are degrees of susceptibility. English Elm is very susceptible. In elmscapes with English Elm mixed with Narrow-leaved Elm or Huntingdon Elm, it has been noticeable that the English Elm has been far more severely affected. This may, in part, be an indirect consequence of its earlier flushing in spring.

Wilt of elm trees is occasionally caused by the fungus *Verticillium alboatrum*. This fungus is not exclusive to elms or even to trees and the damage it causes is only slight. Many fungi grow on dead elm branches. Five appear to be exclusive to elm. These are all, like the agent of Dutch elm disease, ascomycetes, that is, they reproduce by spores (ascospores) which develop inside a special cell, the ascus. The asci are produced in small, flask-shaped bodies of diverse structure. The term perithecium will be used here to cover all of these. The ascospores provide the major distinguishing character of the five fungi. In the medieval jargon of the fungus taxonomist, the stage in the life history of an ascomycete fungus that produces ascospores is called the perfect state. Many ascomycetes produce other spores, called conidiospores, from the ends of special mycelial threads. This stage is called an imperfect state and there may be more than one such. It is a convention of fungal botanical nomenclature that the Latin name of the perfect state is also that of the fungus as a whole. However, the imperfect state may also have its own Latin name, and if there are several imperfect states, each of these may receive a distinguishing Latin name.

In *Crystosporella hypodermia*,[8] the first of the five species, the ascospores are single, colourless, pointed cells. The perithecia are arranged in small clusters. The conidiospores of the imperfect state are single, colourless, sausage-shaped cells. This species is recorded from both Europe and North America.

Eutypella stellulata, the next species, produces single-celled, pale brown, sausage-shaped ascospores. The perithecia are densely clustered. This species occurs both in Europe and North America. Other *Eutypella* species are known on elm. *E. exigua* occurs in Europe and North America. *E. innumerabilis*, *E. longirostris*, *E. scoparia* and *E. tumida* are present in North America.

Quaternaria dissepta[9] has similar shaped ascospores to *E. stellulata* though they are larger. The perithecia are in groups of from two to six. The spores of the imperfect state are single, narrow cells. This species is known both from Europe and North America.

Massaria foedens has distinctive brown ascospores with two cross walls at one end, the whole spore being enclosed in a thick colourless coat. The perithecia are closely set. This species is found throughout Europe. Another *Massaria* species on elm, *M. ulmi*,[10] occurs in central Europe.

The last of the five is *Cucurbitaria naucosa*, distinguished by its brown spores having both transverse and longitudinal cross walls. The perithecia are densely aggregated. The spores of the imperfect state are brown with a single cross wall. This species is known from Europe and North America. Other *Cucurbitaria* species on elm branches occur in Europe, namely *C. ulmicola* in central Europe and *C. ulmea* in northern Europe. The imperfect state of *C. naucosa* is called *Diplodia melaena*. It is often the case that the imperfect state of a fungus occurs much more frequently in nature than the perfect state. In some cases the perfect state has only been obtained under laboratory conditions. There are also many fungi which have an imperfect state but no known perfect state. It is presumed that these have arisen in the course of evolution from ancestors with a perfect state and that this stage in the life history has been suppressed. When the characteristics of the imperfect state corresponding to a particular category of perfect state is known, it may be possible to infer, though with a considerable margin of uncertainty, what sort of perfect state has given rise to an imperfect state for which no perfect state is known. A number of *Diplodia* species are known to be imperfect states of *Cucurbitaria* species, as *D. melaena* is of *C. naucosa*. It is therefore possible that some of the other elm-inhabited *Diplodia* species, such as *D. tephrostoma* of Europe and *D. clavispora* of North America, originated in the course of evolution from extinct *Cucurbitaria* species occurring on elm.

The most conspicuous of the British fungi on elm leaves is *Systremma ulmi*.[11] This becomes apparent in late summer as black pustules, 2–3 mm across, on the upper surface of the leaf. The numerous perithecia are embedded in the pustule. This leaf-spot fungus is a European species

In North America there is a different elm leaf spot. The fungus responsible is *Lambro ulmea*. A closely related species, *L. oharana*, causes leaf spotting on elms in the Himalayas and the Far East. No *Lambro* infecting elm is known in Europe. *L. ulmea*, however, has an imperfect state, *Cylindrosporella ulmea*. The small colourless spores of this state are produced in brown spots on the lower side of the leaf. In both Europe and North America, there is a closely similar imperfect state, *C. inconspicua* for which no perfect state is known. Possibly *C. inconspicua* was derived from an elm *Lambro* which is now extinct.

An elm leaf fungus which has retained its perfect state but only rarely produces it is *Mycosphaerella ulmi*. In the perfect state of this fungus, the ascospores are produced in minute isolated perithecia. This state is seldom seen. The imperfect state, *Septogloeum ulmi*, is very common. It appears as a sprinkling of spore masses, rather like miniature worm casts, on the leaf surface. The spores are narrow and usually have three cross walls. The fungus is of world-wide distribution. Another *Septogloeum*, *S. profusum*, of which a perfect state is not known, occurs on elm in North America. This may have been derived from an elm *Mycosphaerella*. There are some other elm leaf fungi with spores similar to those of *S. ulmi* and which may be connected with undiscovered or extinct elm *Mycosphaerella* species. These are *Cylindrosporium tenuisporium* in North America, *Phleospora ulmea* in Greece and *Septoria jokokawai* in Japan. Another but rare elm leaf fungus is *Mycosphaerella oedema* which is widely distributed in Europe.

A very different leaf spot of elm is caused by the fungus *Taphrina ulmi*. Unlike the leaf fungi just mentioned, *T. ulmi* is usually found, not on the leaves of mature short shoots, but on those of suckers or in regularly trimmed hedges. The mycelial threads of the fungus ramify underneath the leaf cuticle. The fungus first appears as a patch of whitish bloom. This is the spore-bearing area. Later the bloom disappears and the patch turns brown and is no longer identifiable with certainty. The fungus occurs throughout Europe and in North America.

The powdery mildews differ from the leaf fungi so far described in that the mycelial threads are on the surface of the plant they infect rather than internal. No elm mildews occur in the British Isles, probably because the summer temperatures are too low. The mildew *Uncinula clandestina*, which forms grey patches on the upper surfaces of elm leaves, is frequently encountered in southern Europe. The ascospores are produced in black globular structures visible under a hand lens. These are ornamented with characteristic appendages like miniature walking sticks. Elm mildews of the genus *Uncinula* are widespread throughout the Northern Hemisphere. In the Far East the species is *U. kenjiana*. In North America it is *U. macrospora*.

There is another group of fungi, allocated to the genus *Phyllosticta*, which is also responsible for the development of spots on elm leaves. These fungi have no known perfect state and produce spores in small black bodies scattered over the spots. Various species peculiar to elm have been distinguished but it is completely uncertain how distinct these are from one another.[12] The elm *Phyllosticta* species occur throughout the Northern Hemisphere, including Great Britain.

On the whole, the elm fungi do not throw as much light on the evolution and history of the elms as do various other groups of associated organisms. This is partly because they appear not to be restricted as a rule to particular kinds of elm. It is also partly because they tend to be of wide distribution and the spores by which they are disseminated can travel great distances. Distribution patterns consequently do not reveal much about the elms infected by the fungi. It

Fig. 29. *Taphrina ulmi* (× 900).

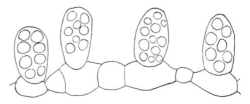

is noteworthy how many elm fungi are present throughout the Northern Hemisphere. This suggests that they evolved early in the history of the elm, and thereafter changed very little.

Other plants

This section is largely dismissive. A number of plants other than the fungi proper are most frequently found associated with elm. The relationships however are never obligatory.

The lichens will be considered first. These are fungi growing in such intimate connection with algae that the association is treated as a plant entity in its own right. Since the alga supplies the fungus with the organic material it cannot synthesize for itself, the lichen does not withdraw any from the substrate to which it is attached. The two principal lichen substrates are rocks and tree trunks. Though lichens attached to tree trunks do not extract organic material from their substrate, the nature of the latter does influence what kinds of lichen are present. The principal determining factor appears to be the degree of acidity of the bark. Elm bark is alkaline. Oak bark, in contrast, is acid.

There are a number of lichens most often found on elm but none are exclusive to it. The other British trees most utilized by the elm lichens are ash and sycamore. Most elm lichens, on occasion, occur on rock surfaces. The lichens favouring elm are mostly grey or greenish-grey encrusting species with circular fruiting bodies. The latter vary in colour. They are orange in *Bacidia rubella*, *Caloplaca luteoalba* and *C. ulcerosa*, *Gyalecta carneolutea*, *G. flotowii* and *G. truncigena*, brown in *Bacidia chlorococca*, and blackish in *Bacidia incompta*, *Buellia punctata* and *Catillaria prasina*. *Opegrapha lichenoides* and *O. vermicillifera* are grey encrusting species with slit-like fruiting bodies. The surface of the second species is covered by a distinctive white bloom. Three of the lichens favouring elm are much-lobed foliose species. *Candelaria concolor* is composed of very small lobes and has yellow fruiting bodies. In *Parmelia borreri*, the lobes are larger, grey above and black below. They seldom fruit. In *Collema fragrans* the dark green lobes, when damp, have a gelatinous consistency.

Mosses, liverworts and ferns are frequently to be found growing on the trunks and larger branches of trees, including elm, especially in damp situations. These types of plant synthesize their own organic material and draw none from any plant to which they might be attached. Slight differences in preference for particular trees have been noted for some mosses and liverworts. These are so slight and so greatly overridden by such major determining factors as light and humidity, that they can be ignored.

A few higher plants parasitize trees, either withdrawing organic material from their roots or, as with the mistletoes, obtaining it from aerial branches. No higher-plant parasite is known to depend exclusively on elm. One root parasite, *Lathraea squamaria*, the toothwort, has elm as one of its preferred hosts. Its whitish shoots bearing dingy purple flowers are found from time to time growing at the foot of elm boles. It occurs throughout Europe and into western Asia.

There appears to be some antagonistic mechanism preventing the infection of elm by the common west European mistletoe, *Viscum album*.[13] Occasionally the mechanism breaks down and mistletoe bunches do develop on elm branches. Elms so infected have been noted at Eastnor (HW), Saltford (So), Wrest Park (Bd) and Yazor (HW).[14]

Protozoa and nematodes

The associated organisms discussed in this and the following sections belong to the animal kingdom. Those in the present section are all secondary associates. They all occur in insects that are primary elm associates.

The protozoa are single-cell organisms, many of which parasitize animals of more complex structure. Those of concern here are parasites of elm bark beetle which are considered below. Two species are known. *Nosema scolyti* infects the blood cells and excretory organs of the beetles.[15] *Stempellia scolyti* goes for the gut.[16] The host range at present known comprises the elm beetles *Scolytus multistriatus* and *S. scolytus*, which occur in the British Isles, and *S. ensifer* and *S. pygmaeus*, present on the European mainland. Both protozoa appear to have a wide distribution in Europe. They have not been reported for the British Isles but neither have they been sought for. Whether they are confined to elm bark beetles is also unknown. Nothing is known about protozoa on other elm insects because there has been no motivation to search for them. The bark beetles were investigated because they distribute the spores responsible for Dutch elm disease. There have been hopes that any disease severely affecting the beetles might retard the spread of Dutch elm disease. Such hopes have been illusory.

The nematodes, or thread worms, like the protozoa, only come into the present discussion as secondary elm associates. Like the protozoa, those relevant here are all associates of elm beetles. Most are associates of bark beetles, though one, *Diplogasteritus labiatus*, parasitizes the North American beetle *Saperda tridentata*, which is an elm wood borer.

Two nematodes are internal parasites of the British elm bark beetles *Scolytus multistriatus* and *S. scolytus*. These are *Parasitaphelenchus oldhami* and *Parasitylenchus scolyti*.[17] Both occur also on the

Continent and *P. oldhami* has been introduced into North America with the host beetle *S. multistriatus*. Another *Parasitylenchus* species, *P. secundus*, closely related to *P. scolyti*, has been found associated with elm bark beetles in Europe.

A considerable number of nematode species have been found either attached externally to elm bark beetles or inhabiting their tunnels. It is uncertain to what extent they are restricted to elm insects. *Cryptaphelenchoides scolyti* is closely attached to the beetle cuticle at one stage of its development but is free living at another. It may be exclusive to elm bark beetles.[18] Two species of *Goodeyus* are reported from bark beetle tunnels in Europe: *G. ulmi* and *G. scolytus*. *G. scolytus* occurs in England where it was first described. Two species of *Synchnotylenchus* are likewise known from bark beetle tunnels, *S. ulmi* in Europe and *S. scolyti* in Colorado. Another nematode, *Cylindrocorpus erectus*, has been found in bark beetle tunnels in New Mexico. The American nematodes both occur in the tunnels of the bark beetle *Scolytus multistriatus* which is an introduced species from Europe. Its nematode associates must either have been brought from Europe, where they are as yet unknown, or they must have existed in association with some American insect before transferring to *S. multistriatus*. Other cases are known, where a European elm associate has first been described from America.

30 Fig. 30. *Parasitaphelenchus oldhami* (×900).

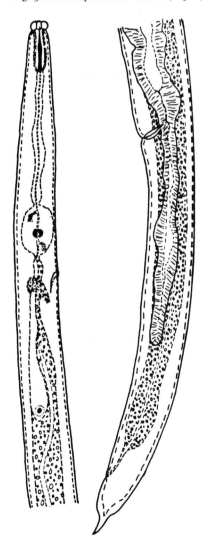

Fig. 31. *Parasitylenchus scolyti* (×200): O, ovary; ov, oviduct; R.S., seminal receptacle; Ut, uterus.

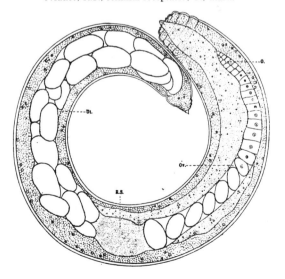

As with the elm protozoa, elm nematodes have hardly been searched for except in association with bark beetles. Again, the motivation has been largely to find some means of controlling the beetles that spread the fungal spores responsible for Dutch elm disease. Much too little is known of the host range or geographical distribution of any of the elm nematodes to throw light on the evolution of the elms on which they ultimately depend. What nematodes may be associated with other elm insects is almost completely unknown.

Bugs, aphids and related insects

The insect order Hemiptera, which is considered in this section, includes a number of the most interesting elm associates. These fall into the five following categories: leaf feeders, stem feeders, shoot gall formers, root gall formers and predators of other elm insects.

Three British bugs are feeders on elm leaves. Two are species of *Orthotylus*, *O. viridinervis* and *O. prasinus*. Both are slender green bugs, differing in the structure of their antennae. The other, *Asciodema fieberi*, is a slender downy bug, the females being greenish and the males orange.

O. viridinervis is widespread in England and also occurs in Wales, Scotland and Ireland. *A. fieberi* is also widespread in England but much less common than *O. viridinervis*. *O. prasinus* is only likely to be found in southern England. The two *Orthotylus* species are widespread in Europe. *A. fieberi* is a rare European species, principally occurring in central Europe and absent from France. Both *A. fieberi* and *O. viridinervis* normally feed on Wych Elm, not on Field Elm.

A. fieberi is of particular interest since its overall distribution and feeding preference suggest that it came into England with its host plant, the Wych Elm, across the bed of the North Sea, in the Boreal period.

Several leaf hoppers are also elm leaf feeders in England. The commonest of these, *Typhlocyba ulmi*, is a small greenish-yellow insect 2.5–4.0 mm long. It is likely to impress its presence on anyone handling elm foliage by its leaping prowess. Its empty casts are commonly to be seen on elm leaves. Some other *Typhlocyba* species seem to have a preference for elm but are also found on other plants. They can only be distinguished by microscopic characters.[19]

The other elm leaf hopper is *Macropsis glandacea*. It is 4.5–5.0 mm long and of a dingy colour.

T. ulmi occurs throughout the British Isles wherever elm is present. *M. glandacea* is confined to southern England. *T. ulmi* is widespread on the European mainland. *M. glandacea*, like *Asciodema fieberi* is a central European species, very rare in France.

The North American elm leaf hoppers are not related to the European species. One of them, *Scaphoideus luteolus*, is important as the vector of the elm disease phloem necrosis.

A white-fly species of the genus *Asterochiton* has been reported as feeding on elm. It has not been named and *Asterochiton* species are not well defined.[20]

Two European leaf feeders, the jumping plant louse *Psylla ulmi* and the aphid *Tinocallis platani*, are both normally restricted to *Ulmus laevis*, the European elm species absent, except for a few recent plantings, from the British Isles. They have been found however on some of the few *U. laevis* trees that have been introduced, as for example at Kew. They have not migrated from this elm to other species growing adjacent.

The European elm leaf feeders are not, in general, found in the Far East or North America, nor have they, usually, related species in these regions.

The aphid genus *Tinocallis* is, however, worldwide on elm.[21] Of its eight species, four are on elm, two on *Zelkova*, the genus nearest related to *Ulmus*, and two are on other species. Clearly, this genus of insects has mainly evolved in association with elm. It presumably entered into association with elm or perhaps the common ancestor of elm and *Zelkova* before continental drift separated the main land masses.

The stem feeder, *Eriococcus spurius*, a scale insect, can be dealt with quickly. The most conspicuous stage is that of the mature female insect. This is 2.5 mm long, purplish brown, with a white cottony fringe. The females are sedentary and congregate in large numbers along the smaller branches of elm trees. It is a major pest of elm in Europe. It has occasionally entered the British Isles on nursery stock but appears not to maintain itself under British climatic conditions.

Five aphids form galls on elm shoots in England. They all belong to a single tribe, the Eriosomatidae, whose evolution has principally taken place in association with elm and *Zelkova*, which are host plants for most of the species. These gall aphids have a complex life cycle. Typically they form galls on elm at one stage and live on the roots of some quite different plant at the other, the two stages alternating 32 regularly. Sometimes, however, when climatic circumstances are suitable, the aphids may persist indefinitely in the root stage on the alternate host. They can thus exist in the absence of elm though they will normally induce gall formation on it when they come in contact with it.

The simplest of the galls is that formed by *Eriosoma ulmi*.[22] This brings about a distinctive crinkling and rolling up of single leaves. The alternate stage of the life cycle is spent on gooseberry roots. Trees heavily galled by this species have been responsible for 'silver rain', the drops glistening from the presence of the white waxy substance secreted by the aphids. The next species, *E. patchiae*, causes puckering on all the leaves of a single shoot. Unlike *E. ulmi* which usually galls leaves on mature shoots, *E. patchiae* goes for juvenile long shoots such as suckers or shoots developing secondarily from the trunk. The alternate stage is on roots of ragwort (*Senecio jacobaea*) and some other *Senecio* species. The third *Eriosoma*, *E. lanuginosum*,[23] forms the most conspicuous gall of all. The development of the shoot on which the aphid has settled is drastically modified and a large bladder of plant tissue, green or purplish in colour, and up to 8 cm across is produced. This contains a sticky fluid whose medicinal uses are mentioned in chapter 8. The alternate stage is on pear roots.

Fig. 32. Galls; *a*, *Eriosoma ulmi* (× 1); *b*, *Eriosoma lanuginosum* (× 1); *c*, *Kaltenbachiella pallida* (× 1); *d*, *Tetraneura ulmi* (× 1).

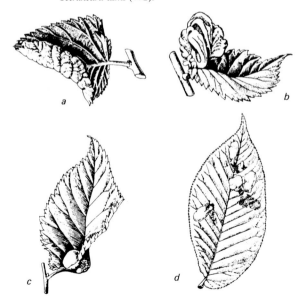

The *Eriosoma* species occur throughout the British Isles except *E. lanuginosum*, which seldom occurs outside the southern half of England. All are widespread in Europe. *E. patchiae* and *E. lanuginosum* do not occur naturally outside Europe. *E. lanuginosum* has however been introduced on pear roots to many localities where elm does not grow. Under favourable conditions it can maintain itself on pear without completing the full life cycle.

Eriosoma species related to *E. ulmi* occur in the Himalayas and the Far East.

There are several *Eriosoma* species in North America, all associated with elm. One of these, *E. lanigerum*, forms rosette leaf galls on *Ulmus americana*. Apple is an alternate host. This is the well-known apple pest known as woolly aphid. It has been introduced into Europe where its elm host is very seldom encountered. Like *E. lanuginosum*, however, it can maintain itself by cutting out the elm stage in its life cycle. It does not gall European elms.

Kaltenbachiella pallida is an aphid which causes the formation of a leaf gall looking like a pale green egg lying on the upper surface of the midrib. The leaf curves around the gall so as partially to encompass it. The alternate stage is on roots of peppermint (*Mentha* species). The aphid occurs throughout the British Isles though it is nowhere common. It is found throughout Europe and in the Himalayas. Other species of *Kaltenbachiella* forming galls on elm occur in the Far East and in North America.

The most elaborate of the British elm leaf galls is that formed by *Tetraneura ulmi*. The galls are stalked, smooth, ovoid bodies on the upper surface of the leaf. When fully developed, the aphids leave from an exit pore that appears near the base of the gall. At this stage it looks much like a small helmet. The leaf surrounding the gall becomes mottled and dense patches of white hairs develop on the under side of the leaf. The gall may be green or red. If it is red, the leaf mottling is likely to be red too.

T. ulmi occurs throughout the British Isles and over most of Europe. The alternate stage, as in all known *Tetraneura* species, is on grass roots, and if cereals are affected some economic damage may result.

Two other *Tetraneura* elm galls occur in Europe. *T. caerulescens* produces bright red globular hairy galls. *T. akinire* produces hairy galls of very variable shape. *T. yezoensis* and *T. sorini*, which produce smooth galls

of its associated organisms migrated to England over the dry bed of the North Sea from central Europe, and not from northern France.

Some of the moths feeding on both Wych and Field Elm may have come in with Wych Elm and transferred to Field Elms when these were introduced. The moths feeding exclusively on Field Elm and of pronounced south-east distribution probably flew over the English Channel and established themselves subsequent to the introduction of Field Elm.

Wasps and related insects

The elm insects in this assemblage (Hymenoptera) comprise leaf eaters, leaf miners, wood borers and secondary associates of other elm insects. Wasps specializing in boring elm wood occur in North America but not in Europe.

The British leaf eaters comprise two closely related species, *Priophorus laevifrons* and *P. ulmi*,[53] the former having a brown spot on the abdomen which the latter does not. The two species have been confused and their relative distribution both in the British Isles and in Europe is uncertain. Another species, *P. pallipes*, feeds on elm in south Europe. An undescribed species feeds on *Ulmus wallichiana* in the Himalayas.

The leaf miner is *Fenusa ulmi*, which excavates a large blotch mine. This differs from that of the beetle *Rhynchaenus alni* in not having a narrow initial section commencing on a vein. *F. ulmi* occurs throughout England and Scotland and over most of Europe. An undescribed *Fenusa* species mines elms in the Far East.[54]

A great many insects of this order parasitize elm insects in Europe. In many cases the parasites are not exclusive to elm insects. Several of the parasites which do appear to be exclusive confine their attention to elm insects such as the scale insect *Eriococcus spurius* and the beetle *Pyrrhalta luteola* which are not true members of the British elm fauna. One parasite however, the ichneumon *Areopraon lepelleyi*,[55] occurs in England and is widespread in Europe. It is exclusive to elm and feeds on gall aphids of the genus *Eriosoma*.

As with the beetles and moths, the European elm insects in this group sometimes have related species in the Far East or the Himalayas. They do not have related species in North America. As in other groups, then, it seems that relationships between elm and the

wasps were entered into before the elm population of Europe and the Far East had become separated.

Mites

One of the most interesting groups of elm associates is that of the eriophyid mites. These differ from other mites in having four legs only, at the fore end of the body, which is elongate and rather worm-like. They are just visible under a hand lens. They fall into three categories: vagrant mites, rust mites and gall formers. The rust mites cause the edges of the leaves to turn brown. They have not been found on elm in the British Isles.

The vagrants are inconspicuous organisms that appear on the under surfaces of elm leaves about the beginning of July and withdraw, presumably to the twigs, early in August.[56] They have no visible effect on the leaves. The commonest of the elm vagrant mites is *Rhyncaphytoptus ulmivagrans*.[57] The distinguishing feature of the genus *Rhyncaphytoptus* is the form of the head. This terminates in a relatively large beak-like structure at right angles to the rest of the body. *Rh. ulmivagrans* occurs throughout England and Europe and is also present in the Far East. It is replaced by a related species, *Rh. atlanticus*,[58] in North America. *Rh. ulmivagrans* shows no restriction to particular kinds of elm. Another less common European species, *Rh. ulmivora*,[59] has been found in southern England. It differs from *Rh. ulmivagrans* in having a conspicuous groove behind the head. An apparently related species, *Peralox insolita*,[60] replaces it in North America.

Two other elm vagrant mites have been discovered in England. *Oxypleurites ulmi*[59] is distinguished by the sharp projections on the segments of the body. This species occurs both in Europe and the Far East. *Aculops calulmi*[61] differs from the other vagrants in that the two bristles behind the head are much longer. When on the march it looks rather like a miniature lobster. Two other species of *Aculops* occur as vagrant mites on elm leaves, *A. neokonoella*[62] was found on *Ulmus pumila* in central Asia. *A. verapasi*[63] occurs on *U. mexicana* in Guatemala. Neither is closely related to *A. calulmi*.

There are two species that form galls on elm leaves in Great Britain. *Eriophyes campestricola* produces easily recognizable pimple-like galls. This species is widespread in the southern half of England. It occurs throughout Europe and is also present in the Far East.

It is exclusive in its diet. In England, it is almost always confined to English Elm and some forms of Narrow-leaved Elm. There appear to be two races of the mite, each exclusive to its own kind of elm. It is a matter of frequent observation that the mites do not pass from one kind of elm to another, even when the branches of the two kinds are touching. It is argued in chapter 3 that the forms of Narrow-leaved Elm that are galled by *E. campestricola* are of French origin, while those that are not are of central European provenance. It is presumed that the former elms carried the mite when they were introduced, whereas the introduced material of the latter happened to be free of the race of the mite present in central Europe.

A related species, *Eriophes brevipunctatus*,[64] is probably responsible for the small pouch-like galls on the European elm species *Ulmus laevis*. In North America, *E. ulmi*, which is very similar to *E. campestricola*, produces galls on *U. americana*. Two more distantly related species also gall elms in the New World. *E. parulmi* occurs in the USA.[65] *E. richensi* galls *U. mexicana* in Central America.[63]

The other British gall mite on elm is *Eriophyes filiformis*. This forms brown scab-like spots on the lower surface of the leaf. The opposite upper surface turns first yellow and finally black. This species is widespread in Great Britain and in Europe. There appear to be at least two races, one on Wych Elm and another on English Elm. It also occurs, though not commonly, on Narrow-leaved Elm.

A related species, *E. multistriatus*, occurs on galled leaves of *Ulmus laevis* on the Continent, but its role in gall causation is uncertain.[66] *E. wallichianae*,[67] which is very similar to *E. filiformis*, causes pustule-like galls on the under surface of the leaves of *U. wallichiana* in the Himalayas. Another related species causes galls on elm in North America but this has not yet been described.[68]

In general, it seems that the association between elms and eriophyid mites was entered into before continental drift separated the Old and New Worlds. Rather different evolutionary relations developed in respect of the vagrant mites and the gall mites. The vagrant mites do not seem to discriminate between different kinds of elm. Different but related mite species occur on either side of the Atlantic. Two vagrant species occur both in Europe and the Far East and presumably antedate the separation of the elm floras of the two regions.

The gall mites are much more restricted in their

Fig. 40. Vagrant mites: *a, Rhyncophytoptus ulmivagrans* (× 300); *b, Rhynchophytoptus ulmivora* (× 300); *c, Oxypleurites ulmi* (× 300); *d, Aculops calulmi* (× 300).

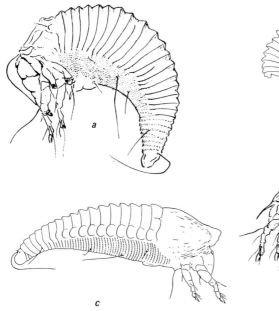

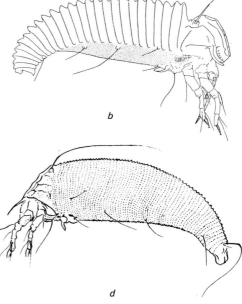

feeding. Particular species or races seem unable to feed on any but one kind of elm. Of the assemblage of mite species related to *E. campestricola*, the American species *E. parulmi* and *E. richensi* may be nearest to the ancestral form.[69] They occur, though not exclusively, on the more primitive elms of the southern USA and Central America. The Eurasiatic species, *E. campestricola*, has not differentiated into different species in Europe and the Far East but it appears to exist in a number of races confined to particular kinds of elm. The eriophyid gall mites have a rather low capacity for aerial dispersal and it seems that their existing distribution in England has been determined by the history of distribution of the elms harbouring them. The combination of limited dispersal capability and highly exclusive feeding preference makes them the most useful of all elm associates for elucidating its history.

Birds

This section is also mainly dismissive. There are some mildly preferential associations between birds and elms but these are very far from exclusive. Preference is based on two diverse roles which elms can play. They may provide a preferred source of nutriment. They may also provide a preferred nesting site.

Elm buds, for example, are sought after by bullfinches (*Pyrrhula pyrrhula*) in spring.[70] As fruit growers know well, other tree buds are just as attractive. There appears to be no British bird that feeds so voraciously on elm buds and fruits as the purple finch (*Carpodacus purpureus*) of North America. The two British birds which have shown a preferential interest in elm for nesting are the rook (*Corvus frugilegus*) and raven (*Corvus corax*). In both cases, it is because elms are normally the tallest trees to hand that they are chosen. In landscapes with little elm, other trees are utilized. The association between British elms and ravens had practically come to an end by the middle of the last century with the decline in the raven population of southern England. It is interesting to recall, however, that ravens' nests in elms were a feature of the coastal marshes of Essex. There was an elm called the Raven Tree on Northey Island. There was another tree similarly designated at Prested Hall, Feering, in the same county.[71] Another lowland site where ravens nested in elms till recently was Croome Park (HW).

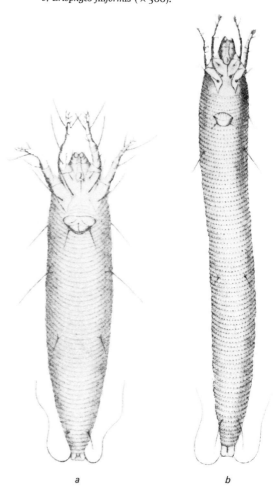

Fig. 41. Gall mites: *a*, *Eriophyes campestricola* (× 500); *b*, *Eriophyes filiformis* (× 500).

a   b

7
Botanical classification

Elm classification is notorious for its difficulty. Two quite different causes are involved. The first concerns the nature of the material. Elms are highly and complexly variable. It is not easy to describe this variability clearly and accurately. Nor is it easy to devise a system of nomenclature for the principal variants that both reflects their real relationships and is convenient to apply. The second cause is quite different. It arises from the International Rules for the Nomenclature of Plants,[1] which govern the assignment of Latin names to plants, and from the difficulties in applying them to recalcitrant material. These difficulties botanists have created for themselves.

In plants such as the elms, which have been studied continuously by botanists since the Renaissance, difficulties arise because the concepts applied in botanical classification have themselves changed between then and now. The nomenclatural rules have likewise changed very significantly over the years. Till recently, the tendency amongst systematic botanists was to give a Latin name to every variant they distinguished. With present-day biometrical methods, the amount of variation uncovered is so large that to apply a Latin name to every recognizable variant would be to plant a thicket of names in which everyone would be irremediably lost. Biometrically minded botanists consequently are inclined to apply Latin names to the broader entities they recognize and to go over to mathematical description of the finer variants.

Classification is hierarchical. At the Renaissance,

the idea that organisms should be named at the two lowest levels in the hierarchy had begun to crystallize. A level of classification that is also a nomenclatural category has subsequently become known as a rank. The two ranks around which the Renaissance botanists were fumbling are those known today as genus, the higher, and species, the lower.

The convention was early established that an organism should be named by stating its genus first, usually a single Latin word, and then its species, a Latin word or phrase qualifying the genus. Thus the elm described in the seventeenth century as *Ulmus folio glabro* has *Ulmus* for its genus and *folio glabro* as the qualifying phrase indicating the species.

In the eighteenth century, a new rank below the species was introduced. This was the variety. Subsequently intermediate ranks made their appearance. Subgenus and section were interposed between genus and species. Subspecies came between species and variety. Subvariety and forma lay below variety. Botanists were now in a position, even when they agreed on the pattern of variation of some material, to differ on the ranks they assigned to it. Those who tend to name variants at the higher ranks, and so recognize many species, are known as splitters. Those who name at the lower ranks and consider many variants are best not assigned Latin names at all are known as lumpers.

The nomenclatural rules favour neither splitter nor lumper. Provided the rules are followed faithfully, the species of the splitter are as nomenclaturally valid as those of the lumper. A nomenclaturally correct name can, however, be bad classification. This happens, for example, when two variants are treated as varieties of one species when they ought rather to have been treated as respective varieties of two different species. This has happened several times, as will be shown later with the elms.

One respect in which the naming rules involve a different conceptual framework from that of the biometrician is in respect of so-called type specimen that fixes the application of a name. The biometrician analyses the variability of a population and arrives at a typical specimen that is most representative of the whole assemblage. For the nomenclatural systematist, the type specimen is the specimen selected for initial description, which may be and sometimes is quite unrepresentative of the population as a whole. It is well known that collectors are liable to pick the most conspicuous plant in a stand, not an average specimen.

A weakness of traditional nomenclature is that the Latin names it establishes provide no clue as to the range of variation to be associated with them. A botanist, on describing a new species, will include under his conception of it, not only the type specimen but various other plants differing from the type specimen in relatively minor respects. Another botanist is entitled to detach some of the variant plants and place them in a new species as long as he leaves the type specimen alone. A third botanist may attach to the first species a second, more recently described species, which he considers not worth separating at species level from the first. The three botanists have accepted species of very different extents. Since all include the type specimen, all are given the same name.

A long-persistent source of error in elm systematics has been the assumption that the kinds of elm present in one region are likely to recur wherever else elm is present. Acting on this preconception, names applied to particular elm populations in one country have often been attached to those in another. Very often the populations in the two countries have been entirely different.

Another proceeding that aggravates confusion in elm systematics is the citation of synonyms. Occasionally a satisfactory description is given of an elm directly observed by an author. He then proceeds to queer his account by listing a series of descriptions of other elms which he supposes to be the same. Often they are nothing of the kind. When an unsatisfactory description is followed by a misleading list of synonyms, the reader might well despair.

In the next two sections, the history of elm systematics in England·will be sketched. Attention will be confined as far as possible to the history of discovery of new kinds. Much of the literature on elm classification is only about name changing. Consideration of this is banished to the notes.

Before Linnaeus

The history of elm classification effectively starts with the Greek author Theophrastus.[2] He described two elms: πτελέα [ptelea] and ὀροπτελέα [oroptelea], literally elm and mountain elm. From his descriptions and from what is known of Greek elms,

it is likely that both elms were forms[3] of Field Elm (*Ulmus minor*).[4]

The next contribution comes from classical Rome. The agricultural author Columella[5] also described two elms, a native elm, *ulmus vernacula* or *ulmus nostras,* and an elm from Cisalpine Gaul, which he called *ulmus gallica,* known alternatively as *atinia.* As with Theophrastus, both elms appear to be forms of Field Elm (*U. minor*). The first confusion was created, as in so many areas of knowledge, by Pliny.[6] He multiplied Columella's two elms into four by treating his alternative designations as different elms. For *ulmus vernacula* he substituted the designation *ulmus sylvestris.* He mentioned Theophrastus' two elms also, designating πτελέα as *ulmus campestris* and ὀροπτελέα as *ulmus montosa.* Here the situation rested till the Renaissance.

It was Theodore of Gaza who first translated Theophrastus' botanical works into Latin, his translation was published in 1483. Theodore rendered πτελέα as *ulmus* and ὀροπτελέα as *montiulmus.*[7] The Renaissance translators usually referred to πτελέα, as *ulmus.* Sometimes they used Pliny's *ulmus campestris.* For ὀροπτελέα, *montiulmus* did not catch on. Pliny's *ulmus montosa* was used by some. The term that won the day was *ulmus montana* which was popularized by Ruellius.[8]

The herbalists then took up the threads. Their principal motivation was to identify what the classical authors had mentioned. Any desire to observe and record the elms around them was quite secondary.

Their first problem was to determine the relationships between Theophrastus' two elms, Columella's two elms and Pliny's four. Then, having sorted this out, they essayed to relate their findings to the elms with which they were familiar. There was a presumption, unjustified, that detailed correspondences existed between the elms of Greece, Italy and central and north-west Europe.

It was generally agreed that Pliny had boobed and that *ulmus gallica* was the same as *atinia.* Opinions were divided on the relations between the Greek and Roman elms. The Flemish herbalist R. Dodoens decided, to begin with, that πτελέα was *ulmus gallica* and ὀροπτελέα was *ulmus nostras.*[9] In the second, French, edition of his herbal,[10] he switched around, and was followed by H. Lyte,[11] whose new English herbal was based on Dodoens' French herbal. Further reflection brought Dodoens back to his first view[12] and he was followed in this by the English herbalist J. Gerard.[13] Dodoens also supposed that Pliny's *ulmus sylvestris* was a third kind of elm about whose nature he could hardly have been less definite. Hornbeam was one of the possibilities with which he toyed, for as shown in chapter 4, elm and hornbeam were much confused at this time. J. Dalechamps,[14] the French herbalist, identified πτελέα with *ulmus nostras* and ὀροπτελέα with *ulmus gallica*. His view was adopted by C. Bauhin,[15] with long-term nomenclatural consequences that will emerge in due course. Whatever view was taken of the relation between Greek and Roman elms, it was assumed in all areas that the local elm corresponded to Theophrastus' πτελέα. There was much uncertainty whether ὀροπτελέα was present and if so, what it was.

It is now necessary to go back a little. The first of the Renaissance herbalists to record variation in the elms they observed was the German botanist H. Bock.[16] He distinguished between an elm with spreading habit, *Ulmus lata*, and an elm with a more erect habit, *U. procera*. The latter must not be confused with the *U. procera* of recent botanists, which is a name for English Elm (*U. minor* var. *vulgaris*). No later botanist has been able to point to two reasonably well distinguished kinds of elms which might have been those which Bock wished to distinguish.[17] They soon disappear from botanical literature.

It was the Flemish herbalists who had the most information to record. Dodoens was confident in his 1554 herbal[9] that Theophrastus' πτελέα, which he equated with Columella's *ulmus gallica*, was the common elm of Brabant. Later, in 1557,[10] he decided that ὀροπτελέα was present in western Flanders. He also thought that there was a third elm, perhaps Pliny's *ulmus sylvestris*, known as *herseleer* or *herenteer*, but this was either hornbeam or confused with it. In successive editions of his works, his opinions vacillated widely. Sometimes ὀροπτελέα was equated with *herseleer*, whatever that was. Confusion over hornbeam persisted.

The other Flemish herbalist, M. de Lobel,[18] knew of two elms. One was the 'common' elm. The other was a broad-leaved elm present in western Flanders, presumably the same as the Flemish elm mentioned above. Dodoens had managed to confuse elm and hornbeam. De Lobel confused elm and lime, since the

tree he called *Tilia mas*, literally male lime, is clearly an elm. He said that this was known by the Flemish name *ypeline*, which possibly means elm-lime.

The sixteenth century English herbalists largely copied Dodoens. Lyte[11] based himself on the 1557 edition of Dodoens' herbal. He did, however, add some important new details. He said that ὀροπτελέα also occurred in Picardy where it was known as *ypreau*. It is shown in chapter 5 that this elm is what is now known in England as the Dutch Elm (*U. × hollandica* nm. *hollandica*). He concluded that Theophrastus' πτελέα is frequent in England, presumably meaning the English Elm (*U. minor* var. *vulgaris*). He also thought it possible that Dodoens' enigmatic third elm was what was known in England as *wych* or *wych hazel*. This could mean that Lyte was familiar with Wych Elm (*U. glabra*).

Gerard[13] also based his herbal on Dodoens' work. For him too, the English Elm (*U. minor* var. *vulgaris*), which he called *Ulmus* without qualification, is Theophrastus' πτελέα. He took over the view that ὀροπτελέα is the same as *herseleer*, whatever this may be. He gave it the name *Ulmus latifolia* and said it grew in Theobalds Park (GL) and Southfleet (Ke). His description is far too sketchy to be able to determine what he had in mind, especially as he, too, confuses elm and hornbeam.

Clearly, elm nomenclature, even at this early stage in its history, was becoming chaotic. Bauhin in 1623 was one of the first to attempt a clean up.[15] He concluded that the elms of which he was aware could be classified under two species. The first, *Ulmus campestris & Theophrasti*, is the earliest example of an elm species designated by an extended phrase, a practice that was to persist until the time of Linnaeus. The name was obviously derived from Pliny's *ulmus campestris*. It corresponded to what most of the herbalists had called *Ulmus* without qualification. The name of the second species, *U. montana*, was taken over from Ruellius. It was meant to cover Theophrastus' ὀροπτελέα, Bock's *U. lata*, Dodoens' second elm, and Gerard's *U. latifolia*. It was, in fact, a rag bag.

Superficially, Bauhin seems to present a classification of the elms of western Europe that corresponds to what is accepted today, that is, a lowland elm, now called the Field Elm (*U. minor*), and a montane species, the Wych Elm (*U. glabra*). His concept of *U. montana*, however, did not correspond to that of Wych Elm (*U. glabra*) as now understood. It was the amalgam described in the previous section, mainly Field Elm (*U. minor*) with a dash of hornbeam.

The year 1633 was a notable one for elm classification. After the confusion that had hitherto reigned, a clear, sober account was published of four kinds of elm growing in England. They were described in sufficient detail and located with sufficient precision for them to be identified with populations existing today. The author was T. Johnson and his contribution appeared in the second edition of Gerard's herbal of which he was editor.[19] The like would not be repeated.

The four elms are as follows. *U. vulgatissima folio lato scabro*, called by Johnson *common elme*, is English Elm (*U. minor* var. *vulgaris*). *U. folio latissimo scabro*, for which the vernacular name *witch hazell* was given, is Wych Elm (*U. glabra*). *U. minor folio angusto scabro* is a form of Narrow-leaved Elm (*U. minor* var. *minor*). Johnson said that it grew between Lymington (Ha) and Christchurch (Do). It still does. *U. folio glabro*, the last of the four, called *witch elm* locally, is also a form of Narrow-leaved Elm (*U. minor* var. *minor*). It grew at and to the north-west of North Ockendon (GL). Today it is still there.

J. Parkinson, the next author to write on elms, heads the long line of nomenclatural revisers.[20] He had nothing new to say of any substance but clearly considered that Johnson's names were too much of a mouthful. He consequently shortened and amended three of them. *Ulmus vulgatissima folio scabro*, English Elm (*U. minor* var. *vulgaris*), became *U. vulgaris*. *U. folio latissimo scabro*, Wych Elm (*U. glabra*), became *U. latiore folio*. *U. minor folio angusto scabro* became *U. minor*.

It is possible to be too hard on Parkinson. The inconvenience of name swapping was not obvious in his time and the botanical rules that eventually outlawed most of it were only formulated much later. Latin names also meant something then to those who handled them. The idea that an unsuitable Latin name once coined had to stick would have struck people in Parkinson's time as odd. As will become evident later, the botanical rules enforce some highly inappropriate names on the elms. Parkinson's names could have been ignored here were it not that some of them were taken up by the post-Linnaean botanists and so passed into current nomenclature.

Two other seventeenth century authors made positive contributions to elm systematics. The Oxford

antiquary R. Plot discovered an elm at Hanwell (Ox) which he named *U. folio angusto glabro*.[21] Elms answering to Plot's description still occur, although their distribution is rather scattered, in northern Oxfordshire. They are yet another form of Narrow-leaved Elm (*U. minor* var. *minor*). Botanists writing on elm in later years ignored Plot's discovery. It was not brought to notice again till the present century when G. C. Druce, also of Oxford, mounted a treasure hunt for the elm that Plot discovered. The paradoxical outcome is described in due course.

L. Plukenet, Botanist to Queen Mary II, described two new elms.[22] The first he called *U. pumila foliis parvis cortice fungosa*. This was probably a small-leaved form of Narrow-leaved Elm (*U. minor* var. *minor*). Plukenet does not give sufficient detail wherewith to identify it. The name disappeared from the English scene. The history of botanical classification abounds in strange quirks, however, and this elm provided Linnaeus with the name *U. pumila*. This he transferred to a very different elm from central Asia and the Far East. Linnaeus' name is still used. Plukenet's other elm, *U. major Hollandica angustis & acuminatis samarris folio latissimo scabro*, is what in England is called the Dutch Elm (*U. × hollandica* nm. *hollandica*). It is not, as pointed out in chapter 2, the *hollandse iep* of the Netherlands. The Dutch Elm is a hybrid between Wych Elm (*U. glabra*) and Field Elm (*U. minor*) though this was not realized till many years later. Plukenet's long qualifying phrase supplied the

Fig. 42. *Ulmus folio angusto glabro*, by Michaell Burghers (× $\frac{1}{2}$).[21]

adjective *hollandica* which survives in contemporary use as described later.

The French botanist J. P. de Tournefort[23] needs a mention at this point. He did not publish any fresh observations, but his choice of elm names guided Linnaeus in the names he chose, and so determined much subsequent usage. Tournefort accepted Johnson's four elm species and Johnson's names for three of them. For *U. vulgatissima folio lato scabro*, the English Elm (*U. minor* var. *vulgaris*), he substituted Bauhin's *U. campestris & Theophrasti*.

Few botanists have managed to complicate elm nomenclature more than P. Miller, Gardener of the Chelsea Botanical Garden. His writings on elms are contained in the successive editions of his *Gardener's Dictionary*, the 'best horticultural book in the world' according to its admirers. In the first edition (1731)[24] he described nine kinds. At that time, the distinction now made between naturally growing and cultivated plants was not drawn. They were all treated equivalently. In fact, since a large part of the British elm flora is, as argued in previous chapters, of human introduction, the distinction is not all that sharp. The elm story is however sufficiently complicated. Miller's

Fig. 43. *Ulmus folio glabro*, by Georg Ehret (× ¼).[24]

purely horticultural kinds, all elms with variegated leaves, will therefore be left on one side. Five elms remain.

Four of the five were Johnson's quartet. Miller was hazy, however, about what Johnson's *U. minor folio angusto scabro* might be, since he gave it the vernacular designation of the 'small-leaved or English elm'. The fifth elm was the Dutch Elm (*U. × hollandica* nm. *hollandica*) of Plukenet. A few years later in what was in effect a supplement to the first edition,[25] three new elms appeared. *U. minor folio angusto glabro* was said to be a narrow-leaved elm abundant in parts of Hertfordshire and the old county of Cambridgeshire. It is clearly a population of the Narrow-leaved Elm (*U. minor* var. *minor*). The two others were birds of passage. *U. folio lato scabro cortice cinereo glabro* was said to be the Irish elm. *U. folio lato scabro angustis samarris* was called the French elm. It is not clear what either was. Miller himself was clearly uncertain for he quietly dropped them in his next round of elm thoughts. They disappear henceforward without trace.

The next round was in 1759, in the seventh edition of the *Dictionary*.[26] Having discarded two of his novelties, Miller was back with Johnson's four elms, the Dutch Elm and his own *U. minor folio angusto glabro*. He proceeded to rename every one. This is aggravating but can be endured. Far worse, he decided that Johnson's *U. vulgatissima folio lato scabro*, the English Elm (*U. minor* var. *vulgaris*), renamed *U. foliis oblongis acuminatis duplicato-serratis basi inaequalibus*, was a form of Wych Elm (*U. glabra*). Johnson's Wych Elm (*U. glabra*) was renamed *U. foliis oblongo-ovatis inaequaliter serratis calycibus foliaceis* and appears to be also treated as a Wych Elm (*U. glabra*) but different from the previous one. *U. minor folio angusto scabro* of Johnson was a population of Narrow-leaved Elm (*U. minor* var. *minor*) from southern Hampshire. Miller renamed it *U. foliis ovatis acuminatis duplico-serratis basi inaequalibus* and seems to have decided to his own satisfaction that it was English Elm (*U. minor* var. *vulgaris*). Johnson's *U. folio glabro* escaped with no more than a renaming. It became *U. foliis ovatis glabris acute serratis*. The Dutch Elm escaped likewise. It became *U. foliis ovatis acuminatis rugosis inaequaliter serratis cortice fungoso*. Miller's own *U. minor folio angusto glabro* was renamed *U. foliis oblongo-ovatis glabris acuminatis duplico-serratis*. The unhelpful remark was added that it was sometimes called the

43

Irish elm, presumably connected in some way with the discarded Irish elm.

One discovery was never published. A. Buddle collected a sample of the Hybrid Elm population of Essex which he named *Ulmus folio latissimo glabro*.[27] The name got no further than the herbarium sheet carrying the specimen. Such was the situation when the new ideas of Linnaeus first drew the reluctant attention of British botanists.

Linnaeus and after[28]

Linnaeus' great contribution was clarity of systematization. The fact that what he clearly expressed did not always correspond to the situation in real life only emerged when the first euphoria was over. Linnaeus, like many other systematic botanists attempting to cover the higher plants in general, was a lumper. By making extensive use of the new rank of variety, he was able to reclassify many examples of what had been treated as separate species as subordinate categories of the species he did accept. Linnaeus tackled the elms in 1737 and reduced the lot to a single species which he called *Ulmus fructu membranaceo*. Johnson's four elms, as cited by Tournefort, became varieties of it.[29]

Linnaeus' *Species Plantarum*, which ushered in the Latin single-word species designations still current, came out in 1753.[30] It included three elms. *U. pumila*, the central Asiatic species, has been mentioned already. *U. americana* was the designation of the one North American elm with which Linnaeus was acquainted. All the European elms found themselves under *U. campestris*,[31] the front end of Bauhin's *U. campestris & Theophrasti*, also Pliny's expression for Theophrastus' πτελέα. It was implied by this treatment that there was only one European elm in the Garden of Eden, Johnson's four elms being derived from it. The nomenclatural rules now in force take the publication of the *Species Plantarum* as their starting point. Earlier names have only to be taken into consideration in as far as they are needed to explain post-Linnaean designations.

Linnaeus' ideas did not commend themselves to everyone. It was some time before his system of nomenclature ousted all others. The first English botanist to give a Linnaean treatment to the elms was W. Hudson who made a thoroughly bad job of it.[32] He opted for two species into which he fitted Johnson's four elms. His first species was Linnaeus' *U. campestris*.

He took the English Elm (*U. minor* var. *vulgaris*) to be the typical form of this species and he attached Johnson's Hampshire elm to it as a variety. His second species he called *U. glabra*. He made the typical form of this the Wych Elm (*U. glabra*) and Johnson's *U. folio glabro* was made a variety of this. Hudson's Latin name for the Wych Elm has priority of all post-Linnaean species names for this entity. It was unhappily chosen since the adjective *glabra* normally denotes that the leaf is smooth. Here it means that the bark is smooth on younger branches. The leaf is usually rough.

Hudson's classification is an excellent example of how an arrangement conforming strictly to nomenclatural rules can be bad systematics. The two subordinate varieties are Narrow-leaved Elms (*U. minor* var. *minor*). They are more closely related to one another than they are to either of the typical forms of the two species to which they are attached. One suspects that Hudson was guided by no more than the vernacular name *wych elm*, usually applied to Wych Elm (*U. glabra*), but in Essex, as Gerard recorded, applied to Narrow-leaved Elm (*U. minor* var. *minor*).

Miller came round to the Linnaean system a few years later.[33] In the eighth edition of his *Dictionary*, single adjectives in poor Latin (here corrected) replaced the extended phrases that had designated the six elms he had latterly recognized. The elm based on Johnson's name for the English Elm (*U. minor* var. *vulgaris*) was called *U. campestris*, the name that Hudson had used for this kind. Miller, however, appears still to have thought that it was a kind of Wych Elm (*U. glabra*). Johnson's Wych Elm (*U. glabra*) became *U. scabra*. This was a better name than Hudson's, but Hudson had got in first so *U. scabra* has no standing. The Narrow-leaved Elm (*U. minor* var. *minor*) that Johnson described from Hampshire and which Miller confused with the English Elm, became *U. sativa*.[34] Johnson's other form of Narrow-leaved Elm (*U. minor* var. *minor*) became *U. glabra*. This would have been an appropriate name for this elm. Unfortunately Hudson had used this name for the Wych Elm (*U. glabra*) so Miller's designation falls victim to the rules concerning priority that were later and retrospectively made obligatory. The Dutch Elm (*U.* × *hollandica* nm. *hollandica*) presented no problems and became simply *U. hollandica*. Miller's own Narrow-leaved Elm (*U. minor* var. *minor*) became *U. minor*.

Two of Miller's names survive in contemporary use. *Ulmus minor* is the earliest name for a Field Elm (*U. minor*) that conforms with all the nomenclatural rules. It is therefore of present obligation for this general category of elm even though used in a much wider sense than Miller envisaged. *U.* × *hollandica* with the addition of the multiplication sign to denote a hybrid involving different speces, is used for all elms resulting from crossing between Wych Elm (*U. glabra*) and Field Elm (*U. minor*).

The next English botanist to introduce a new elm was R. Weston.[35] He used a highly vulnerable way of expressing his thoughts on classification. Species he represented in roman type. Varieties were indicated by italics. This put him and later readers at the mercy of his printer, who sometimes got it wrong. His new elm was *U. campestris* var. *cornubiensis*. This is the Cornish Elm (*U. minor* var. *cornubiensis*). He also described an elm under the name *U. hollandica* var. *angustifolia*. This is just possibly the *hollandse iep* of the Netherlands.

Attention must now be paid to a designation that kept confusion seething steadily for most of the nineteenth century. In 1785, the German botanist C. Moench described an elm which he named *Ulmus suberosa*.[36] There is little doubt that what Moench meant by this name, and what Continental botanists in general thought he meant by it, was an elm similar to Field Elm (*U. minor*) but with conspicuous corky branches. There are, as mentioned in chapter 2, differences in the tendency to corkiness shown by different elms, but these are not now regarded as sufficient to warrant recognition of a distinct species or even variety.[37] The only recent publication of major status to retain *U. suberosa* is that hallowed shrine of conservative botanical thought, the *Flora of the USSR*. Even in the Soviet Union, its abolition has been urged.[38]

In England, *Ulmus suberosa* had a very different fate. In the first edition of J. Sowerby and J. E. Smith's *English Botany*,[39] it was used for the English Elm (*U. minor* var. *vulgaris*). In the third edition of the same work it was used as virtually equivalent to Field Elm (*U. minor*) in general.[40] Fortunately it faded out of responsible British publications thereafter.

The botanist with the keenest eye for elm in the early nineteenth century was J. Lindley, Professor of Botany at London University. He describes four new elms.[41] Three were forms of Narrow-leaved Elm (*U. minor* var. *minor*). One, a single tree in southern Warwickshire, he called *U. carpinifolia*. This name was current in Germany for some form of Narrow-leaved Elm (*U. minor* var. *minor*). The other two he described as varieties of *U. glabra* as understood by Miller. Of these, var. *glandulosa* came from near Ludlow (Sa) and var. *latifolia* from Essex. The fourth elm was a small-leaved Cornish Elm (*U. minor* var. *cornubiensis*). He named it *U. stricta* var. *parvifolia*. None of Lindley's new elms were sufficiently distinct to be accepted by later botanists.

J. C. Loudon, whose enormous output of published material extends over botany, horticulture, landscape gardening and architecture, provided a full account of the range of elms in the 1830s. His publications are particularly important on account of his close contacts with nurserymen. Masters of Canterbury and Loddiges of Hackney, for example, were collecting and distributing new elms. Their catalogues name but do not describe their novelties. Loudon provided further information on some of these.[42] Unfortunately his own descriptions tended to be perfunctory and do not help as much as one would have hoped. Loddiges had apparently obtained the Guernsey Elm (*U. minor* var. *sarniensis*) from the Channel Islands.[43] Loudon, with the bare bones of a description, introduced it into scientific currency under the designation *U. campestris* var. *sarniensis*.

A lull in elm investigation followed Loudon's work. Nothing new was reported till the present century when G. C. Druce set out to find the elm that Plot had described. He failed. He found instead another very distinctive elm, Lock's Elm (*U. minor* var. *lockii*), which he described as *U. sativa* var. *lockii*.[44] Then, with mistaken confidence that this was indeed the elm described by Plot, he upgraded his discovery from variety to species and renamed it *Ulmus plotii*.[45] The nomenclatural rules are indifferent to false insinuations in the names they regulate. Druce's singularly inappropriate designation is correct nomenclature.

Several other English elms were described about this time. None of them were sufficiently well defined to be taken up by later botanists. A. Henry described a Cornish variety of Dutch Elm (*U.* × *hollandica* nm. *hollandica*) under the name *U. major* var. *daveyi*.[46] C. E. Moss, at Cambridge, named two East Anglian varieties of Narrow-leaved Elm (*U. minor* var. *minor*) in the sumptuous and never finished *Cambridge British Flora*. These he called *U. nitens* var. *hunnybunii* and *U. nitens* var. *sowerbyi*. The former variety was named in

honour of Edward Hunnybun, the illustrator of the *Cambridge British Flora*. A further variant, subvar. *pseudostricta*, was attached to var. *hunnybunii*.[47]

Another lull in elm research ensued. It ended in the 1930s. A Leicestershire population of *U. minor* var. *minor* was then graced with species rank under the name *U. elegantissima* by A. R. Horwood.[48] The first epidemic of Dutch Elm disease had struck meanwhile. There was an urgent need for a satisfactory account of elm variation as an adjunct to the investigations mounted on the disease and on the possible existence of resistant elms. The first studies, largely conservative in outlook and revisionary in content, were by H. Bancroft.[49]

The next contribution was a long series of papers by R. Melville of Kew,[50] who subjected the whole of the existing elm classification to close scrutiny. He proposed the recognition of one old, *U. carpinifolia*, and two new species, *U. diversifolia* and *U. coritana*, within what is here called the Narrow-leaved Elm (*U. minor* var. *minor*), and he described three varieties, var. *angustifolia*, var. *media*[51] and var. *rotundifolia* of *U. coritana*. *U. diversifolia* was subsequently discarded by its author who, on second thoughts, decided it was a species hybrid.

The end of the classical era of classification came quietly in the 1950s when radically different ways of handling organic variation came into use. In these, organisms were conceived as points in an *n*-dimensional space, each dimension being a character or group of characters that varied together. Classification became very much a matter of mapping these points, and identifying aggregates of points that could be interpreted as corresponding, to some extent, to the old classification categories. The mathematics required for handling this approach was that of multivariate analysis. Soon computers were enlisted to deal with the heavy load of computation involved.[52] It is possible to exaggerate the differences between classical and numerical approaches to classification. In general, the traditional procedure of allocating Latin names to well-defined entities is accepted by the numerical systematist at what may be called the broad species level, that is, the sort of species a lumper would accept. Beneath this level, instead of an assemblage of Latin names, the numerical systematist uses mathematical description, adapted to the material, in an attempt to mirror as exactly as possible the complex patterns of variation that occur in real life. It is not surprising that this development has been the cause of some tension between systematic botanists continuing to use traditional procedures and the biometricians.[53]

World array

Although the elms of the British Isles are remarkable in the degree of variation they show, they are not a cross section of the elms of the world.[54] The number of satisfactorily defined species is about 30. They are usually divided into five groups, about which there is remarkably little difference of opinion. Only one group is represented in England and only two in Europe. The part of the world showing the greatest range in variation is not, as so often is the case, China, but North America.

The groups are usually designated sections. One botanist, J.-E. Planchon, called his groups subgenera, but they correspond in all essential respects with everybody else's sections. The sectional arrangement accepted here is as follows.

Section *Microptelea*[55] includes autumn-flowering elms with a well-developed wing to the hairless fruit. The section is confined to south-east Asia. *Ulmus parvifolia*[56] is the best known species.

Section *Trichoptelea*[57] comprises autumn-flowering elms with fruits inconspicuously winged and hairy over their entire surface. These elms are confined to the southern United States.

Section *Chaetoptelea*[58] includes spring-flowering elms with fruits inconspicuously winged or with no wing and hairy over their entire surface. It includes three species from North and Central America and one species, *U. villosa*,[59] from the western Himalayas.

The next assemblage, section *Blepharocarpus*,[60] is well defined. It includes spring-flowering elms with the fruit hairy only around the edges of the wing, which is moderately well developed. The flowers have long stalks. There are three elms in this group, *U. americana*[61] of North America, *U. divaricata*[62] of Mexico, and *U. laevis*[63] in eastern and central Europe.

The remaining elms are spring-flowering with well developed wings to the fruits. The latter may be hairy all over, hairy only over the seed, or, as in the English representatives, devoid of hairs. The current nomenclatural rules require that this group be called section *Ulmus*.[64]

Some botanists have subdivided this section into subgroups with the designation series. One botanist interposed an additional rank, subsection, between section and series, so that the Field Elm, for example, is allocated to series *Nitentes* of subsection *Foliaceae* of section *Ulmus* of *Ulmus*.[65] For a relatively small genus such as *Ulmus*, this Byzantine hierarchy of botanical ranks is excessive. One rank, the section, between genus and species will do.

Section *Ulmus* is distributed over three major regions. The first is the northern part of the circumpolar belt of temperate forest. This comprises three segments, in Europe, the Far East and eastern North America, respectively. In the European and Far Eastern segments, the elm present is Wych Elm (*U. glabra*). In North America this is replaced by *U. rubra*,[66] the slippery elm. There are outliers of this region in the mountainous regions further south. In one of these, the Himalayas, there is a distinctive elm, *U. wallichiana*.[67]

The second region is in the southern part of the temperate forest zone. It comprises segments in Europe and the Far East and also an outlier in the Himalayas. The European representative is the Field Elm (*U. minor*). The Himalayan counterpart is *U. chumlia*.[68] The corresponding Far Eastern elms mostly go under the names *U. davidiana* or *U. wilsoniana*.[69] These are further discussed below.

The third region is the east Asian steppe which supports two elms: *Ulmus pumila*, most closely related to the elms of the second region, and *U. macrocarpa*,[70] whose affinities are closer with the elms of the first region.

The problem has now to be faced as to how many botanical ranks beneath the species are best taken up to designate British elms. The lower ranks so far used for naming elms comprise subspecies, variety, subvariety and forma. Again one is faced with the prospect of having to handle such names as *Ulmus nitens* subsp. *nitens* var. *hunnybunii* subvar. *pseudostricta*. This is also excessive. The view is strongly held here that one botanical rank beneath the species, the variety, is sufficient for naming purposes. Should further distinction be required for any particular purpose, then it should be achieved outside the realm of Latin nomenclature.

The remaining decision to be made is the number of species to recognize. If *Ulmus laevis*, which does not occur wild in this country, is excluded, similar considerations apply to Europe in general as to England in particular. Only three possibilities need consideration. The simplest solution is the two-species concept. In this, two species are accepted, the northern and montane Wych Elm (*U. glabra*) and the southern and lowland Field Elm (*U. minor*). There is one interspecific hybrid, called here the Hybrid Elm (*U.* × *hollandica*).

The four-species concept gives specific status to the form of Wych Elm without hairs on its fruits (*Ulmus glabra*), Wych Elm with hairs over the seed (*U. elliptica*), Field Elm excluding English Elm (*U. minor* in a more restricted sense) and English Elm (*U. procera*). There are six possible two-species hybrids, of which all but one, *U. elliptica* × *U. procera*, occur in the wild. There are four possible three-species hybrids. It would be possible to accept *U. elliptica* but not *U. procera*. To do the opposite would be rather perverse.

Lastly, there is the many-species concept. By further splitting of *U. minor*, up to seven British elm species can be recognized. If the same approach were applied to the European elms as a whole, it would be difficult, without inconsistency of treatment, to get away with less than 20 species. The number of possible two-species, three-species and higher hybrids goes up to correspond.

There are three criteria for deciding how many species to recognize within a biological population. A barrier to crossing is, in general, a good indication of differentiation at the species level, but there is none within the elms now under consideration. Discontinuity in structural and physiological characters is a second indication of differentiation at species level. This justifies the separation of Wych Elm (*U. glabra*) from Field Elm (*U. minor*). It is doubtful whether any other discontinuity of this magnitude is discoverable among the British elms. The third criterion is pragmatic. There is little point in erecting species that no one, not even non-specialist botanists, can identify. With the many-species concept, identification becomes so difficult and the number of possible hybrids so large, that, as a means of communicating information, the system breaks down. The two-species concept is adopted in this book.[71]

A key to the species and varieties of elms recognized is given in Appendix 1.

Wych Elm (*Ulmus glabra*)[72]

This elm, when standing alone, grows into a large tree, up to 40 m high. The branches, which remain smooth for several years, spread out fan-wise, giving a rounded silhouette. The bark of the trunk is grooved. The roots do not throw up suckers. The leaves average 10–11 cm in length and usually taper abruptly. Sometimes the shoulders of the leaf run out into cusps. The stalk in English specimens is very short but not always so on the Continent. Usually the leaves are very rough above but occasional samples completely smooth above are encountered. Total tooth number is usually over 130.

The fruit is about 2 cm long with a central seed. The fruit clusters, which contain fertile seed, remain longer on the tree than in the Field Elm (*U. minor*). They turn brown eventually so that the entire tree in early June appears from a distance brown rather than green.

Chemically, this species differs from the Field Elm (*U. minor*) in its great variability in isoperoxidase content. This is associated, as explained in chapter 2, with its sexual rather than vegetative reproduction.

There are several organisms which occur on this species but not on other elms. They include the plant bugs *Asciodema fieberi* and *Orthotylus viridinervis*, the butterfly *Strymonidea w-album*, the white-letter hair streak, and the moths *Abraxas sylvata*, the clouded magpie and *Discoloxia blomeri*, Blomer's rivulet.

It is also likely that some organisms which occur on other elms too have Wych Elm strains. These include the elm gall mite *Eriophyes filiformis* and the gall fly *Physemocesis ulmi*.

Wych Elm is adapted to a colder climate than Field Elm (*U. minor*) and is adversely affected by summer drought. Reproduction, as stated above is by seed.

As regards distribution, Wych Elm is pre-eminently a tree of the north of England and the Welsh border. In the south it is mainly found on higher ground, as the Cotswolds, the Chilterns and the Mendips. In Wales, it is mainly to be found in the river valleys in the east and south. It is widespread in Scotland, extending as far north on the west side as Assynt (Sutherland) only 50 km south of Cape Wrath. Except where recently planted, Wych Elm is rare in Ireland. It appears to be indigenous in some of the Antrim valleys.

In Europe, Wych Elm is again mainly a northern tree. Its Continental distribution is described in chapter 3. The west Asian stations of Wych Elm include Turkey, the Alborz mountains in northern Iran, and Afghanistan. There is a major discontinuity in distribution involving Siberia and central Asia. Wych

44

17

16

1, 120, 122

Elm reappears in the Soviet Far East, northern China, Korea and northern Japan.

This species is the only elm certainly or even probably native to England. It entered the country most likely over the dry bed of the North Sea in the Boreal period.

The species most closely related to it are as follows. *Ulmus bergmanniana*[73] occurs on mountains in central China. It may not be distinct at species level. *U. wallichiana* replaces Wych Elm in the Himalayas. Less closely related are *U. uyematsui*[74] on mountainous terrain in Taiwan, *U. changii*[75] from Chekiang, *U. macrocarpa*, a small bush-like tree in Chinese steppe country, *U. gaussenii*,[76] a Chinese elm closely similar to *U. macrocarpa*, and *U. rubra*, the slippery elm of North America.

Fig. 44. Wych Elm (*Ulmus glabra*), by Edward Hunnybun ($\times \frac{2}{3}$).[72]

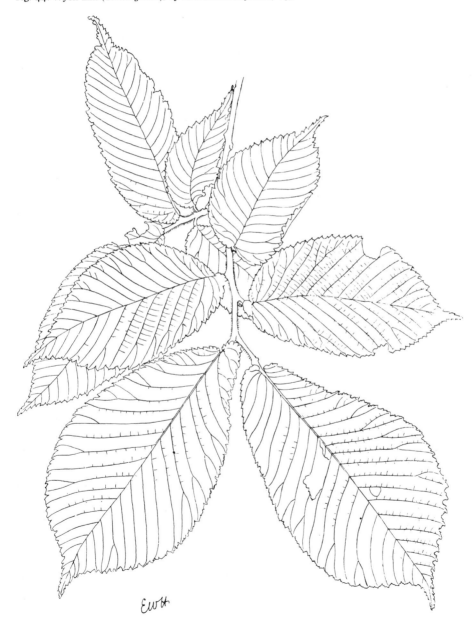

Wych Elm is moderately variable in the British Isles but there are no British variants thought worthy of recognition as named varieties.[77] The existence of smooth-leaved variants has been noted already. One variant character abroad is the leaf stalk. This may be relatively long in Spain. The presence of hair over the seeds on the fruit characterizes *U. glabra* var. *trautvetteri*,[78] which occurs mainly in the USSR.

Field Elm (*Ulmus minor*)[79]

This extremely variable species has diverse branching habits which are described under its varieties. All varieties throw up suckers from the roots. The leaf is also variable. It is usually less than 7 cm long and the total number of teeth is usually below 110. The stalk is relatively long. The fruit is 1–2 cm long and the seed is displaced towards the apical notch.

It appears, as mentioned in chapter 2, that Field Elm differs from Wych Elm (*U. glabra*) in two chemical respects. Its leaves contain the flavonoid quercitin 3-glucuronide which Wych Elm does not. And with regard to their isoperoxidases, the English representatives of Field Elm differ from Wych Elm in showing little variation. The majority show the same combination of two particular isoperoxidases.[80]

Several of the leaf-mining moths are normally restricted to this elm. They include *Buccalatrix albedinella*, *Phyllonorycter schreberella*, *Stigmella marginicollela* and *S. viscerella*. The gall mite *Eriophyes campestricola* also has Field Elm as its host.

This species is adapted to warmer conditions than Wych Elm (*U. glabra*) and is more tolerant of summer drought. Reproduction is normally vegetative by root suckers. Seed is only set in rare, especially favourable, seasons.

The detailed distribution of this species is given under its varieties. In general it is a species of the Midlands and southern England. It is present in southern Wales. It is absent, except for a few relatively recent plantations, in Scotland. In Ireland, where its status is dubious, it has a scattered distribution, with a concentration in the south-east. Most elm in Ireland goes back no further than to eighteenth century plantings. Possibly some of the Field Elm in the south-east is older, perhaps much older.

The Field Elm has an extensive distribution in 16 Europe. This is outlined in chapter 3. It is characteristically a tree of lowlands in central and southern Europe, particularly in association with rivers. In Africa it is only of long-standing occurrence in Algeria.

In Asia the Field Elm occurs in Turkey, and there are scattered locations in Lebanon, Syria and Israel. It is present in the forest of the Alborz mountains in northern Iran. Elsewhere it is only present as one of the ball-headed horticultural elms which are planted widely in central Asia.

The Field Elm appears not to be a native of northern Europe. It is believed to have been introduced into the British Isles on a number of separate occasions from the Later Bronze Age onwards. The evidence for this is presented in chapter 3.

The species most clearly related to the Field Elm are the Far Eastern elms *Ulmus davidiana* and *U. wilsoniana*, which are doubtfully separate from one another. More distantly related are *U. castaneifolia*[81] of south-west China, *U. pumila*, the elm of the central Asiatic deserts, and *U. chumlia* of the western Himalayas.

The British varieties of Field Elm regarded as sufficiently distinct to merit recognition are described in the sections immediately following. Other varieties are distinguishable on the European mainland.[82] However, the extensive mass collection of material necessary for ascertaining whether well characterized and reliably separable varieties exist has only been undertaken in Yugoslavia,[83] northern France[84] and northern Spain. It is not at present possible to present an overall breakdown of the European Field Elm into regional varieties.

A character whose variation is associated on the Continent with geographical situation is hairiness of the leaf blade and leaf stalk. Elms with densely hairy leaves tend to be more frequent as one proceeds southward towards the Mediterranean. Several such have been given varietal or even specific names, but it is highly doubtful whether all the elms with densely hairy leaves are closely related.[85] They vary greatly in other leaf characters. The view taken here is that it is premature to erect a variety on leaf hairiness until more material has been collected and analysed for other characters.

Narrow-leaved Elm (*Ulmus minor* var. *minor*)[86]

This elm grows up to 30 m. The trunk may be straight or crooked. The canopy is always rather open, and the head often umbrella shaped. The bark is usually grooved.

Leaf breadth is usually less than two-thirds the length. The leaves may be light or dark green. They are usually smooth above, but sometimes matt.

It is noteworthy that only some populations of Narrow-leaved Elm are infected by the gall mite *Eriophyes campestricola*. The strain that occurs on this variety in England appears to be different from that occurring on English Elm (*U. minor* var. *vulgaris*).

Fig. 45. Narrow-leaved Elm (*Ulmus minor* var. *minor*), by James Sowerby (× 1).[86]

Fig. 49. Dutch Elm (*Ulmus × hollandica* nm. *hollandica*), by Edward Hunnybun (× ¼).[97]

7 cm long, but with a relatively longer stalk than in Wych Elm (*U. glabra*). The upper surface is often smooth or matt. Most sucker readily from the root.

Hybrid Elms are likely to be purchased sporadically wherever the two parental species occur together. In three areas they occur extensively and form a major component of the landscape. The largest area is a band extending across Essex from the Hertfordshire border to Suffolk and in an extensive area of southern Suffolk. The next largest is in northern Bedfordshire and adjoining parts of Northamptonshire. The third is a small area to the southwest of Lincoln.

Comparable zones of hybrids between Wych Elm (*U. glabra*) and Field Elm (*U. minor*) occur in Europe, notably in Picardy and Cotentin in northern France.[95] The Cotentin hybridization zone also extends over the Channel Islands. Such areas appear usually to have arisen when Field Elm has been taken by man outside its area of native distribution into locations where Wych Elm is indigenous. The two principal notomorphs of Hybrid Elm (*U. × hollandica*) are described in the next two sections.[96]

Dutch Elm (*Ulmus × hollandica* nm. hollandica)[97]

Dutch Elm must be clearly distinguished from the *hollandse iep* of the Netherlands. The latter is sometimes called Belgica. It is another notomorph of Hybrid Elm and is the elm principally planted throughout the Netherlands in recent times. Dutch Elm as here understood is a tree growing to 30 m. The silhouette is distinctive amongst elms in being open and scraggy. The main branches are crooked. From a distance Dutch Elm can be confused with oak. The smaller branches are often marked by conspicuous cork flanges. The leaves are relatively broad and in trimmed hedges can be confused with those of English Elm (*U. minor* var. *vulgaris*). The marginal tooth number is normally higher, over 120.

In isoperoxidase content, Dutch Elm is indistinguishable from Field Elm (*U. minor*). The gall mite *Eriophyes campestricola* does not gall the Dutch Elm in England but may do so in the Netherlands. Probably a different race of mite occurs there which is absent from England.

Dutch Elm is of scattered occurrence in England but is a conspicuous component of the landscape in exposed coastal areas in the west, such as the Land's

End peninsula, where it characteristically grows in company with sycamore (*Acer pseudoplatanus*). In Wales it also occurs in exposed coastal areas, dominating much of the landscape behind Carmarthen Bay. It is quite widely planted in Scotland and Ireland.

In Europe, Dutch Elm is a frequent tree in Picardy, where it was formerly known as *ypereau*. It was introduced to England via the Netherlands in the seventeenth century, as described in chapter 5. It still occurs, though only rarely, in the Netherlands.

It shows little variability in the British Isles.[98]

Huntingdon Elm (*Ulmus × hollandica* nm. vegeta)[99]

This is a rapidly growing tree which may reach a height of 40 m. The trunk usually divides into a number of major branches which spread out fan-wise, the whole tree having a rounded silhouette. It is the European elm nearest in habit to *Ulmus americana*, the American elm.

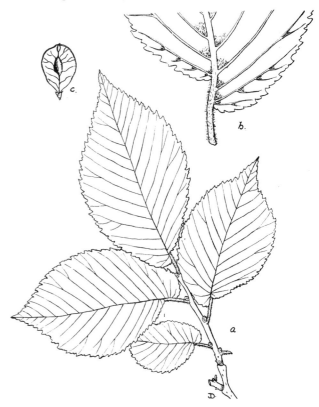

Fig. 50. Huntingdon Elm (*Ulmus × hollandica* nm. *vegeta*), by Jean Dickson.[99] *a*, shoot ($\times \frac{1}{2}$); *b*, leaf base ($\times 1\frac{1}{2}$); *c*, fruit ($\times \frac{1}{2}$).

8

Utilization I[1]

Elm is not always utilized because of one or other of its particular qualities. Under some circumstances, any tree will do. Such general applications are not considered further. What is of interest here is the large number of cases where elm is used preferentially. Preference is a relative concept. A tree will be preferred in one place for a particular application when there is no other tree locally available that is equally satisfactory. Elsewhere, another tree may replace it. Elm was preferred for archery bows in parts of northern Europe where yew was absent. When yew was available, elm fell to second place. In North America, yew and elm both give way to Osage orange (*Maclura pomifera*), the *bois d'arc* of the French Canadian settlers.

During the historic period, some uses to which elm has been put have come down from prehistoric times. Some early uses have died out. New applications, unknown to prehistoric man have come in. To obtain a complete picture of elm utilization in Europe, it is therefore necessary to investigate elm uses elsewhere, particularly in North America and the Far East. Such uses give the range of likely utilization in Europe in early times.

Earlier general uses can be inferred from certain restricted uses still surviving. Emergency use, for example as a famine food, is likely to indicate a general use in earlier and harder times. Children's uses, similarly, may reflect earlier adult use. Lastly, what is now done for animals may formerly have been done for man. Elm leaves have been boiled for pigs in quite

recent times. It is recorded below how they were formerly used for human sustenance.

Pollarding and trimming

Depending on the uses to which they are to be put, elms are grown in one of four different ways. The simplest is to leave the tree untrimmed so that it achieves its natural shape. In recent times, this has been the most usual treatment. It was not the most usual treatment over large areas of England in earlier times.

Elm lends itself readily to pollarding. The main stem is cut off about 3–4 m from ground level and the branches that then issue from the crown are lopped regularly every so many years.[2] Formerly, lopped branches, whether from pollards or from trees trimmed in other ways, were used as animal fodder as described below. In more recent times, lopped branches were more likely to be used as stakes or firewood. Though old elm pollards are recognizable over a wide area in eastern England, the practice of regular pollarding has been largely discontinued. It was practised till very recently in several Cambridgeshire parishes.[3]

In parts of Italy from Classical times till the present century, elms were trained as live supports for vines. The tree habit required was a straight main stem 3–4 m high with a few stout, spreading, but relatively short branches at the head. These hold aloft the vine shoots and their bunches. The technique of trimming the elm and tending the vine is described in detail by the classical Latin agricultural writers.[4] It was called marriage and provided an important literary theme referred to in chapter 10. Although vines may be seen growing up elms elsewhere, for example in Spain and Portugal, the technique of marriage seems not to have spread beyond Italy. Even there it was limited to certain areas, notably central Italy and Emilia. In some parts of Italy, as around Naples, elms and other trees were trained as posts between which wires were run to support vines. This is not marriage.

The heavy demand in seventeenth–eighteenth century England for elm water mains, described later in this chapter, led to a particular form of trimming. A visitor from Scotland[5] described the method used around London. 'The Custom there is to cut off all the Side branches, Close by the Body of the Tree, and only to leave a small head, so that in the Winter they look

52

in a Manner like a very tall Hedge, and in the Spring are as Bare as May poles except the small Head'. Trees so treated are conspicuous in early topographical paintings.[6] The custom persisted into the present century, long after the demand for water mains had ceased, especially in the Thames valley.

Timber I

The traditional date for felling elm for timber utilization was All Saints' Day (1 November).[7] The special mechanical properties of elm timber are toughness, resistance to splitting and elasticity. These result from the interweaving of the wood fibres. They 'lock' and the timber is described as 'cross-grained'. The grain of the wood, when seen at a polished surface, is decoratively patterned. This pattern is termed 'partridge breast', 'crazy' grain, 'grain and flower' or 'bale and bastard' in England[8] and *pied d'cat* [cat's claw] in the Channel Islands.[9] It results from the wavy disposition of the lesser wood vessels mentioned in chapter 2. Old elm wood develops a characteristic bloom, reminiscent of plum skin.

The special uses of elm timber are so diverse that it is most convenient to deal with them chronologically. The earliest known elm artefacts are archers' bows. They have been found in deposits of Mesolithic age in Denmark.[10] Elm remained a bowyer's timber as long as archery was of any practical significance. It was the timber used by the medieval Welsh archers.[11] It features, as mentioned in the next chapter, in the various statutes enacted to regulate the manufacture of archery bows.

In Europe generally, elm took second place to yew for making bows. The importance of yew and elm in archery reveals itself in several ways. The two Celtic tribal names in prehistoric France derived from tree names were the Eburovices [yew warriors] and Lemovices [elm warriors]. Sometimes the name of the tree became that of the bow. The Indoeuropean linguistic root that gave *taxus*, yew, in Latin, became

Fig. 52. Wicken (Ca) elms II, by Anthony Day.

τόξον [toxon], bow, in Greek.[12] In early Scandinavian literature, *almr*, originally elm, became a standard metaphor for bow. In North America, as already mentioned, other timbers outclassed both yew and elm for archery.

In prehistoric times, wooden weapons for slashing and thrusting were used. Elm swords, javelins and spears have been excavated, also elm shields.[13] As metal became increasingly available, these types of wooden weapon disappeared.

A prehistoric use that has lasted into the present century is the use of elm timber in cart, waggon and chariot construction.[14] The part most particularly requiring elm timber is the hub of the spoked wheel. Whenever elm has been available, it has been the preferred timber for this purpose. It resists splitting better than other timbers both when the spokes are driven in and subsequently. Resistance to splitting is

so important that wheelwrights have long sought out sources of supply of particularly tough elm wood. Toughness depends partly on kind of elm and partly on terrain, light stony soils giving a tougher timber. The elms at Dovercourt (Ex), mentioned further below, were renowned for their toughness. In northern France, wheelwrights from long distances around had recourse to the elms at Vareddes (Seine-et-Marne).[15]

While elm was almost always used for the hubs of wheels and oak by preference for the spokes, the felloes that compose the rim could be either elm or ash. The boards that formed the bed of a cart or waggon were frequently of elm, especially if they had to stand up to shovelling. A long-boarded construction was preferable for this purpose. Under such conditions of hard usage, elm *duffs* or dents, where other timbers split.[16] The wheelbarrow has a long pedigree too. The hub of its complexly constructed four-spoked wheel

Fig. 53. Landscape with rainbow, Henley on Thames (Ox), by Jan Siberechts.[6]

was also elm by preference, and so also were its boards.

In early times, it seems that elm was a preferred wood for turnery, again because of its resistance to splitting. Elm bowls turned with a pole lathe were produced in a workshop at Bucklebury (Br) till the present century.[17] This was probably a survival of an industry once widespread. The medieval Devizes (Wi) skippet, used for holding documents, was turned from elm.[18]

Two reasons account for the eclipse of elm turned ware. Sycamore was introduced into England during the Middle Ages and its timber proved smoother and easier to work than elm. It was also believed in some quarters that elm turned ware gave an offtaste to the liquor it might hold.[19] This was not the view of the elm turners who maintained that elm bowls were the only sort that did not discolour with use and did not stain the fluids they contained.[20] Recently there has been a revival of interest in elm for turnery. John Thompson has specialized in the turnery of burr elm, which gives a particular beautiful surface effect.

Elm has also been used in the construction of ploughs. Its use for the plough beam was recorded by Hesiod and Virgil[21] but it was not so used in later times. It has been used for the mouldboard in England up to the present century.[22] In Hampshire, this was sawn from a single block. In Cambridgeshire it was shaped by adze. Red lead was painted on the surface. Elm mouldboards were preferred to metal mouldboards for working on heavy soils. Clay stuck to a metal surface. The elm surface shed adherent soil better.

In early Roman times, the use of elm that would have occurred first to most people would have been for flogging slaves. It is a pervasive theme in the writings of Plautus.[23] The fasces carried by the lictors were originally bundles of elm rods holding the lictors' axes.[24] The elm rods were used for beating those convicted, the axe for beheading. Later birch rods were substituted for elm rods.[25]

Many uses of elm timber date from medieval times. Timber-framed houses were usually of oak, but elm framing was used where oak was in short supply, as in Cambridgeshire and the Sussex coastal plain. Elm was frequently used for weatherboards before deal became available. Elm weatherboards were often cut with their lower edges retaining the original wavy outline of the surface of the bole.

Till the recent introduction of metal headstocks, elm, whenever available, had been used exclusively for the headstock of a bell. The method of hanging bells till recently was of medieval invention and involved the attachment of the bell by a variety of metal furnishings to an elm block, the headstock. This was so mounted in the bell frame that it could be swung, together with its bell, under the control of the bell rope. Only elm, of timbers generally available in England, had been found capable of standing up to the severe mechanical strains generated by a swinging bell. Many bells of medieval age are still attached to medieval elm headstocks.[26]

The use of elm for coffin boards goes back to the Middle Ages, Elm can be sawn thin, an obvious advantage for the function now under consideration. How widely it was used has varied. In medieval times, elm was used for coffins at any social level, from royalty downwards,[27] where it could be afforded.

For Queen Elizabeth I, however, while elm provided the coffin boards, these were surmounted by ornamentation in oak.[28] In the centuries that

Fig. 54. Fourteenth century bell at Kingston Lisle (Ox) with original elm headstock and iron work. From F. Sharpe. The church bells of Berkshire. VII. *Berks. Arch. J.* (1942) 46, 60.

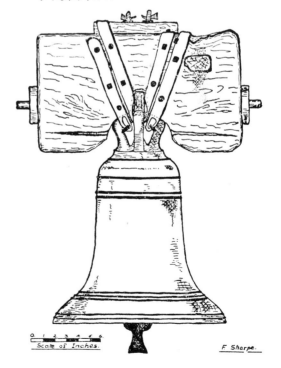

Scale of Inches.

F Sharpe.

followed, elm lost out at both ends of the social spectrum. The rich opted for oak as more stylish and easier to carve. The poor, if able to afford a coffin, were likely to have to make do with imported deal. In the nineteenth century, the *Illustrated London News* printed with relish the undertakers' descriptions of their receptacles for the remains of the exalted. Most usually three concentric coffins were used, each of different material and none of elm. For those of moderate means, elm remained the coffin board generally used. The association of elm with burial has become irrevocable. Innovations in coffin construction were few. One such was the fish-tail coffin where the sides were curved after nicking the boards by shallow saw cuts.[29] A tightly sealed thirteenth century elm coffin within an outer lead casing at Danbury (Ex), opened in the eighteenth century, caused some surprise. The well-preserved body within it was immersed in what the reported of the opening called 'catchup' in which were floating feathers, flowers and herbs.[30]

A medieval use of elm that continued into the present century was for artillery.[31] It was found in late medieval times that an elm gunstock took the recoil after firing better than other timbers. So, the provision of elm for artillery became a standard item in military supply. During Tudor times, the scarcity of elm in Ireland led to importation of elm timber from England to meet the needs of the English army.[32] The planting of elms along trunk roads in France created a highly distinctive landscape feature. In origin, it was a measure to supply the royal army with a necessary item of ordnance.[33]

Fig. 56. Statue of Sir William Walworth, in elm, by Edward Pierce.

Brave WALWORTH Knight Lord Major y Slew
Rebellious TYLER in his Alarmes
The KING therefore did give in Liew
The Dagger, to the Cityes Armes
In the 4th Year of Rich. 2nd Anno Domini 1381

Fig. 55. Elm flanged wheel from Brosely (Sa).

Fig. 57. Heraldic panel in elm behind the choir stalls of King's College Chapel, Cambridge, by William Fels.

Fig. 58. Reclining figure in elm, by Henry Moore.[39]

Elm was used from medieval times onwards for floor boards and stair treads, particularly in situations where rough usage was to be expected. For both uses, as with coffins, it lost ground later. The opulent preferred oak. The less well off used deal.

Pulley blocks were needed by builders and at sea. A specimen of Roman age was of ash.[34] Latterly elm was preferred, no doubt, again, because of its resistance to splitting. It was probably used in earlier times for the pulley sheaves as well, but lignum vitae has replaced it for this purpose since it became available.[35]

The use of elm for beetle heads is attested in the sixteenth century. The elms at Dovercourt (Ex), mentioned above, were renowned for this purpose. They were: 'a strong, knurly, and knotted and crooked sort of Elms, famous for their several uses in Husbandry, which with using wear like Iron'.[36] Beetle heads in Suffolk were often made from an elm *tod* or burr.[37] Related uses were the employment of elm for the heads of some types of masons' mallets and for flax-beating bills. Elm was sometimes used for the cogs on the wheels in mill gearing, but for this purpose apple wood was first preference. Some other uses for elm in mills are considered below.

The ability of elm to stand up to rough treatment without splitting resulted in its use for the construction of mine and quarry trucks. Before iron wheels became readily available, these trucks would have been fitted with wooden wheels. Elm was especially suitable. The flanged wheel for holding trucks to the rails they ran over appears to have been an English invention. One of the earliest to survive is 55 of elm.[38]

The cross-grained texture of elm timber presents problems to the wood carver. It is difficult for a stroke to follow the grain. In early times elm was seldom used for this purpose. There was, however, some outstanding exceptions. These include the elm statue of the fishmonger Lord Mayor, Sir William Walworth, who 56 struck down Wat Tyler. This was carved by Edward Pierce and stands today in the Fishmongers' Hall, London. Equally notable are the carved heraldic panels behind the fellows' stalls in King's College Chapel, 57 Cambridge. These were executed by William Fels.

In the present century, with improved technical facilities, elm has become a favoured medium for sculpture, partly because its grain can be exploited to emphasize or intersect the contours of the object sculpted. The best known of elm sculptures are those by Henry Moore[39] and Barbara Hepworth.[40] Other less 58, 59 abstract elm sculptures have been wrought by Ernst Blensdorf,[41] Roger Maycox and Bernard Reynolds.

Elm has been used for chairs and tables for

Fig. 59. Hollow form with white in elm, by Barbara Hepworth.[40]

many years, often as a substitute for more highly esteemed timbers. In some chair designs it played an essential role. These were designs involving a dished seat, that is a seat moulded to the contours of the sitter's bottom. The best known of these designs is the 60 Windsor chair. In this, the legs and back sticks are of beech. The bow is steamed ash. The dished seat, into which the legs and back are fitted, is of elm.[42] It was formerly rough shaped by adze by a craftsman appropriately named a bottomer.

Another design with a dished seat is the 61 Mendlesham chair, made originally by Daniel Day. This has an elm dished seat. The back is a light and elegant composition, usually in cherry wood. It normally incorporates a horizontal row of three wooden balls held between rails. This was a common motif in the older Suffolk chair, which had a simply curved, not a dished seat.[43] The traditional Hereford-

shire chair also had elm for the seat. This was severely flat.[44]

Other forms of seating in which elm was preferentially used include the settle, usually entirely of elm, and the three-legged milking stool whose seat was best of elm. Since milking stools have the legs fitted into the seat in the same way as a Windsor chair, they have been regarded as one of its ancestors. Saddle trees, which may also be treated as a form of seating, were preferentially of smoked elm.[45]

Various situations in which appliances are subjected to considerable mechanical strain have been dealt with by use of elm. The drumhead of ship and quarry capstans is one example. Specification of elm for this purpose is to be found in the fifteenth century steward's books of Southampton.[46] The stone nut in the wooden gearing of water-mill and windmills is another.[47] Elm bosses have been used for the trundle head of mill-wheels.[48]

Circumstances where knocking or kicking were inevitable also provided grounds for preferential use of elm. It was used, for example, for stable and cowshed

Fig. 60. Windsor chair with elm dished bottom.

Fig. 61. Mendlesham chair with elm dished bottom.

partitions, for mangers and for cattle cribs. Elm blocks were occasionally laid for street paving. The preferred timber for threshing floors varied. In some counties it was oak. Elm was used in Suffolk[49] and in the coastal plain of Sussex.

Dough troughs were customarily of elm. The choice seems to have been influenced by the folklore of bread making. Dough was thought to rise better in a cross-grained container.

Of elm appliances used in the recent past, space can only be given to a selection. Blacksmiths' and domestic bellows boards were usually of elm. So, often, was the rope maker's top. The mealing table used for grinding firework materials was either of elm or walnut.[50] Stricks for sharpening scythes were made by pasting an abrasive mixture on an elm wood base. Their manufacture was a local industry at Humberston (Hu).[51]

Later still, elm is found employed in dartboard construction, doubtless because it holds the darts. It has been used also for cable drums. For ice hockey sticks, *Ulmus thomasii*, the rock elm of North America and a particularly tough kind, was preferred.[52]

Elm withes have few special uses. They were used for binding whin faggots (*Sarothamnus scoparius*) for heating bread ovens in Suffolk cottages.[53]

Timber II

Elm timber is exceptional among British timbers in resisting decay when in continuous contact with water. This property has led to a number of uses. For these, unseasoned green timber is normally taken.[54]

Underwater piles have long been preferentially of elm. Roman pilework is mainly of oak and Roman writers do not mention elm for this purpose. An elm pile was excavated, however, from beneath the Roman wall at Gloucester.[55] The extensive use of elm for piles started in the Middle Ages. Vast quantities of elm piles supported the superstructure of London and Rochester bridges.[56] They were also used under Bedford bridge. This relatively late use of elm may have been a consequence of earlier shortage of supply. It may also have been due to earlier ignorance of the resistance to decay of elm timber in contact with water.

The pilework along the London waterfront in Roman times was, as elsewhere, largely of oak. No elm

Fig. 62. Elm water pipes at Clerkenwell (GL).[59]

has been detected. In medieval times, the piles were largely of elm.[57] The use of elm for groynes, lock gates and mud scoops also depends on its resistance to decay in contact with water.

The wooden water wheel of a water mill was usually constructed with a framework, axis and arms, of oak. The parts in continual contact with water were preferentially of elm.[58] These comprised the two 'rings' which formed the rims. Let into the rings were the 'starts' which supported the bucket boards. Each bucket consisted of a 'float' and a 'sole'. It was often necessary to bring the water to the wheel by a wooden channel, the flume, the flow being regulated by a penstock. All these were best made of elm.

Elm water mains were used extensively in England in the seventeenth and eighteenth centuries and created a major demand for elm trunks. These were trimmed for straightness of bole as mentioned earlier in this chapter. The trunks were formerly bored by hand using a large auger. The sound of the boring, audible over great distances, was a distinctive countryside sound. Later, machine bored pipes became available. The trunks were cut into sections about 2 m long. The pipes were tapered at one end so that the narrow end of one pipe could be driven into the wide end of another.[59] When laid, holes were drilled at intervals and plugged for fire emergencies, the plugs being drawn when water was needed.

Towns known to have had elm water mains are mostly in or near a region where English Elm is frequent. It may be, in some cases, that it was the demand for water pipes that led to elm being grown locally.

Elm pipes were not only used for carrying water. At Droitwich (HW) they were used to convey brine in the salt works.[60]

Wooden water pumps were also mainly made of elm. Both the pump casing and the pump plunger would be of this timber. Such pumps were widely used for lifting water from wells for domestic use.[61] They were also used to keep mines dry and to dispose of ship bilge.

In the recent past, when lavatory seats were of wood, elm was a preferred timber. A precedent is supplied by the medieval gardrobe of Ashbury (Ox) manor for which a substantial elm board provided the seat.[62] Tolerance of dampness may also have led to the choice of elm for cheese vats.

A number of different and sometimes conflicting requirements dictates the choice of timber in boat building. Resistance to decay when wet is only one requirement. Many parts of a boat do not come into continual or even regular contact with water. Mechanical properties are likely to be the principal consideration in choosing internal timbers. For the outer planking, other qualities, such as resistance to boring animals, have also to be considered.

Elm is relatively seldom encountered in the construction of early boats. This does not necessarily mean that other timbers were preferred. In many cases it is likely that boat builders lacked sufficient supplies of elm timber. In later times, it is clear that the choice of elm for local craft depended on availability of elm in quantity in the hinterland of the boat yards.

For large craft, oak was the preferred timber for the frame. This was due to its strength and the ease with which the requisite curved pieces could be obtained from naturally growing branches. Elm was preferred for two items of the frame, the keel and the kelson. The likely reason for this preference was not, however, resistance to decay under wet conditions but the ease of obtaining long straight pieces.

For planking, especially underwater planking, elm was much sought after. It was used for this purpose in such craft as the Deal (Ke), Rye (ES), Hastings (ES) and Cornwall luggers, the Bideford (Dv) polacker, the Thames, Devon and Cornwall barges, the Bridgwater (So) and Weston super Mare (So) flatners, the Somerset turf boat, the Bude tub boat, the Exe salmon boat, the Bristol pilot cutter, the Fenland lighter and the canal narrow boat.[63] It is significant that all these craft are of southern provenance, reflecting local availability of elm in quantity. Elm was also used for the paddles of paddle steamboats.

Elm was used in various boat accessories. In some cases its use was consequent on its mechanical properties rather than its tolerance of wet conditions. This is particularly so with pulley blocks and capstans which have been considered in the previous section. Other elm items include dead eyes in rigging, the caps joining mast spars and the bobbins and rollers of trawler nets. Two local nautical accessories deserve mention. One is the wooden bailer,[64] made of elm. The other is the killick, or stone anchor, used by Cornish fishing craft. For this, Cornish Elm, rather than English Elm, was preferred.[65] Of recent industrial uses based on resistance of elm to decay under wet conditions, the most conspicuous was the incorporation of elm baffle boards in water-cooling towers.

All the uses considered so far in this section have been in respect of water in the liquid phase. One use depended on the capacity of elm to tolerate contact with steam. Malleable timbers such as ash or elm itself can be bent when steamed. For this operation, they were put in a steaming chest. This was constructed of elm.

Combustion

It has been noted throughout the Northern hemisphere that elm wood catches alight with some difficulty. Once alight, it burns slowly and steadily without crackling and throwing sparks. These observations provide the burden of the various fuel rhymes:

> Elm wood burns like churchyard mould.
> Even the very flames are cold.

Elsewhere:

> Elm logs like to smouldering flax
> No flame is seen.[66]

Slow and steady burning is sometimes advantageous. The Arikara Indians of North America use a fire bed of elm wood for firing their pottery.[67] In parts of Norfolk it was the custom to dispense cider as long as the Yule log remained alight during Christmas night.[68] Revellers found that they did best when the log was of elm.

Slowness to ignite is a desideratum in the operation of the braking machinery of windmills. Braking is effected by pulling a brake band on to the rim of the brake wheel. The brake band sometimes consisted of a series of wooden blocks threaded along a flexible metal rod. The blocks and the rim of the brake wheel against which they pressed had to be of a slow-igniting timber. Poplar was best. The more generally available elm was second best.

It was probably the toughness of the wood that led to elm being used for production of fire by drilling. This method is widespread in primitive cultures. In Europe it only survived into recent times as a ritual performance for the generation of need fire, believed to be efficacious in such calamities as disease epidemics in cattle. Any association it might have had with elm had by then been lost. The Ainu of northern Japan and Sakhalim used an elm drill for fire production. In fact the Ainu name for elm, *chikisani*,

is derived from *chickisa*, which means fire drilling. The Ainu also used an even more ancient method for producing fire, namely by rubbing two pieces of elm root wood together.[69] In North America, the Iroquois used elm for fire production both by rubbing two pieces together and by drilling.[70]

Elm was used for preference for curing English home-grown tobacco during the short period during which this was being urged.[20] No part of elm is alkaloidal but the roots have been smoked as a tobacco substitute.[71]

The two utilizable end products of elm combustion are charcoal and wood ash. Elm charcoal is of the second degree. Wood ash from elm is relatively rich in potassium. It was, in consequence, a preferred source of potash and hence a starting point in soap manufacture. The usefulness of elm for soap making was recognized as far afield as Tibet.[72] One other use to which elm ash was put was by saltpetre boilers.[73] They used it to remove the scum from their lye.

Bark

As in other trees, the bark of elm is composed of an outer corky layer and a more delicate inner bark. Neither is much used now in Europe. In earlier times in Europe and till recently in the Far East and North America both had manifold applications.

The outer bark will be taken first. This had two main uses. The first was for covering houses. This was a practice among Indians of the eastern side of North America where elms are native. It was used for various types of dwelling, in particular the rectangular long houses of the Iroquois and the hemispherical houses of some other tribes.[74] Amongst the Huron, elm bark was also used, but cedar bark was more highly esteemed.[75]

The second main use of elm bark was for covering canoes, again amongst North American Indians. In the more northerly part of North America, as by the Huron and Algonquin, birch bark was used. Their neighbours to the south, especially the Iroquois, used elm bark.[76] Green elm bark was sometimes used for food containers by the Algonquin.[77]

The inner bark has a greater range of use. It contains bast fibres utilizable for textiles. In Europe the principal product was bast rope, which was very widely used. In France it served, in particular, for well ropes.[78] Elm bast however took second place to lime bast. The ancient Welsh laws refer to ropes of *llwyf*.[79]

As explained in chapter 4, this word was used indiscriminately for elm and lime in medieval Welsh. It is likely that elm would have been largely used. Elm was also used for cordage in northern India.[80]

Elm bark fibres were woven into coarse cloth for garments by the Ainu of northern Japan and Sakhalin.[81] Amongst the North American Indians, the inner bark fibre preferred for weaving was that of mulberry (*Morus rubra*). This gave a fine cloth. Elm was sometimes substituted, especially by the Creek Indians.[82]

The inner bark of elm has a relatively high content of nutritive carbohydrate. It has been utilized over most of the area of distribution of elm for food.[83] Elm bark bread has been used as a famine food in Norway and the USSR in quite recent times.[84] It is probable that this late emergency use indicates an earlier general use. At least it is more palatable than elm sawdust, used as a famine food in Germany in World War I.[85]

There is considerable variation between elm species in the content of nutritive material in the inner bark. It is particularly high in the inner bark of the American species *Ulmus rubra*, the slippery elm. This was used as a food but its principal application was in medicine. It is considered further below.

The inner bark extract of elm has adhesive properties. These were utilized in China and Japan. The extract was used in these countries as a tile cement, as a lubricant, for confecting incense sticks, and for setting hair.[86]

A use of elm bark recorded only for the North American Indians is in rendering bison fat.[87]

Foliage

The feeding of elm leaves to animals was one of the earliest, most widespread, and most important uses of the tree. It probably provided the motivation for the planting and importation of elm by man in prehistoric times, as discussed in chapter 3. The basis of this preferred use of elm leaves for animal feed seems to be less any high concentration of nutritive material than the absence of noxious substances or unwanted side effects.

In ancient Greece, use of elm foliage for animal feed was recorded by Aristotle.[88] All the Roman agricultural writers from Cato onwards mention it.[89] Amongst the early Irish, elm was *cara ceathra* [friend of cattle], as explained in more detail in the next

chapter. The designation presumably refers to the same application. Elm foliage was fed to animals in China and Japan. It is still an important animal feed in northern India.[90]

Elm leaves were fed to animals over much of Europe till recent times, perhaps more widely in France, Switzerland and Scandinavia than elsewhere. They were so used in England up to the present century.

The mode of utilization varied. Young leafy shoots might be lopped, made into bundles, and dried. Alternatively, the leaves might be stripped from the branches and dried. In either case, the leaves would have been fed in the dried state in earlier times.[91]

In more recent times, elm leaves have been boiled. This has been done for pigs in England up to the present century.[92] Possibly elm leaves were processed as a sauerkraut in the central European area.

Elm leaves were eaten by children in France. It was suggested in the first section of this chapter that what pigs and children may have eaten in recent times is likely to have been a food of general consumption in earlier times. In fact Dioscorides refers to boiling the leaves as a vegetable dish.[93] In the ninth century, the Japanese traveller Ennin described how he was regaled with elm leaf soup when passing through Hopeh.[94] Nearer to home, the monks of Fountains Abbey, in the difficult early days of their new foundation, lived off cooked leaves from the large elm that gave them shelter.[95] Hugh de Kirkstall records this as a feat of endurance. It was a feast of fat things compared with the diet of the founding monks of Clairvaux who had to subsist on beech leaves. In recent times, elm leaf meal has been used in bread making under famine conditions in central Europe.

Elm leaves have also been tried as a substitute for, or an adulterant, of tea. Neither in Russia nor England, where they were tried, was this experiment successful.[96]

Flowers and fruits

Utilization of flowers and fruits can be covered quickly. The flowers provide a pollen source for bees in early spring. The only direct use which has been noted is by the Chinese in Kansu. They roll the flowers in flour and steam them as a springtime delicacy.[97]

The fruits were fed to goat kids in Roman times.[98] They are eaten by children in France.[99] Again, what animals and children have eaten in historic times could well have been a generally consumed food in prehistoric Europe. This is rendered all the more likely by finding that the Chinese have eaten elm fruits up to the present century.[100]

Medicine

Elms contain little of any physiologically active or medically useful substances. What there is occurs mainly in the inner bark. This contains a certain amount of tannins and of mucilaginous carbohydrates. The concentrations vary with the kind of elm. When the concentration of tannins is sufficiently high, the inner bark can be classified as mildly astringent. This is the situation in the European species. If there is sufficient mucilage, the inner bark can be applied for soothing raw surfaces. It can be taken internally as a bland food in cases of diarrhea or dysentery. This is the case in *Ulmus rubra*, the slippery elm, of North America.

The history of pharmacological use is largely a record of delusion. Fortunately, as regards the elms, though they may have contributed little to any patient's recovery, they are likely to have killed very few. It will not be possible in what follows to consider the medicinal qualities of elm pharmaceuticals in isolation from the medical folklore that went with them.

The classical pharmacological tradition will be taken first. This is enshrined in the writings of Dioscorides.[93] Inner bark, leaf, decayed wood and root are all used. External applications of one or other of these parts were recommended for skin eruptions, healing of wounds and promotion of union of fractures. Taken internally, the inner bark in wine or water was said to bring up phlegm. The fluid in the galls caused by aphids was used as a face lotion to clear the skin. Later Greek medical writers added little to this inventory.[101] Pliny took over Dioscorides acount much as it was.[102]

Dioscorides had been translated into Latin by the fifth century and his work was widely used in western Europe right through the Middle Ages and later. His works were translated into Syriac in the ninth century and then into Arabic. However the Arabic pharmacology of elm became completely confused.[103] Greek πτελέα [ptelea], elm, was translated into Syriac as ﭏﯹﯹﯹ [daddara], in turn rendered into Arabic as داردر [dardar]. The newly coined Arabic word was not

generally understood and was eventually misinterpreted to mean ash (*Fraxinus*). So elm dropped out of later Arabic pharmacology in as far as this was based on Dioscorides. The Greek text of Dioscorides became available again in western Europe in the fifteenth century, and several fresh translations into Latin were made during the Renaissance period.

One of the contributions of the earlier Arabic medical writers that passed into European medicine was their classification of drugs. Elm was classified by Ibn Sina (Avicenna) as cold in the first degree and dry.[104]

Decoction or extract of inner bark of elm featured in the pharmacopoeias of most European countries. It disappeared from the *British Pharmacopoeia* towards the end of the last century. One bold sceptic had, as early as 1868, breathed the suspicion that the preparation was practically inert.[105] It had meanwhile been attributed with an enlarged spectrum of usefulness; as a sore throat gargle,[106] for increasing perspiration, as a diuretic, for reducing fever, to improve muscle tone,[107] and for alleviation of vaginal discharges.[108]

The fluid in the leaf galls was not taken over by the pharmacopoeias but it continued in esteem in popular use. It was usually known as St John's oil, presumably because the galls reached full development on the Feast of the Nativity of St John the Baptist (24 June). In England, it was known more prosaically as elm water. It acquired various new uses. In Italy it was recommended for relief of rheumatic pains and bruises.[109] In Bulgaria, it was used to arrest bleeding.[110]

Parallel with the main classical pharmacological tradition, there were other more arcane traditions. Pliny cited some of these. He said the pounded leaves relieved sore feet. The exuded sap received credit as a hair restorative.

Marcellus[111] recommended external application of elm inner bark in sea water for relief of gout. An elm wreath around the temples countered dizziness. A draught of Falernian wine with elm leaves and an odd number of pepper grains relieved cough. This remedy had to be taken on a fasting stomach. It presumably worked equally well without the elm leaves.

More mysterious still are the remedies of Sextus Placitus.[112] For vaginal discharge he recommended that the patient squat over hot coals upon which had been sprinkled the powdered dung of a bull that had been feeding on elm leaves. Fortunately for patients, neither Marcellus nor Sextus Placitus were much read in later ages.

The Anglo-Saxons were acquainted with the pharmacological tradition of Dioscorides. They had another of their own. One of its favourite devices was the boiling of the bark of some ten to twenty different trees in urine or milk.[113] Amongst the barks specified were those of *elm* or *wice*. Such concoctions were recommended for skin eruptions, healing of wounds and shingles.

In the seventeenth century there was a vogue in England for the medical use of exuded tree sap. Elm sap was first recommended for liver complaints. Later it was thought to reduce fever.[114] In ancient Ireland there was a celebrated elm at the monastery at Lothra. The sap of this tree had such wonderful properties that monks from elsewhere forsook their own monasteries to imbibe it. The sap was said to be a sweet-tasting fluid with the flavour of wine. This one could guess. More remarkable, it was sufficient food and drink for those who partook of it. Even more remarkable still, everyone who drank it experienced the flavour that he liked best of all. There are suggestive parallels between this fluid and the slime flux discussed in chapter 6.[115]

The only elm pharmaceutical that has survived modern scrutiny is not a European product at all but the inner bark of *Ulmus rubra*, slippery elm, mentioned above. This was widely used by North American Indians for relief of diarrhea and dysenteries, and after childbirth. It was also used externally to soothe raw surfaces. Its later use, boiled in water with gunpowder to expedite labour in childbirth, crosses the border between medicine and sympathetic magic.[87, 116] The European settlers in North America took over slippery elm bark from the Indians and it was admitted to and remains in the American pharmacopoeia. It is imported into Europe as a bland food for costive patients. It also entered the domain of underground medicine both in North America and England. Pieces of the inner bark inserted into the neck of the womb absorb moisture and swell, so acting as an abortifacient.[117] To circumvent this, it is now sold in granular form rather than in sizable pieces.

Entirely distinct from all the foregoing is the role of elm as an allergen. Elm pollen causes an allergy of the upper respiratory tract, manifested as hay fever, in some individuals.[118] In some cases, allergic reaction to pollen is coupled with allergic reaction to a number of other tree pollens. A rather unexpected association has been found between allergic reaction to elm pollen and to plane pollen. However, some people are allergic to plane pollen but not to that of elm.

Elm wood has occasionally been implicated as a cause of wood cutter's eczema. It now seems that this condition, for which a number of tree species have been blamed as causes, is induced by liverworts or lichens on the surface of the bark.[119] The hairs of elm leaves may penetrate the skin to cause a mildly irritant contact dermatitis. A few individuals, in addition, develop an allergic contact dermatitis on handling elm foliage, as, for example, when hedge trimming.[120] Patch tests in Denmark revealed that some 25% of individuals were sensitive.

9
Utilization II

The preceding chapter was concerned with direct utilization of elm. In this chapter a survey is presented of the various other ways in which elm has penetrated into human affairs.

Amenity[1]

Elm planting, up to the end of the Middle Ages, was almost entirely utilitarian. In open-field country, it was largely confined to the closes that lay behind the tenements of nucleate villages. In some areas elm was the principal tree so planted and created a screen hiding the houses from view from anyone approaching. Such villages were, till recently, readily identifiable, more so than in Victorian times when the post-enclosure hedges were still thick on the ground. Many examples existed in southern Cambridgeshire, as at Boxworth, Elsworth, Grantchester and Teversham.

With dispersed settlement and early enclosure, as in much of Essex, elms were likely to be much more generally scattered over the entire terrain of a parish.

Specimen elm trees, very often on village greens or at road junctions, were frequent in medieval times and many such trees survived into recent times or were replaced by new ones.

The history of amenity planting of elm is continuous from the late Middle Ages onwards. Such notable figures as Richard III and Cardinal Wolsey were reputed elm planters.[2] Rewley Abbey outside Oxford had a formal planting of elms that was supposed, falsely, to represent the abbot in chapter.[3] The medieval Church passed from the scene with Abbot Feckenham planting elms at Westminster Abbey. When summoned by Elizabeth I, he insisted on completing his planting. Proving equally inflexible over matters of religion, he soon found himself in the Tower of London.[4]

The French practice of planting roadside elms to provide timber for gunstocks was noted in the previous chapter. It gave rise, in France, to one of the most distinctive landscapes in Europe. In Champagne, for example, long, straight, double rows of elms shading the major trunk roads cut across the immense treeless arable plain. Roadside planting, often with elm, became frequent in England from the eighteenth century onwards, divested however of military function.

Though the planting of elm avenues is commonly regarded as a French fashion introduced in the seventeenth century by the landscape designer Mollet, there are examples that are earlier. John Evelyn notes the remains of the elm avenue at Nonsuch Palace (Su), already demolished in his time.[5] The elm avenue at Longford Castle (Wi) may have gone back to the late sixteenth century.[6]

Mollett was however a potent influence. His plan was simple.[7] A grand avenue of *orme femelle* or lime was to be centred on the front facade of the grand house. This conception may have been borrowed from the prevalent layout of the grounds of Palladian villas in the hinterland of Venice. In England, English Elm was most usually substituted for *orme femelle*, which, as pointed out in chapter 4, was a designation of the group of Hybrid Elms widespread in northern France. The approach avenue could be supplemented by avenues centred on other facades of the grand house. Further avenues might also be laid out in the surrounding grounds. Sometimes these would radiate from a suitable object, usually a piece of statuary. The outward vistas would frequently also close on a suitable object, perhaps another piece of statuary, but often a distant church spire. A rare device, said to have been used at Canons in Little Stanmore (GL), was a layout of diagonal elm avenues centred on the corners of the grand house.[8] Elm avenues were sometimes laid out as rows of trees flanking a rectilinear canal as at Boreham House (Ex).

Le Nôtre, who laid out the grounds at Versailles, and made extensive use of elm,[9] has had attributed to him the layouts of St James's Park[10] and Greenwich Park.[11] Both were laid out in the seventeenth century formal French style but Le Nôtre appears not to have been directly responsible. St James's Park was planted with elm and lime and had a rectilinear canal. It was entirely re-landscaped in the nineteenth century. Greenwich Park was planted with elm and chestnut.

The foremost of the early landscape gardeners was Henry Wise, who was responsible for the principal elm avenue at Blenheim Palace near Woodstock (Ox).[12]

The supreme artist in formal elm planting was Charles Bridgman.[13] Perhaps his best approach avenue was the Narrow-leaved Elm avenue of Wimpole Hall (Ca),[14] so useful for locating Bassingbourn airfield from the air. Bridgman's masterpiece was Kensington Palace Gardens,[15] regarded by some as London's foremost ornament. Its landscaping had been begun by Wise. Bridgman recast it completely. His plan was a vast expanse of spaced trees of English Elm. Radiating avenues centred on the palace cut through this. Other avenues intersected the radiating set. It had matured by the nineteenth century. The one change made before the elms began to die was the re-aligning of one of the intersecting avenues, the Lancaster Walk, on the Albert Memorial. Some of the eulogies lavished upon the elms of Kensington Gardens are reproduced in the next chapter.

Another French importation was the mall. Originally this was a long narrow pitch, properly floored with powdered oyster shell, required for the ball game pall mall. The pitch was hemmed around by rows of elms planted on either side. Later, both in France and England, the mall became any grove of elms disposed in parallel ranks. In France, it became a shelter for sitters. In the colder climate of England it became a promenade for walkers. The seven-ranked mall at Tours was thought by Evelyn to be the noblest in Europe.[16]

The first English mall, with two rows of elms, and used for the game, was laid down just north of St James's Park.[17] The site is the present Pall Mall. Later the mall was moved into the park and incorporated into the formal planting scheme already mentioned. It served initially as a pitch for the game, latterly as a fashion promenade. The site is still The Mall. It was first planted with rows of elms and of lime. The limes appear not to have lasted long. The elms survived till their eventual replacement by planes.

Other elm malls, for promenade, were subsequently laid down, as at Moorfields, Chiswick and Hammersmith.[18] The last was particularly associated with Catherine of Braganza, consort of Charles II. The principal walks around London where ladies and gentlemen could display their finery were, successively, Moorfield Mall, Gray's Inn walks, The Mall, The Ring in Hyde Park and the Long Walk in Kensington Palace Gardens. All but The Ring were under elm.[19]

The two older universities have two landscape features of their own. The first is the academic grove. This is quite different in function to the village grove mentioned in chapter 5. It exists to provide a setting for thought and contemplation. Its antecedent is the olive grove that sheltered Plato's Academy. A fine overseas example was the Lundagård, the Wych Elm grove of Lund University in Sweden. In Oxford and Cambridge, where the colleges monopolize the university landscape, the academic groves pertain to them rather than to the university as such. The Cambridge Backs, however, though owned by several colleges, became a single harmonious elmscape from enlightened common policy. The college groves were diversely composed. Trees other than elm were usually present but the latter was generally conspicuous.[20] Magdalen College Grove was the finest example in Oxford.

The second university feature is the academic walk. Its function overlaps with that of the academic grove, but it is more to ease discussion between peripatetic scholars. Its antecedent is Aristotle's Lyceum. Many of the colleges have their own walks. These were often of elm. In Cambridge, academic walks of English elm were planted some distance out into the western fields. The line of the walks is preserved in the road plan of the Victorian expansion. Christ Church Broad Walk was the outstanding Oxford example.

Reaction against the French style of formal landscape planting set in half way through the eighteenth century. It has been analysed at length and in Byzantine depth by many writers.[21] In part it was merely anti-French. Disciplined rectilinear planting designs were identified with the autocracy of the French monarchy. For a freedom-respecting nation like England, freer, less constrained layouts were deemed appropriate. Nature was to be the guide.

The Grand Tour terminating in Rome also played a part. Not only did this result in a taste for the scenery of the Campagna, it also led to an interest in the painters who idealized this scenery. Foremost amongst these was Claude whose serene landscape backgrounds were typically constructed on the following formula. The view point is elevated. The viewer is confronted with alternating planes of light and dark as he looks into the depth of the picture. A soft golden light illuminates the horizon. This ideal landscape coincided in sentiment with the many prospect poems that were composed in the wake of Sir John Denham's *Cooper's Hill*.[22] These had rather little of topographical detail but much moralizing, uttered from a viewpoint similar to that in Claude's paintings. It seemed to be assumed that viewing such landscapes would promote virtuous reflections. The cult of melancholic meditation, also fashionable at the time, was readily assimilated into the new attitude to landscape.

A last component of this mix was chinoiserie. Starting with the Jesuits, reports were trickling into Europe of the ways in which the Chinese dealt with landscape. It was gathered that symmetry and straight lines were 'out' in China and that what were 'in' were studied asymmetry and layouts based on curves. The new package was designated by the mistransliterated and probably misunderstood term *sharawadgi*.[23]

So, since in England Nature was to be the guide, the next step was to decide what mood of Nature should be the source of inspiration. The seminal thinking about this was presented to English readers in Burke's *Essay on the sublime*.[24] He established two categories of scenery. The sublime was distinguished by such attributes as vastness, ruggedness, abrupt deviations in contour, darkness, gloom. It was linked with the human emotion of pain. The beautiful, on the other hand, was smaller in scale. Changes in contour were gentle. Overall effects were of mildness, lightness and delicacy. The associated human emotion was one of pleasure.

Perhaps implicit in Burke, but certainly explicit in the writing of Uvedale Price,[25] is the recognition that what identifies the beautiful is the set of characteristics most admired by men in the female body. It was easy to establish equations between these categories and particular painters. Salvator Rosa represented the sublime. Claude was the exemplar of the beautiful. The analysis of these two categories became a fashionable pursuit in the eighteenth century and has been revived in recent decades. The situation was complicated by the addition of a third category, not accepted by everyone, the picturesque.[26] This originally meant, as its name implies, of such a quality that it would make

a good picture. The ingredients of this were 'roughness and sudden variation joined to irregularity'. It did not, however, include such attributes as vastness, darkness and gloom, which belonged to the sublime.

The impact of this ferment on the practical business of tree planting was diverse. Burke had expressed the view that tall trees such as elm were awesome and majestic. This put them into the category of the sublime rather than the beautiful. The general opinion however was that though Wych Elm could, on occasion, be sublime, the English Elm was definitely beautiful, and not sublime. It might, however, by losing its leader, be promoted to the category of the picturesque.[27]

The landscape planters of the late eighteenth and early nineteenth centuries were agreed that Nature must be followed, as against the old French formalism. They differed in the value they attached to its three aspects.

William Kent was the first major exponent of the new school of landscape planters.[28] Only one of his designs, at Rousham (Ox), has survived.[29] This is so constructed that the visitor, pursuing a set route, meets a series of viewpoints, beautiful or picturesque, at which he may pause to reflect. One of these is at the end of an elm avenue, left over from an earlier layout by Bridgman. Kent did not grub up his predecessors' plantings as resolutely as his successor Lancelot Brown.

Kent favoured mixed plantings of tree species. They were thought to be more natural. So it comes about that it is usually easy to pick out late eighteenth and early nineteenth century landscape planting by the ecologically monstrous combinations of unlikely tree species that the planters threw together. The boundaries of the plantations were curves such as Nature seldom conceived.

Lancelot Brown, better known as Capability Brown, is the best known of the eighteenth century landscape designers.[30] His achievement, though great, has been exaggerated. He did not change the face of England. He landscaped several hundred estates, but it would have been easy, as it is still is, to travel widely in England without setting eyes on any of them. Brown did not scruple to destroy what he found on arrival. Elm avenues, in particular, were felled without flinching. If the trees flanked a rectilinear canal, so much the worse for both. The elms in Savernake Broad

Walk (Wi) went at his command. He reduced the number at Castle Ashby (Np).[31] He did have moments of compunction. The main elm avenue in Blenheim Park survived his visits, though not the lesser elm avenues.[32] The avenues in Althorp Park (Np), Corsham Court (Wi), Longleat (Wi) and the south avenue at Wimpole Hall (Ca) escaped. Brown was not the only destroyer of elm avenues. Gales could wreak havoc. Many avenues that survived the vogue of naturalistic landscaping fell before the gale of March 1916.[33]

Brown's landscapes come under the category of the beautiful. The plantation boundaries are soft flowing lines. The atmosphere created is of serenity and repose. An artificial lake of curved perimeter is usual. The design usually includes a peripheral tree belt and scattered tree clumps.

Fig. 63. Croome Park (HW), landscaped with English Elm by Capability Brown.[35]

Like other planters of the naturalistic school, Brown favoured mixed plantings of tree species. His favourite ingredients were beech, elm, oak and conifers.[34] He was well disposed to elm, informally planted, and some of his landscapes where he used elm as his principal tree are delightful. Croome Park (HW) 63 is an example.[35]

Humphrey Repton,[36] the third of the principal eighteenth century landscape planters, was less ruthless than Brown in destroying earlier work. He would make openings, for instance, in an elm avenue, to create lateral views, but also so as to preserve the vistas from either end. He preferred mass plantings to narrow belts. Brown's hill top clusters had been criticized as too small to be dignified. Repton planted generously on elevations.

Repton planted mixed tree species like the others. He recommended round-headed trees to serve as a foil to Gothic buildings and either round-headed or conical trees to go with buildings in Grecian style.

sun and moon, and the days of the week is well known. Parallel associations were also made with metals and with plants. In earlier times only some plants were linked with planets. Elm was not amongst them. It was only quite late, in Tudor times, that an attempt was made to devise planetary associations for all plants, usually based on their supposed medicinal properties.[73] Elm was placed under Saturn. Plants were also associated with fixed stars, and with the zodiacal constellations and the decans into which they were divided. Elm had no such role.

The witch cult also requires a mention. It is not surprising to find that particular elms were, or were believed to be, the sites of witch-cult proceedings. The Tubney Tree (Ox), mentioned above, before it was whitewashed by Matthew Arnold, appears to have had such a reputation. It probably needs only a little Gothic fancy to endow any picturesquely ancient elm in this way. To hear the Devil roaring in hell, it was only necessary to circle three elms near Mylor church (Co) nine times and sunwise and then put an ear to the trunk of one of them.[74] This Cornish recipe is however in common form. Various other objects can be circled with like effect.

The prophylactic use of elm in neutralizing witchcraft is a consequence of confusion between Anglo-Saxon *wice*, *wicce* and *cwic* (*beam*), later, respectively, *wych*, *witch* and *wicken*. Rowan (*Sorbus aucuparia*), known as *wicken* over a large area of the northern half of England, was renowned as an antidote against witchcraft. Elm was also known as *wicken* in part of the same area, as mentioned in chapter 4. Latterly *wych* and *witch* were pronounced alike. So elm, designated as *wych*, *wych elm* or *wicken* took over some of the magical attributes of rowan. Bewitched pigs, for instance, in Lincolnshire were wreathed in elm to remove their spells.[75] Elm twigs, sometimes bound up with rowan, were equally effective against witches in Hereford and Worcester.[76]

The Bible

What elm there is in the Bible is confined to the Old Testament. The best candidate is represented by the Hebrew word תדהר [tidhar]. This is not known outside the Bible and there occurs but twice, in Isaiah 41. 19 and Isaiah 60. 13. The latter passage clearly recalls the former, so, in effect, the occurrence is unique. The Dead Sea Scrolls have the variant תרהר [tirhar]. This is unlikely to be correct in view of the

second century transliteration θαδαάρ [thadaar] in the Greek versions of the Bible made by Aquila and Theodotion.

In the first passage, Isaiah 41. 19, תדהר is the second of a sequence of three trees, this sequence following a previous sequence of four trees earlier in the verse. In the second passage, Isaiah 60. 13, the set of three trees is repeated, and in the same order as before. The Greek translators responsible for the Septuagint version of the Old Testament were clearly uncertain about their botany. The difficulties in the first passage were nearly waltzed around by telescoping the seven trees in the two sequences into a single sequence of five trees and discreetly omitting any equivalent for תדהר. In the second passage, the difficulties were faced a little more resolutely. תדהר was translated as πεύκη [peuke], a Greek word for pine or spruce.

The more cautious of the later translators of the Old Testament into Greek left תדהר severely alone, merely transliterating as indicated above. The second century translator Symmachus, however, stuck his neck out. In the second Isaiah passage, he followed the Septuagint and translated it as πεύκη. But in the first passage, whether from information to hand or out of his head, he translated it as πτελέα [ptelea], the usual Greek word for elm. Since elm occurs today in Samaria, though only rarely, its earlier occurrence in the area can be taken for granted. Symmachus' rendering is not impossible but is not otherwise supported.

The Jewish glosses on תדהר in the targums and Talmud make it clear that no one, when these were written, had a clue what it meant. Guesses ranged over ash, maple, oak or various conifers.

The Old Latin versions of Isaiah 41. 19 follow the Septuagint in having a single sequence of trees. The number of trees varies, however, and the text used by Cyprian of Carthage,[77] with six trees, has *ulmus* as the penultimate term. The redactor responsible for this text presumably knew of Symmachus' translation. For Isaiah 60. 13, the Old Latin versions render תדהר as *pinus*, an obvious translation of Septuagint πεύκη. In the Vulgate Latin version translated by Jerome direct from the Hebrew, he lapsed sadly from his usual principle of providing a single rendering for each of his source terms. For Isaiah 41. 19, he followed Symmachus and rendered תדהר as *ulmus*. But for Isaiah 41. 13 he took the Septuagint as his guide and rendered it as *pinus*. Moreover, though the Vulgate preserves the order of the trees in Isaiah 41. 19, it swops the second and third in Isaiah 60. 13, *pinus* for תדהר coming last.

So, in Europe, where for many centuries the Vulgate was the exclusive biblical text, תדהר lived on partly as *ulmus* and partly as *pinus*. When Wycliffe translated the Bible into English, the Vulgate was his source. He did not translate the Vulgate *ulmus* as *elm* as might have been expected, but transliterated it as *ulm*.[78] It would be interesting to know why he felt unable to provide an English equivalent.

Later translations of the Vulgate, such as the Douai Bible and the Knox version gave *elm* for תדהר in Isaiah 41. 19 and *pine* in Isaiah 60. 13.

The Reformation brought with it a spate of new Old Testament translations from the Hebrew and much vacillation as to what to do with תדהר. Luther appeared to think that few botanists would read the Bible and that when a tree appeared there that would be unfamiliar to the average north European reader, it was to be rendered by some tree with which he was familiar. So he solved the problem, presumably to his own satisfaction, by translating תדהר as *buche* [beech], which is the last thing it could possibly have meant.

Coverdale, guided by Jerome, rendered תדהר as *elme* in Isaiah 41. 19. It eluded him completely in Isaiah 60. 13 since, not realizing that the Vulgate had reversed the order of the last two trees, he took Jerome's second tree as his second and the Septuagint third tree as his third. Both, the latter inaccurately, rendered the same third tree of the Hebrew text which is box.

The translators of the Geneva Bible realized that something had gone wrong. They achieved consistency, at least, by rendering תדהר as *elme* in both passages.

The translators of the Authorized Version thought again. They elected to follow the Septuagint. So, תדהר was translated in both passages as *pine*. The Revised Version followed suit.

Of recent versions, the Jerusalem Bible is the worst botanically. It manages to find different translations both for the first and the third tree in the two Isaiah passages. For reasons that are not apparent תדהר is translated in both as *plane*. The Revised Standard Version follows the Jerusalem Bible with *plane* for תדהר but is at least consistent in its translation of its companion trees. The translators of the New English Bible decided to go back to a conifer for תדהר

but chose *fir* instead of *pine*, perhaps because they preferred a biblical tree to be of Germanic etymology. To protect themselves they added a footnote 'or elm' for Isaiah 60. 13 but not, for some mysterious reason, for Isaiah 41. 19. Theirs seems to represent fairly exactly the state of mind of biblical translators two thousand years ago.

The other Hebrew word that has fathered elms is אלה [elah]. Septuagint translators usually render it as τερέβινθος [terebinthos], which is *Pistacia terebinthus* or as δρῦς [drus], which can mean either tree in general or oak in particular. The Old Latin and Vulgate versions normally follow the Septuagint, translating δρῦς as *quercus* [oak] and transliterating τερέβινθος as *terebinthus*.

At the Reformation, two passages with אלה raised problems. Both mentioned more than one tree. Isaiah 6. 13 referred to אלה and a second tree which the Septuagint translator rendered as βάλανος [balanos] another word for oak. The two trees appeared in the Vulgate as *terebinthus* and *quercus*. This did not bother Coverdale whose *terebinth* and *oak* merely reflected the Vulgate. The later Reformation translators, imbued with Luther's principle that familiar trees were to be substituted for unfamiliar ones, could have no truck with terebinth. Nor could they translate אלה as *oak* if they retained this word for the second tree. The solution of the Geneva Bible translators was to render אלה as *elme*. This gave them the opportunity to append an improving footnote on the remnant that was to survive the wrath of the Lord:

> For the fewnesse they shall seeme to be eaten up: yet they shall florishe as a tree which in winter loseth its leaves, and seemeth to bee deade, yet in sommer is fresh, and greene.

It appears that subsequent translators thought that the intrusion of *elme* here was unwarranted. Accordingly for *elme* they substituted *teil*, an alternative word for lime (*Tilia*), now long disused. The choice of this tree may have been influenced by Luther's translation of this verse, where he has *eiche* and *linde*, in that order. The translation *teil* first appeared in the Bishops' Bible and was copied from there into the Authorized Version. In the Revised Version, it was thrown out and *terebinth* was reintroduced.

The second problem passage with אלה is Hosea 4. 13. The Septuagint translator of this book appears to have been one of the rare category of translators

there, the proverbs would be misleading in East Anglia and Cornwall where the elms flush up to four weeks later.

Other proverbs are concerned with soil fertility. Examples are:

A good elm never grew on bad ground.[99]

and

Tall elms and fat cows.[100]

Greater precision is implied in:

Good elm, good barley;
Good oak, good wheat.[101]

A Welsh proverb appears to attempt a long-range weather forecast:

Pan fo llwyf ei blodau, blwyddyn y saith ffrwythlondeb

When the elms flower it is the year of sevenfold fruitfulness.[102]

Some proverbs assess the chances of lightning strike:

Under oak there comes a stroke,
Under elm there comes a calm
And under ash there comes a crash.[103]

The propensity of elm, mainly English Elm, to drop large branches without warning, was noted in chapter 2. Since people have, on occasion, been injured or killed as a consequence, it is not surprising that warning proverbs are current of the type:

Every elm has its man.[104]

Proverbial expressions of moralizing tendency seldom refer to elm. The expression 'straight as an elm', is common to the British Isles and France.[105] It usually refers to physical uprightness rather than moral. The Spanish proverb

es pedir peras al olmo

it is asking for pears off the elm

is enshrined in Don Quixote, Sancho quoting it in the hope of escaping an uncongenial adventure. From this source it entered English proverbial lore.[106]

The justice elm is pre-eminently a French institution. It has given rise to the expression:

attendez-moi sous l'orme

meet me under the elm,

uttered when the party addressed is not expected to fulfil his stated intentions. Justice under a village elm gained a reputation for inefficiency and procrastina-

tion. A corresponding expression was not coined in England.

Emblem and device

Elm was part of the repertoire of devices in the emblem books from the start. In the first emblem book proper, that of the Italian Andrea Alciati,[107] there is an illustration of a vine supported by the branches of *73* a dead elm. The theme is the victory of friendship over death. The sources of the emblem are twofold. One is the Italian viticultural practice of marrying vines to elms, described in the preceding chapter. The other is an epigram of Antipater of Thessalonica in which a

Fig. 73. Emblems of friendship: *above*, sixteenth century: *below*, eighteenth century.[111]

vine covering a dead plane tree personifies the mistress whose love outlasts her lover's life.[108]

In the first English emblem book, compiled by Geoffrey Whitney,[109] the same illustration and the same theme, though with less supporting text, appears. Like most emblem books, Whitney's ran through many editions.

Another Italian emblem book quarried by English emblematists was that of Cesare Ripa.[110] He presented several themes illustrated by elms. Friendship was represented, as with Alciati, by a vine supported by the branches of a dead elm. Campania Felix, the Italian province, had a vine supported by an elm as a background feature. Several English versions of Ripa's emblem book were published in the eighteenth century. The illustrations of the 1709 edition were newly done.[111] The theme of Calabria also received a background elm in the English editions, but the original had *ornello*, which is *Fraxinus ornus*.

As a morally neutral device, elm identifies a number of English inns and public houses. The commonest device is a single tree designated The Elm or The Elm Tree. Inns so called are found as far apart as Bingley (WY) in the north, Chiseldon (Wi) in the west, and Cambridge (Ca) in the east. The trees may be replicated. For reasons not immediately apparent, the number is usually uneven, One Elm at Stratford upon Avon (Wa), Three Elms at Chignall St James (Ex) and Dartmouth (Dv), and Nine Elms at Lydiard Tregose (Wi).

The designation White Elm is characteristic of East Anglia. Inns so called are located at Copdock (Sf) and Woolpit (Sf). The significance of this name is discussed below. Elms seldom consort with other devices in inn signs. The Fox and Elm at Tuffley (Gl) is an exception.

Elm only appears rarely as a charge in heraldry, whether on the Continent or in England.[112] Usually, but not always, it is borne on the arms of families whose surnames are derived from elm.

The entire tree may be the charge. This is the rule in Continental bearings. It is exceptional in England. The most common continental charge is a single elm on a mount or terrace, as for de Lolme of Geneva.[113] There may be more than one tree. Van Holmen of Middelburg has three.[114] Elms may be combined with other charges or imposed upon them. Lormier of Rotterdam, for example, has a chevron between three elms. Olmius bears a mullet and elm on either side of an embattled fesse.[115] An English example is provided by the arms of Nelmes, three elms, as with Van Holmen.

The leaf is the other elm charge. It is always replicated. In Continental arms the leaves are usually green. In England they are usually golden, presumably a rendering of the autumnal colour. A typical English example is the bearing of the Elmes family: five golden elm leaves on each of two bars. The exceptional green elm leaf in an English coat of arms features in the Elmy arms: three green leaves on a chief.

The oldest monumental inscriptions in the British Isles are in Ogham characters. Stones carrying these are frequent in Ireland and also occur in those parts of Great Britain where there was early settlement by Irish immigrants. Knowledge of the Ogham alphabet and much curious and arcane lore about it persisted in Ireland till the Middle Ages when it was written down. The order of the letters differed from that in the Greek or Latin alphabets, and the names of the letters were derived from Irish tree names whose initial consonants they were.[116] The alphabet was called *beth-luis-nion*. The first and last terms present no difficulty: *beth* meant birch; *nion* was ash. The middle term *luis* normally meant bright. The medieval Irish scholars who record all this explained that *luis*, in this context, meant rowan (*Sorbus aucuparia*) whose berries are brightly coloured. It appears that a tree word beginning with *l* had to be found to replace an earlier tree name that was no longer understood. This word had been *lem*, elm, which had become a very rare tree in Ireland in the Middle Ages. That this is what happened is confirmed by one of the descriptive designations given to Ogham *l*, namely *cara ceathra* [friend of cattle]. Almost certainly, this would be an ancient title of elm.

Planting schemes based on the initial letters of tree names are the least happy of alphabetical exercises. The French seem to have thought of this first.[117] Some relatively recent English examples were laid out in Eynsford (Ke) with elm for *e*.[118] Since there is no correspondence between alphabetic value and landscape effect, the results of this kind of operation are likely to be lamentable.

Those tempted in this direction should sublimate their impulses by constructing chronograms. Although these were much more popular in France than in

England, as the mural tablets in French churches testify, a few were devised north of the Channel. One, by Michael Sparkes, was in honour of the Great Hollow Tree at Hampstead, referred to in chapter 14.[119] It runs:

> esto saCrata Deo MIrabILIs arbor!
>
> *O marvellous tree, may you be dedicated to God.*

The numerical letters MDCLIII provide the date of the little exercise, 1653.

An elucidation of the white elm can fittingly conclude this section. The White Oak of Whiteoak Farm, Craswell (HW) was so called because it bore white leaves, in addition to green and variegated leaves.[120] However, trees were sometimes whitewashed as landmarks. There was such a whitewashed oak at Tong (Sa).[121] The White Tree on Bristol Downs was an elm also whitened annually as a landmark.[122] There was a White Ash at Clayton le Moors (La)[123] but in what way it was white does not appear. It can hardly be supposed that the White Elm inns of East Anglia were named after white-leaved elms. It is more likely that they had whitewashed trees before them as just described.

Commemorative trees[124]

Specimen elms often commemorate persons of note. Sometimes the person commemorated is the planter. More often he or she will have performed some act beneath the elm. The acts most usually remembered in this way are shooting a deer, directing a battle, ordering someone else to be hung or beheaded, or being hung or beheaded oneself. The French even commemorate an elm that served as urinal to Napoleon I.[44]

Some care has to be exercised in accepting an association. Some are genuine. Others are transferences effected in folk imagination. Henry VIII, for instance, was reputed to have listened to the preaching of Bishop Latimer under Latimer's Elm near Monken Hadley (GL). He did not. The elm was originally named after a local man called Latimer. Later generations supposed that the Latimer in whose name the elm rejoiced was the bishop. Henry VIII was then brought out into the country and sat under the tree for a full length sermon. The episode is not even plausible.

To earn an elm, it is not necessary to be very good, very bad or even of historic importance. It is usually necessary to be vivid. Elms associated with saints have been considered already. Royalty comes next.

In France, monarchs of the Old Regime who earned most elms were Louis IX and Henri IV.[125] Louis IX was vivid from sanctity. Henry IV was vivid from carnality. Napoleon I, the most vivid ruler of the New Regime, is mentioned above.

The earliest English monarch to earn an elm was Stephen with whom an elm at Chequers (Bu) was most improbably associated.[126] Next was John, who reputedly met the hunt in Cranborne Chase under the Larmer Tree near Tollard Royal (Wi), mentioned earlier in this chapter. There is no early evidence to substantiate this story either. Richard III's Trees at Sandal Castle (WY) were elms.[127] They could not have been contemporary with the king but could have replaced others that were. Since Richard's credits are few, one is tempted to concede this one.

Queen Elizabeth I has more elms associated with her than any other sovereign. Usually she shoots at deer from beneath them. There are two reasons for the association. First, her royal progresses took her into a large number of deer parks. Secondly, she was built up into a cult figure even in her lifetime and remained so, to some extent, in later centuries.

Both Charles II and his consort had elms associated with them. With the former it was the curiously shaped elm in Hampton Court Park (GL) called King Charles' Swing. With the latter, it was Hammersmith Mall (GL). Popular imagination, however, was caught more by Charles' cousin Rupert, the most colourful of the royalist commanders in the Civil War. An elm beneath which he dined was pointed out at Bolton Abbey (NY).[128] An elm from which he hung a man was an object of regard at Henley on Thames (Ox.)[129] Rupert's Walk was beneath elms in Woolwich (GL).[130]

Of heads of state outside the circle of royalty, few have planted more elms than the Presidents of the United States.[131] Oliver Cromwell was recalled by an elm at Lowestoft (Sf).[132] The association was the basis of an unsuccessful conservation plea. Minor politicians in general do not engrave themselves on folk memory. Cardinal Wolsey, however, was recalled as a reputed planter of elms in St James's Park.[133] Neville Chamberlain planted an elm at the Prime Minister's country retreat at Chequers (Bu). Disraeli was a great tree planter, though not particularly of elm. Gladstone felled trees to relax.

Abroad, elms were not infrequently associated with military leaders. The Washington Elm, near Cambridge, Massachusetts, beneath which George Washington reviewed the army that was to secure the independence of the United States of America, became a national monument. The Duke of Wellington followed the course of the Battle of Waterloo from beneath an elm tree that was subsequently carved to pieces by relic hunters.[134] There are no English examples of comparable fame. It is not, however, always easy to distinguish between elms commemorating military leaders and those associated with particular military encounters. The latter are dealt with below.

Religious innovators take to the fields when churches are closed to them. They are often encountered preaching beneath elms. Luther did so, Calvin likewise.[44] The outdoor preachers who made the greatest impact on English folk memory were John Bunyan and John Wesley. Bunyan tended to preach under oak. Wesley was fond of elm. Elms under which he is said to have preached or upon which he commented were pointed out at Corston (Av),[135] Stony Stratford (Bu),[136] on the road between Northampton and Towcester,[137] near Portsmouth[138] (Ha) and in Camborne in Cornwall.[139] He planted a double row of

elms at Kingswood (Av) and preached under the shade of the trees some 40 years later.[140] One other preacher whose colourful delivery drew crowds was Edward Irving. An elm under which he preached stood on Hampstead Heath (GL).[141]

Literary achievement is seldom rewarded by a niche in English folk memory. Among the few elms associated with writers were Milton's Elm at Chalfont St Giles (Bu), and the Nine Elms on Hampstead Heath, frequented by Alexander Pope and his circle of friends.[142]

Visitors from abroad paid homage to Byron at the elm in Harrow churchyard where, as described in chapter 10, his romantic strivings took shape. English admirers, less interested in literary achievement, sought out the elm in Devil's Wood in the grounds at Newstead Priory carved with the initials of Byron and his beloved half-sister.[143]

Elms are so integral a part of many English landscapes that it is not surprising that they feature as the backcloth to momentous decisions. Sir E. B. Chain recalls a decision of great consequence reached with H. W. Florey. This was to initiate the research programme that uncovered the antibiotic properties of penicillin. The decision was made under an elm in the New Parks, Oxford.[144] Lady Ottoline

Fig. 74. Rupert's Elm, Henley on Thames (Ox).[129]

Morrell and her husband finally agreed that she should give herself to Bertrand Russell. Their decision was made, so she recalls, looking out to the elms around Newington (Ox).[145]

Fig. 75. Albert! Spare those trees, from *Punch*.[150]

Other elms commemorate events rather than persons. Treaty elms are an instance. The *ormeteau ferré* of Gisors has already been mentioned. A number of elms marked the sites of treaties between English immigrants and North American Indians. They include the Logan Elm in Ohio, the Council Elm, Johnston, and Fort Stanwix Elm, Rome, both in New York State, and the Penn Treaty Elm near Philadelphia. In recent centuries in Europe, treaty making has been an indoor matter and so the treaty elm does not feature.

Sites of battles were sometimes marked by elms. The Battle of Evesham (HW), as mentioned in chapter 13, was marked by the elms around Battle Well. An old elm was also reputed to mark the field of the Battle of Barnet (GL) during the Wars of the Roses. Sometimes elms were planted locally to celebrate military victories elsewhere. Perhaps more often, popular legends that they were so planted became attached to elms whose origins had nothing to do with their subsequent reputations. The defeat of Charles Stuart, the Young Pretender, at the Battle of Culloden

Fig. 76. Closing of the Great Exhibition, by H. Bibby.[151]

was kept in popular memory by the three elms called Faith, Hope and Charity on Southampton Common (Ha), and by the elm on the green at South Collingham (Ng).[146]

The Gunpowder Plot was 'never to be forgot'. It was commemorated by an elm outside Deane's Almshouses in Basingstoke (Ha).[147]

The other event that was commemorated by several elms was the Revolution Settlement of 1688. Examples of elms performing this role existed at Bisham (Br),[148] Kempsey (HW)[64] and Gunnerside (NY).[149] To put his loyalty beyond all question, the person responsible for the last planted a Dutch Elm.

Trees of liberty, often elms, were mainly a French Revolutionary institution, though possibly inspired by the Liberty Elm at Boston, USA. They were planted in other countries too, when revolutionary fervour reached them. As a long-term sylvicultural investment they were a failure. Change of government in a politically sensitive country involved the felling of such trees of liberty as the previous government had planted.

Elms otherwise devoid of particular associations sometimes had notoriety thrust upon them. The elms in Hyde Park (GL) are a case in point. The initial plans for the erection of the original Crystal Palace to house the Great Exhibition of 1851 involved the felling of a certain number of elms. Those opposed to the exhibition seized their opportunity. They became conservationists. There was an outcry in the Press, debate in Parliament and even a cartoon in *Punch*.[150] *75* Consequently, though a few elms were quietly felled, several large ones were carefully preserved inside the Palace which was constructed around them. They picture conspicuously in the numerous published *76* illustrations of its interior.[151]

The most recent incident of this sort occurred in Stockholm. Some elms in Kungsträdgården were threatened by plans to construct an underground station entrance. Conservationists arose in protest. This time they failed to move the authorities, even though the singers of the Stockholm Opera, which overlooked the doomed trees, came out on to the balcony and invoked the elms in harmony.[152]

10
The writer's reaction

Elms have been a major component of the English landscape for many centuries. They have necessarily been frequent objects of perception. They have also been literary symbols as far back as literature extends. In this chapter, their place in English literature, whether as perceived object or literary symbol, is explored.

Object of perception

Over large areas of England, particularly southern England, elms have been so ubiquitous that they cannot have been overlooked. They may, however, not have been knowingly perceived. Distance, in some cases, prevents recognition of what kind of tree is under observation. Normally, there are always enough trees sufficiently close for the kinds to be identifiable. Identification does, indeed, require some knowledge, and lack or defect of this accounts for some failures to identify correctly. Elm can generally be recognized for what it is at moderate distances. Its principal kinds are also, in general, recognizable if sufficiently close, as, for example, when seen a field's distance away from a train window.

Errors in identification are, however, to be expected. Several are enshrined in the works of standard authors. William Cowper described a much loved view over the Ouse:

There, fast rooted in his bank
Stand, never overlook'd, our favourite elms.[1]

The scene has been engraved.[2] It presumably inspired Betjeman's stanza:

O God the Olney Hymns abound
With words of Grace which Thou didst choose,
And wet the elm above the hedge
Reflected in the winding Ouse.[3]

The elms were, as Sir George Throckmorton took delight in informing Cowper, poplars.[4]

Another misidentification, which generations of Oxford men have tried to explain away is

The signal-elm, that looks on Ilsley Downs,

consecrated by Matthew Arnold to the memory of Arthur Clough. This was an oak.[5]

Truth may also be compromised by citing elms in places where they do not occur. This is exemplified by the botanically slapdash Virginia Woolf whose intrusion of elms into Skye earned a reproof that wounded her deeply.[6]

Clearcut error is less common than distortion. The relative frequency of trees correctly identified may, for instance, be manipulated into improbability. A writer may also observe at one level of discrimination and write at another. More usually the change is in the direction of greater precision. Virginia Woolf, whose diaries and correspondence invite such comparisons, has numerous allusions to trees in general in these productions but few to particular kinds of tree. In her novels, particular kinds of tree are frequent. It would appear that Mrs Woolf observed at a general level of perception but considered that particularization was a literary desideratum. The converse situation, where an author observes precisely but writes generally is much rarer. Jane Austen is an instance. Her letters reveal precise recognition of kinds of tree. Particularization is largely absent from her novels.

One pervasive cause of distortion is the incubus of earlier writings with which an author is familiar. This interposes a veil between the writer and his objects of perception. The thickest of such veils was woven by the classical authors. Much of the botanical sterility of the seventeenth and eighteenth century writings results from the compulsion to follow Classical precedents, rather than to react directly to what is perceived. As regards elm, it is the Classical motif of the marriage of vine to elm, to be considered later, that dominates post-Renaissance English verse. Vines so treated would have been objects of perception only for those who kept their eyes skinned during the right part of the Grand Tour.

The prospect poetry introduced into English

literature by Sir John Denham,[7] with minimum arboreal detail and much moralizing, had a similar effect. Prospect versifiers followed their precedents slavishly. It is only by eliminating what previous authors had said of their prospects that one can have any confidence that a later writer has made an original observation. Take even Coleridge's description of the landscape at Nether Stowey (So):

> Ah! what a luxury of landscape meets
> My gaze! proud towers, and lots more dear to me,
> Elm-shadow'd Fields, and prospect-bounding Sea![8]

Elms do abound here, but was not Coleridge perhaps reworking Willian Bowles' quite different Wiltshire prospect of

> villages
> Thick scattered, 'mid the intermingling elms?[9]

The presumption in the foregoing paragraph is that error or distortion is largely unconscious. This may not be so. On his English tour, Nathaniel Hawthorne passed through the village of Whitnash (Wa). He recorded the appearance of the large elm by the church in his note book.[10] Yet in his travelogue *Our old home*,[11] the elm, still at Whitnash, has become a yew, to the perpetual danger of all such cattle as might roam the pages of this publication. It persists as a yew, still at Whitnash, in his novel *Septimius*.[12]

The rhyming position can seduce the most moral of authors. Excused by her youth, George Eliot wrote:

> the rookery elms
> Whose tall old trunks had each a grassy mound,
> So rich for us, we counted them as realms
> With varied products.[13]

The rookery trees were oaks.[14]

Reality may also be sacrificed for various literary devices. The conceit of having Nature reflect the feelings of the principal characters of a novel is well established, though Nature is indifferent to these. So, in *David Copperfield*, Dickens changes the appearance of the elms at Blundeston (Sf) to mark the progression of the narrative.[15] George Eliot manipulates the elms around Milby (Nuneaton, Wa) to amplify the emotions of the characters in *Janet's repentance*.[16] And in *The woodlanders*, Hardy fells the elm that threatens John Smith when the latter dies.[17]

Symmetry may also seduce. Early in *The excursion*, Wordsworth introduces

> a brotherhood of lofty elms,[18]

as observed near Racedown (Do).[19] In his manuscript, the poem culminates round a joyful oak in a valley in the north of England. But as printed the poem reads:

> A wide-spread elm
> Stands in our valley, named the JOYFUL
> TREE.[20]

Clearly Wordsworth's instinct for symmetry had been dissatisfied. In the published version it is a climactic elm that balances the opening brotherhood.

What has just been described above is what an author is liable to do with elms that he has knowingly perceived. He may, however, concern himself with elms someone else has perceived. When Hardy wrote

> They are blithely breakfasting all –
> Men and maidens – yea,
> Under the summer tree,[21]

he describes an elm he had never perceived. He lighted upon it in Emma Hardy's diary after her death. It cannot be always assumed, either, that the apparent observer was the actual observer. Coleridge's lines on the landscape at Nether Stowey (So) have already been quoted. But was Coleridge the actual observer of the elms, assuming that it was the local elms that the poem described? Coleridge roamed the countryside with William Wordsworth, a more acute observer than Coleridge, and with Dorothy Wordsworth, a more acute observer still. Either could have made the observations that Coleridge elaborated. Finally, Crabbe's poem *On a drawing of the elm tree under which the Duke of Wellington stood several times during the Battle of Waterloo*,[22] must, as its title states, be at least one remove from the original tree.

It is not implied that literary composition is to be criticized for recession from reality. It is merely to be noted that such recession is common, as far as the elm is concerned. Having established this, the way is clear to exemplify how elms, knowingly perceived, have been described.

The salient characteristic of the basilica of St Peter's at Rome is size. So it is with elm. On average, elms are markedly taller than other landscape trees and endow a landscape in which they are present with a third dimension. It is the loss of this dimension, of height or depth, that has most brutally transformed so much of the landscape of southern England in the latest Dutch elm disease epidemic. Thus, the characteristic adjectives applied to elms are giant, gigantic, great, high, huge, lofty, tall, towering. Of these, the most graphic term is great, conveying not only height but also an overtone of solemnity. The height of the elm did not much concern the Classical authors. It has been a continuing theme with English writers since at least the time of John Gay, who describes how

> From the tall elm a shower of leaves is borne.[23]

The landscape to the south of Reading (Br) was the literary preserve of Mary Mitford. Here,

> towering high above all, the tall and leafy elms[24]

dominate the countryside. Similarly, in the neighbourhood of Langley Burrell (Wi) Francis Kilvert recorded how

> the great elms shaded the road from the glowing sunshine and everything was still and beautiful and green.[25]

When Tolkien needed to measure his giants, an elm is the yardstick.[26] The association between giants and elms is much older, featuring in the Cornish legend of Tom and Giant Blunderbuss, the West Country version of the East Anglian story of Tom Hickathrift.[27]

The coincidence of height, straightness and hedgerow planting has led writers such as Richard Jefferies to describe elm alignments as cliffs.[28] This simile has been handled most felicitously by Betjeman, with

> Pale corn waves rippling to a shore
> The shadowy cliffs of elm between.[29]

The immediate simile suggested by the skyline silhouette of an English Elm hedgerow has been avoided by most writers, presumably as faintly ludicrous when verbalized. Lady Ottoline Morell expressed it in prose:

> a line of old elm trees like old men ambling across.[30]

Geoffrey Grigson put it into verse:

> Elms, old men with thinned-out hair,
> And mouths down-turned, express
> The oldness of the English scene.[31]

Descriptions of the bole have tended to refer to ancient and hollow elms. It would be hard to better

William Strode's evocation of the Great Hollow Tree at Westwell (Ox):

> How proper, o how fayre
> A seate were this for old Diogenes
> To grumble in and bark out oracles,
> And answer to the Raven's augury
> That builds above.[32]

The configuration of the major branches is described with a combination of charm and local loyalty by Glyn Morgan:

> Around the fields soared tall elms with branches like Gothic tracery writing the unmistakeable signature of Essex across the sky.[33]

The ultimate branches of the elm are distinguished by the fineness of the spray. Gerard Manley Hopkins' line

> Only the beakleaved bows dragonish damask the tool-smooth bleak light[34]

was, it appears, inspired by the sight of elm branches in the evening light at Cumnor (Ox).[35] George Orwell's description is more immediate:

> multitudinous twigs making a sepia-coloured lace against the sky.[36]

The movement of elm branches in the wind admits of great variety. Leigh Hunt describes a gentle breeze:

> a neighbouring wood of elms
> Was moved, and stirred and whispered loftily,
> Much like a pomp of warriors with plumed helms,
> When some great general whom they long to see
> Is heard behind them, coming in swift dignity.[37]

Virginia Woolf invoked a strong blast, causing a

> superb upward rise (like the beak of a ship up a wave) of the elm branches as the wind raised them.[38]

Elm foliage presents a very diverse appearance as the season progresses. Spring is advertised, in Hopkins' words,

> by the thick heaps and armfuls of the wet pellets of young green.[39]

Then, as they swell, the buds throw off their pink scales which D. H. Lawrence used to strew upon his lovers.[40] In the early summer, the leaves, seen through Baron Corvo's eyes, are 'exquisitely lush and vivid'.[41]

Thereafter, the leaves of the English Elm, though not of the Narrow-leaved Elm, darken to dusky green.

Autumn is the season of flamboyance. While the leaves of oak and ash turn their own shades of brown, those of elm turn bright yellow. This is particularly the case with the English Elm. Eleanor Rohde comments:

> The rich and almost countless shades of browns and reds of other trees are suggestive of treasures of earthly splendour, but the pure, clear gold of the elm leaves is ethereal, and turns one's thoughts to the 'transparent gold' of S. John's vision.[42]

William Barnes then describes how the wind brings the

> Feäded leaves, – a-whirlen round,
> Down to ground, in yollow beds.[43]

Wych Elm leaves also colour well. The autumnal scene caught by D. H. Lawrence was

> a sickly gleam of sunlight, brightening on some great elm-leaves near at hand till they looked like ripe lemons hanging.[44]

The Decadents appropriated autumnal yellow. Lionel Johnson's Decadent melancholy was expressed as:

> yellowing elms
> Moan, as the quick gust overwhelms
> Their wintry fellowship of boughs.[45]

Oscar Wilde, in his florilegium of Decadent conceits, *Symphony in yellow*, noted how

> The yellow leaves begin to fade
> And flutter from the Temple elms.[46]

Elm leaves, being of soft texture, do not drop with the rustle or crackle of coarser leaves. Silently, as E. J. Scovell has it, they

> Fall in still air and light unjarred like snow.[47]

As they lie, they give off, as Edward Thomas noted, a smell like tea.[48]

On the most remarkable feature of elm leaves, their extraordinary bilateral asymmetry, the poets are silent.

In European elms, it is not the leaves that appear first, but the flowers.[49] Their coloration is one of the subtlest landscape tones. At close quarters, they appear dingy, but Mary Webb, looking at them in the distance, with the sun behind them, saw how

wildfire runs along the elms as the red buds push out.[50]

The colour of the flowers is elusive. Hopkins defined it as 'smoky claret'.[51] For M. Armfield it was 'a pure Tyrian purple as in Coptic linens'.[52]

Only the flowers of Wych Elm regularly give rise to fertile fruits. These turn greenish-yellow and finally brown while still attached to the tree. In early summer Wych Elm can be picked out from other trees by their brownness. In other elms, the fruits are infertile. They also turn greenish-yellow and may be held thus for some time before they are shed. It is the shedding of the fruits that has drawn the attention of the poets, most particularly those of China. However, in Chinese poetry, elm is very much an extra touch to a landscape of willow. There are large numbers of Chinese poems in which willows are the sole tree, but very few where elm plays such a role. Han Yu provides an example involving both trees.

楊花楡莢無才思
惟解漫天作雪飛

willow catkins and elm fruits scatter unaware over the heavens like snow.[53]

In China the shedding of elm fruits and the dropping of willow catkins carry a note of melancholy. Spring is passing. The English reaction, represented by Barnes, is more robust:

> An' greeny elmsceäles do strew
> The road below the dousty shoe.[54]

With the exception of the single olfactory observation, the writers so far quoted have concerned themselves solely with the visual qualities of elms. It is time to turn to their auditory qualities when animated by movement of the air around them. One of the first English poets to concern himself with these was Milton:

> malus auster
> Miscet cuncta foris, & desuper intonat ulmo.
> *an ill south wind confounds everything outside and thunders aloft in the elm.*[55]

Bloomfield, similarly, describes a storm in whose passage

> Full-leaf'd elms, his dwelling's shade by day,
> With mimic thunder gives its fury way.[56]

More usually, elms are said to roar rather than thunder in a high wind, as with Meredith:

> A roar thro' the tall twin elm-trees
> The mustering storm betrayed.[57]

Reactions to gentler winds are more diverse and more intriguing. John Clare composed two poems to particular elms and of each he describes its sound. Beneath one he would lie down

> And hear the laugh of summer leaves above.

The other, he recalled as an

> Old elm, that murmured in our chimney-top
> The sweetest anthem autumn ever made.[58]

The musical analogy recurs in *The woodlanders*. Hardy, describing the elm that threatens John Smith, notes

> the melancholy Gregorian melodies which the air wrung out of it.[59]

The author came from a family of church musicians and retained a nostalgia for church music. The adjective melancholy serves to underline the doleful role of the tree. It also reveals Hardy's unfamiliarity with a domain of church music then closed to the general run of church musicians.

Morgan Forster takes up the musical analogy in *Howard's End*. The elm this time is a benign symbol:

> The tree rustled. It had made music before they were born, and would continue after their deaths, but its song was of the moment.[60]

A quite different but perhaps more perceptive description of the sound of elms in wind and rain is due to Edmund Blunden, who recalled sitting in church and hearing

> the elms without like waterfalls,
> While the cold arches murmured every prayer.[61]

Environment

What has been considered till now is elm with no more environment than sufficient light to see it or sufficient wind to hear it. Much of what has been written on elm is concerned, not with the tree in isolation, but as sited in its physical environment, or its biological environment, or as used by man.

First then for the physical environment: sun, moon, stars and mist. No one in his senses looks through elms at the sun when it is high in the sky. Just after sunrise or just before sunset is a different matter. It was while

three quarters of the scarlet sum was settling among the branches of the elm in front[62] that D. H. Lawrence opened his earliest short story. Shortly before sunset too, and shortly after sunrise, it is possible to see, as William Barnes loved to recall, the coppery glow of elm foliage in low-shot sunlight.

Elm shadows have something of the monumentality of the elms themselves. Frederick Faber surpassed his better-known verse when he described

> Where elm-cast shadows dimly green
> On the dew beaded pasture sleep.[63]

It is a convention of Arabian poetry to liken rippling water to chain mail. In a poem of Al-Isra'ili of Seville, the elms by the river are fancied as charging at it:

> أعلامٌ خَزٍّ فوق سُمْرِ رماحٍ   وكأنّما الأنشامُ فوق جِنانه
> لـمّا رأتْهُ مدرّعًا لـكـفـاحِ   لا غروَ أن قامتْ عليه أسُطُراً

> *It was as though the elms over its gardens were silken banners over brown lances. No wonder they rose in ranks against it when they saw it mail-clad for battle.*[64]

Crabbe provides a more prosaic English counterpart:

> On a clear stream the vivid sunbeams play'd,
> Through noble elms, and on the surface made
> That moving picture, checker'd light and shade.[65]

On occasion one comes upon a sight extolled by Edward Thomas:

> the glow of elms where an autumn rainbow sets a foot among them.[66]

The moon shining through elms is more often observed. Kilvert is uncomplicated:

> the full moon rose gloriously behind the elms in the East.[67]

More introspective, Edward Lowbury draws on the same perception to heighten –

> The moon rises, maddening, enormous –

then interrupt –

> Suddenly the wind blows, and the moon behind an elm
> Becomes a hundred trembling stars, laughing and laughable –

a stream of night thoughts.[68] For Wilfred Gibson, night terror reigns:

Between the midnight pillars of black elms
The old moon hangs, a thin cold amber flame
Over low ghostly mist.[69]

The shadows cast by elms in moonlight have a sombre quality. For Barnes this aspect is not salient, as in his address to *The Elm in home-ground*:

the moonlight marks anew
Thy murky shadow on the dew,
So slowly o'er the sleeping flow'rs
Onsliding through the nightly hours.[70]

But in Edward Thomas' war-time poem *Liberty*, it is paramount:

The last light has gone out of the world, except
This moonlight lying on the grass like frost
Beyond the brink of the tall elm's shadow.[71]

Milton had noted the imperviousness to starlight of the elm in summer when he proffered his invitation:

Under the shady roof
Of branching Elm Star-proof,
Follow me.[72]

Keats, making his observations during the winter snows, saw

the black elm tops 'mong the freezing stars.[73]

Samuel Palmer's reaction to elms at night, in as far as they influenced his paintings, are considered in the next chapter. A passage in his sketch book is, however, worth reproducing here as it illustrates one of the most significant cross fertilizations between the literary and visual arts:

So exquisite is the glistening of the stars through loopholes in the thick woven canopy of ancient elm trees – of stars differing in glory, and one of prime lustre piercing the gloom – and all dancing with instant change as the leaves play in the wind that I cannot help thinking that Milton intended his 'Shady roof, Of branching elm star proof' as a double stroke – as he tells of the impervious leafy gloom – glancing at its

Fig. 77. Elm avenue at Harefield Place (GL).[72]

77

beautiful opposite – 'Loop holes cut through thickest shade and in them socketed the gems which sparkle on the Ethiopic forehead of the night.[74]

In contrast, Gibson's thoughts on elms in starlight are painful. In his poem *The elm* he describes how it became a

> great funeral-plume of black
> So awful in the cold starlight,[75]

a more chilling evocation, even, than his picture of elms in moonlight.

Occasionally the constellations are particularized.[76] Rupert Brooke writing from Rugby remarks how

Casabianca sits and snores in the elms.[77]

Were he older, no doubt it would have been Cassiopeia. Morgan Forster's astronomy is more assured. Towards the end of *The longest journey*, he assembled two of his characters at nightfall as

> the shoulders of Orion rose behind them over the topmost boughs of the elm,[78]

preparatory to pushing the difficult one under a train.

The height of the elms and their usual siting in hedgerow alignment are together responsible for the distinctive appearance of elms in mist, fading into invisibility both horizontally along the row and upwards. Walter de la Mare sensed some of this in depicting a scene where

Fig. 78. Elms in frost, by H. Barrett.[81]

The towering elms are lost in mist.[79]

Edward Lowbury's encounter with elms is more graphic:

Two mountains by my road suddenly emerge
Like phantoms on the fringes of existence
Yet filling half the sky.[80]

The fineness of the ultimate branches of elms has already been noted. It is responsible for the remarkable appearance of elms in an air frost. In the words of A. M. Dell:

A tall elm coated with rime is surely the most exquisite object in the whole field of nature.[81]

Now for the biological environment. Edward Thomas had a keen eye for the wild flowers characteristically present under elms. Along the road from Market Lavington to Westbury (Wi), where Salisbury Plain drops down suddenly to the clay plain, he encountered, in one place,

high banks of loose earth and elm roots, half draped by arum, dandelion, ground ivy, and parsley [chervil], and the flowers of speedwell and deadnettle,[82]

and in another

a bank of nettles and celandine under elm trees.[83]

Barnes concentrates more on single and more conspicuous species. He describes both

banks wi' primrose-beds bespread,
An' steätely elms over head.

and

Where elems high, in steätely ranks
Do rise vrom yollow cowslip-banks.[84]

In the West Country only does one expect to find elms shading daffodils, as delighted Kilvert at Bredwardine (HW).

Birds on elms have attracted much more attention than the flowers under them. Virgil's line

nec gemere aeria cessabit turtur ab ulmo

nor will the turtle dove cease to moan from the lofty elm[85]

introduces a theme that will run for the next two millenia. Of English poets, Andrew Marvell is amongst the first to write of doves in this context:

Sad pair unto the Elms they moan.[86]

Tennyson, whose line on the same theme is the best

known of all, was presumably familiar with the earlier lines both of Virgil and Marvell. His is discussed in a later section of this chapter. A Decadent dove was observed by Oscar Wilde in Magdalen College grove, Oxford:

the gloom of the wych-elm's hollow is lit with the iris sheen
Of the burnished rainbow throat and the silver breast of a dove.[87]

However, the dove is not the bird principally associated with the elm in English literature. This role is taken by the rook whose preference for the highest tree locally available for its nesting site inevitably links it with elm.

Joseph Addison is one of the earliest writers to reflect on the cawing of rooks in elms:

I am very much delighted with this sort of noise, which I consider as a kind of natural prayer to that Being who supplies the wants of His whole creation, and who, in the beautiful language of the Psalms, feedeth the young ravens that call upon him.[88]

It is Frances Cornford who presents this theme most graphically. A shot is fired, scattering a rook colony,

Then, suavely slow and gradually dumb,
Back in a circling saraband they come
Each to his elm-bough, neither fast nor soon,
Black judges of the golden afternoon.[89]

The rooks in the elms of *Tom Brown's schooldays* call for attention later in this section. Henry James' acclimatization to his adopted country was advertised by the tedious insistence with which rook-haunted elms were planted along his English travelogue.[90]

Other birds associated with elms follow a long way behind. The owl is prominent among these. One is portrayed in *The unsafe elm* of Christopher Hassall:

Hunched like a toy, here sat the vigilant
Occasional owl, frigidly basking on the bough,
Moon-bathing.[91]

Another is depicted even more evocatively by Dylan Thomas:

Only a hoot-owl
Hollows, a grassblade blown in cupped hands, in the looted elms.[92]

The rest of the elm aviary, cuckoo, greenfinch, linnet, redwing, robin, thrush, willow warbler, wren

78

and wryneck, and even the gull, will be found variously in the pages of Richard Jefferies and Edward Thomas.

The associated birds sharply differentiate the elm literature of England from that of North America. Who could strike a more foreign note for an Englishman than Longfellow when he wrote:

> from the stately elm I hear
> The blue-bird prophesying spring?[93]

Just as foreign to English readers is the association between cicada and elm. Alphonse Daudet describes a scene in southern France where

> sur les ormeaux du bord du chemin, tout couverts de poussière blanche, des milliers de cigales se répondent d'un arbre à l'autre.
>
> *on the roadside elms, all covered in white dust, thousands of cicadas trill to each other from tree to tree.*[94]

Antonio Machado brought before his readers summer in Spain:

> Dentro de un olmo sonaba la sempiterna tijera
> de la cigarra cantora, el monorritmo jovial,
> entre metal y madera,
> que es la canción estival.
>
> *Within an elm the everlasting chirp of a cicada singer sounded, a cheerful recurring rhythm as between metal and wood, which is the song of summer.*[95]

Elm insects have not drawn the attention of many English poets. One so attracted was Andrew Young. In his poem *The elm beetle* the sight of the hieroglyphic bark mines of the beetle opened up for him a vision of the gates before the throne of Osiris.[96]

Bark beetles are, as indicated in chapter 6, the carriers of Dutch elm disease. It has been hinted throughout this chapter that no very close correspondence between Nature and its representation in literature is to be expected. The catastrophic effect of Dutch elm disease on the landscape of southern England is a case in point. Writers, in general, have failed to notice it. Edmund Lowbury, Andrew Young's son-in-law, was one of the very few who

> looked away
> From the pandemic shock
> Of sick and ravished boles.[97]

The Elm Decline of the fossil pollen record described in chapter 3 was probably not pathological in the usual sense, though resulting in a marked diminution in the frequency of elm over much of England. It is the subject of a poetic commentary by Norman Nicholson, who drew a gloomy parallel between the Neolithic forest clearances around the time of the Elm Decline and the threats to the environment posed by contemporary man.[98]

Elms are pre-eminently hedgerow trees in England and their consequent disposition in alignment is a characteristic feature. John Scott of Amwell (He) was one of the first English poets to bring this out when he wrote of

> Stansted's farms inclos'd,
> With aged elms in rows dispos'd.[99]

With Barnes, elm ranks became a glorious obsession. They crept into the line quoted above on springtime elms over cowslip banks. Winter also is a time

> When elem stems do rise, in row,
> Dark brown, vrom hangens under snow.[100]

The planting of elms along major roads was, as is shown in the preceding two chapters, a French idea, initially for purely military purposes. It is not surprising that roadside elms feature prominently in French literature. As the poet Léon Duvauchel expressed it:

> Les ormes sont les jardins
> Des vieilles routes de France.
>
> *The elms are the gardens of the old trunk roads of France.*[101]

Neither is it surprising that when an English poet found himself in France, as did the young Wordsworth, his theme was:

> O'er Gallia's wastes of corn my footsteps led;
> Her files of road-elms, high above my head
> In long-drawn vista, rustling in the breeze.[102]

Although the inspiration behind elm avenues was largely French, they were sufficiently widespread in England to encourage the supposition that they might have flourished in English literature as they did in French. This appears not to be so. For the French, an elm avenue was an appropriate place for lovers' meetings. One of Theophile Gautier's lovers tells how it was:

> Nous suivions une grande allée d'ormes d'une hauteur prodigieuse; le soleil descendait sur nous, tiède et blonde, tamisé par les déchiquetures

du feuillage; – des losanges d'outremer scintilla-
ient par places dans des nuages pommelés, de
grandes lignes d'un bleu pâle jonchaient les
bords de l'horizon et se changeaient en un vert
pomme extrêmement tendre, lorsque'elles se
recontraient avec les tons orangés du couchant.

*We followed a great avenue of prodigiously high
elms; the mild, golden sun shone down upon us, its
light filtering through the gaps in the foliage;
lozenges of ultramarine shone here and there
between the dappled clouds, and great lines of pale
blue strewed the verge of the horizon, changing into
a most delicate apple green where they met the
orange hues of the sunset.*[103]

Baron Corvo did indeed provide Pope Hadrian
VII with a two-mile elm avenue but the scene was by
the Nemorensian lake, not in England.[104] One of the
rare elm avenues in English literature appears
suddenly in chapter 37 of George Eliot's *Middlemarch*,
where it leads to Lowick Manor.[105] Since a lime avenue
plays this role elsewhere in the novel, the possibility
that George Eliot was capable of a Freudian slip forces
itself on the incredulous reader.

The elm mall is so specially a French institution
that its celebration must be sought in French
literature. It is hardly possible to picture M. Bergeret
except under the mall whither Anatole France directed
him, where he could forget

sous les arbres classiques, dans la solitude
aimable, sa femme et ses deux filles et sa vie
étroite dans son étroit logis.

*under the classic trees, in the friendly solitude, his
wife and his two daughters and his confined life in
his confined home.*[106]

The association between elm and sacred spring
is a theme of Scottish and Welsh literature, but not of
English. Sir Walter Scott opens *The lady of the lake* by
invocation to the

Harp of the North! that mouldering long hast
hung
On the witch-elm that shades Saint Fillan's
spring.[107]

The finest homage to a sacred spring, however,
is a line on St Winifred's Well at Holywell (Clwyd)
attributed to Iolo Goch.

Iorddanen dan lwyfen deg
Jordan under a fair elm.[108]

Elms and churches are often mentioned together, 3, *80*
because elms are the trees most comparable in height
to church tower or spire. Ebenezer Elliott portrays a
northern church which

thy grey tower and thee
Coeval elms hide from the passer-by.[109]

Arthur Clough describes a complementary
scene in a gentler landscape in the south:

the solemn eve came down
With its blue vapour upon field and wood
And elm-embosomed spire.[110]

Both poets dwell appreciatively on the association. Not
so Richard Jefferies:

I wish the trees, the elms, would grow tall
enough to hide the steeples and towers which
stand up so stiff and stark, and bare and cold.[111]

The two occur together in North America too, even in
Vladimir Nabokov's *Lolita*, where

the slender white church and the enormous
elms[112]

of a New England township provide a necessary
background.

An even more particular link associates elm
with cathedral close. Charles Dickens unfolds *The
mystery of Edwin Drood*[113] as his characters thread their
ways under the elms of Rochester Cathedral Close.
Bram Stoker, in one of the less remarked passages of
Dracula, takes his heroine to Exeter and solaces her *79*
with

the great elms of the cathedral close, with their
great black stems standing out against the old
yellow stone of the cathedral.[114]

Others have portrayed the elms of the cathedral
closes of Lichfield, Salisbury, Wells and Worcester.[115]
The elms of no cathedral close, however, have so
generally recognized a place in English literature as
those of Barchester, Trollope's amalgam of Salisbury,
Wells and Winchester. Thither Trollope directed the
unhappy Mr Harding who

slowly passing beneath the elms of the close,
could scarcely bring himself to believe that the
words which he had heard had proceeded from
the pulpit of Barchester cathedral.[116]

While the justice elms mentioned in the
preceding chapter feature occasionally in the English
scene, they were never so frequent nor so much a part

The bark is peeled from the many-branched...elm tree, and hung to dry in the sun. It is ground, grating, in a foot mortar. It is pounded in a hand mortar.[121]

A high proportion of the schools, universities and places of concourse that have had the greatest formative influence on the English way of life, have been generously furnished with elm. It is not unreasonable to suppose that this elm may have entered into the background of the minds of the denizens of these places and to have remained there when they passed to other walks of life.

Amongst schools, Eton, Harrow[122] and Rugby were all encompassed with elm. The playing fields of Eton, where the outcomes of so many historic events were reputedly determined, were shaded by elms documented back to the seventeenth century. The Brocas, another Eton landmark, was conspicuous by its elm clump.[123] It had become traditional, in the Latin verses composed by Eton scholars in the eighteenth century, to include an invocation to an elm. Thus H. V. Bayley:

80, 81

82

Ulme, precor vivas! nec te Jovis ignea tanget
Ira, neque Aelio turbine verrat hiems!

O elm, may you live long, I pray. And may the fiery wrath of Jove not touch you, nor winter sweep with its whirlwind.[124]

One facet of the Romantic movement was born under an elm in Harrow churchyard. Byron reveals its origin:

Thou drooping Elm! beneath whose bough I lay,
And frequently mus'd the twilight hours away;
Where, as they once were wont, my limbs recline
But, ah! without the thoughts which then were mine.
How do thy branches, mourning to the blast,
Invite the bosom to recall the past,
And seem to whisper, as they gently swell,
'Take, while thou canst, a lingering last farewell'.[125]

Fig. 82. The Brocas, Eton (Br), by David Cox.[123]

The spot was so dear to Byron that he buried his little daughter Allegra nearby.[126] The elm became a tree sacred to the Romantic movement. Chateaubriand made a pilgrimage thereto and invoked it:

> Salut, antique ormeau, au pied duquel Byron enfant s'abondonnait aux caprices de son âge, alors que je rêvois *René* sous ton ombre, sous cette même ombre où plus tard le poète vint à son tour rêver *Childe-Harold*!

> *Hail, venerable elm, at whose feet the child Byron abandoned himself to the whims of his years, where I mused on* René *under your shade, the same shade where, later, the poet came in his turn to muse on* Childe Harold!*[127]

Some doggerel lines attributed to Francis Litchfield, when a pupil at Rugby reveal more about response to an elm than do many more sophisticated verses. He addresses the Treen Tree which was felled in 1819:

> When Avon's banks with hope and fear,
> My blushing childhood ventured near,
> Thou first didst bid its sorrows end,
> And wert unto it as a friend.[128]

For the studied reaction, listen to Matthew Arnold brooding in the school chapel, while

> the elms
> Fade into dimness apace,
> Silent.[129]

The popular image of Rugby was created by Thomas Hughes. As Arthur George recovers from his fever in *Tom Brown's schooldays*,

> he looked out of the windows again, as if he couldn't bear to lose a moment of the sunset, while the tops of the great feathery elms, round which the rooks were circling and clanging, returning in flocks from their evening's foraging parties.[130]

Fig. 83. Oxford, by John Baptist Malchair.[132]

Readers of Hughes' book will have noted how the disposition of the elms on the playing field determined the configuration of the pitch. Rupert Brooke's star-gazing through the Rugby elms has already been mentioned.[131]

83 Those going up to Oxford[132] or Cambridge were assured of an elmscape in either case. Amenity planting at Oxford went back to monastic times. The monks of Oseney did

> expend much in furnishing pleasant Walks by the River's side, and invironing them with Elm Trees.[133]

College groves such as those of Magdalen, Merton and St John's colleges, were most often of elm and Oscar Wilde's reflections in the first have already

84 been noted. The elms that impressed themselves most on Oxford men were those in the Broad Walk in Christ Church Meadow, planted by Dean Fell in 1670.[134] Here

Fig. 84. Broad Walk, Christ Church Meadow, Oxford, by Lonsdale Ragg.[134]

George Whitefield, John Wesley's coworker, wrestled in the spirit.[135] Cardinal Newman described it just past its prime:

> They say there are dons here who recollect when it was unbroken, nay, when you might walk under it in hard rain, and get no wet.[136]

And here Lewis Carroll walked with Alice and the Red Queen.[137]

> Betjeman found charms in north Oxford when

> a Cherwell mist dissolveth on elm-discovering skies.[138]

Cambridge, likewise, is environed with elm. The college groves tended to be mixed plantings but with elm as a major component. Out into the country ran the various elm-shaded walks. The elms that made most impact were those to the west of the Cam, along 85, 8 the Backs.[139]

It was at Cambridge that William Wordsworth discovered that a poet can thrive on level terrain. He did not adjust well to the life of his college and sought refuge regularly outside its walls where

> lofty elms
> Inviting shades of opportune recess,
> Bestowed composure on a neighbourhood
> Unpeaceful in itself.[140]

College fellows were granted right to matrimony in the mid-nineteenth century and Frances Cornford, a granddaughter of Charles Darwin and one of the first generation of academic children, wrote of Cambridge as she had known it:

> The figure of a scholar carrying back
> Books to the library, absorbed, content,
> Seeming as everlasting as the elms
> Bark-wrinkled, puddled round their roots, the bells,
> And the far shouting in the football fields.[141]

Bertrand Russell, casting philosophical detachment aside, is even more enthusiastic:

> It is lovely weather, and the yellow elms are heavenly, and all the people are good and nice, and it is a perfect paradise after the hell of Paris.[142]

The chorus of rapture had its dissident, Virginia Woolf. She began all right:

> The trees are pure gold and orange, and no place in the world can be lovelier than Cambridge.

Fig. 85. The Backs, Cambridge, by William Cook after J. M. Ince.[139]

Fig. 86. The Backs, Cambridge, by Joseph Pennell.[139]

Fig. 87. Kensington Palace Gardens (GL), by Seymour Haden.[144]

Then the worm entered in. It was a detestable place,

> when I think of it, I vomit, that's all – a green
> vomit...in the dusk, in the college garden...we
> have the tragedy from start to finish, and
> Choristers hidden in the Elms.[143]

Excluding places of education, no elmscape has been eulogized so highly or depicted so frequently as Bridgman's masterpiece, Kensington Palace Gardens, in the period before its original elms were felled.[144] Disraeli set the tone:

> In exactly ten minutes it is within the power of
> every man to free himself from all the tumults
> of the world; the pangs of love, the throbs of
> ambition, the wear and tear of play, the
> recriminating boudoir, the conspiring club, the
> rattling hell, and find himself in a sublime
> sylvan solitude superior to the cedars of

Lebanon, and inferior only in extent to the chestnut forests of Anatolia. It is Kensington Gardens that is almost the only place that has realized his idea of the forests of Spenser and Ariosto.[145]

W. H. Hudson lets himself go too, though in a different direction:

> It was like a wood where the trees were
> self-planted, and grew together in charming
> disorder, reaching a height of about one
> hundred feet or more. Of the fine sights of
> London known to me, including the turbid,
> rushing Thames, spanned by its vast stone
> bridges, the cathedral with its sombre cloud-like
> dome, and the endless hurrying procession of
> Cheapside, this impressed me the most.[146]

In Disraeli's case,[147] his adulation of Kensington

Fig. 88. Kensington Palace Gardens (GL), by Arthur Rackham.[148]

Gardens was fuelled further by the fancy, shared by other Victorians, that here was the virgin bower of the young Princess Victoria. The early years of Princess Mary of Teck, later consort to King George V, were also spent in Kensington Palace and allowed a recycling of the same sentiments.

Another denizen of Kensington Gardens was Peter Pan. Barrie had met 'the boys' there and the rest 88 followed.[148]

Contemporary with the metropolitan Elysium was a provincial Parnassus. This was Clifton Grove, an elm plantation alongside the River Trent near Nottingham. It first achieved more than local fame from the poem *Clifton Grove* by the ailing hand of

Henry Kirk White,[149] who left his initials on one of the trees.[150] What later local poets wrote of it is best not quoted.[151] Its finest moments were when D. H. Lawrence led his lovers to where

> on either side stood the elms-trees like pillars along a great aisle, arching over and making high up a roof from which the dead leaves fell.[152]

This roll of particular elmscapes can fittingly conclude with Garsington Manor, where Lady Ottoline 89, Morell fried her mixed grills of Oxford, Cambridge and Bloomsbury. Lady Ottoline's enthusings over the local elms can be taken as read.[153] The reaction of D. H.

Fig. 89. Garsington (Ox), by Mark Gertler.[153]

Lawrence, cut off from his northern roots, is more interesting.

> this house of the Ottoline's – it is England – my God, it breaks my soul – their England, these shafted windows, the elm-trees, the blue distance – the past, the great past, crumbling down, breaking down.[154]

Kinds of elm

Few writers reveal any awareness of the conspicuous differences between the various kinds of elm. Tennyson was an exception. He grew up at Somersby in the Lincolnshire Wolds where the local elms are Narrow-leaved Elm or Hybrid Elms. In the *Ode to memory* he portrays 'the seven elms' around the rectory. Later, when he left Somersby, he found that elm for most people meant English Elm, so, in *In memoriam*, the same trees became

> Witch-elms that counterchange the floor
> Of this flat lawn with dusk and bright.[155]

Robert Bridges specifies a witch-elm in his first eclogue, located in a Somerset landscape.[156] The elm already mentioned that features so conspicuously in *Howard's End* is also a 'wych-elm', based on a tree that grew at Forster's home at Rooksnest (He).[157]

The principal literary comparisons between kinds of elm have resulted from transatlantic travel in both directions. Matthew Arnold, for instance, having arrived in New England, remarked:

> The American elm I cannot prefer to the English, but still I admire it extremely.[158]

Fig. 90. Garsington (Ox), by Gilbert Spencer.[153]

Having travelled in the opposite direction, Nathaniel Hawthorne surveyed the English Elms in Greenwich Park. They were

> scarcely so beautiful, however, nor so stately, as an avenue of American elms, because these English trees have not such tall, columnar trunks, but are John Bullish in their structure – mighty of girth, but short between the ground and the branches, and round-headed. Our trees 'high over-arched with echoing walks between', have the greater resemblance to the Gothic aisle of a cathedral.[159]

Wendell Holmes' comparison is more contrived:

> I have thought that the compactness and the robustness about the English elm, which are replaced by the long, tapering limbs, and willowy grace and far-spreading reach of our own, might find a certain parallelism in the people, especially the females of the two countries.[160]

New England puritanism is liable to erupt in such comparisons, and Henry James, reacting to the lushness of the Warwickshire elmscape, concluded that it

> sins by excess of nutritive suggestion.[161]

Literary artifice

The syllable *elm*, with its exclusively liquid consonants, falls softly on the ear. It seems that this liquid quality, rather than the tree denoted, led to the appearance of the word in some literary contexts. Matthew Arnold's signal elm, already considered, is a case in point. The abruptly guttural *oak*, though accurately designating Arnold's tree, would have disturbed the gently elegiac atmosphere of Arnold's poem. As noted earlier, Virgil and Marvell anticipated Tennyson's line:

> The moan of doves in immemorial elms.[162]

Where Tennyson improved on his precursors was in composing a line where liquid consonants outnumber all others.

Puns and play upon sounds are seldom translatable. Wendell Holmes could not resist juxtaposing Thomas Gray's

> youth at the prow

with

> pleasure at the 'elm.[163]

In Catalan, Provençal and some French dialects, the words for elm and man fall together. The Catalan poet Auzias March exploited this circumstance to produce the cryptic couplet:

> L'om qui n'es menys es arbre menys de fruyt,
> Homs en bell ort son los homens del mon.
>
> *The elm that is least is the tree with few fruits.*
> *Men in the world are those in a fine garden.*[164]

Iranian poets made repeated use of the antithesis between the pomegranate, with succulent fruits, and the elm, fruitless in the popular sense. This antithesis was heightened by the similarities in the Persian words for the two plants, نار [nar] for pomegranate and نارون [narvan] for elm. So when Farrukhi calls down everlasting happiness for the recipient of his verse, it is for

تا نبود نار بر نارون

> *as long as the elm bears no pomegranate.*[165]

Elm seldom enters into metaphorical use in English. Shakespeare's

> Answer, thou dead elm, answer,

is a rare and unflattering exception.[166]

In the Welsh and particularly in the older Scandinavian literatures, elm is a standard metaphor for the archer's bow. An early Welsh example occurs in the poem *Yspeil Taliessin*, with the line

> Prieff lwyvyð, rhein onwyð yw i arveu
>
> *Strong elms [bows] and ashen spears are his weapons.*[167]

The death song attributed to the viking Ragnar Loðbrok had the lines:

> Umdi álmr pá er oddar
> Allhrat slitu skyrtur
>
> *The elm [bow] whined as the points [of the weapons]*
> *swiftly pierced the shirts [of mail].*[168]

More recondite still, such double metaphors as álmsvell, literally elm ice, that is, arrow, abound in the literature of the skalds.[169]

The most contrived literary conceit involving elm emanated not from one of the English metaphysical poets, but from the American author James Russell Lowell, who wrote:

> our tall elm, this hundredth year
> Doge of our leafy Venice here,

Who with an annual ring doth wed
The blue Adriatic overhead,
Shadows with his palatial mass
The deep canals of flowing grass.[170]

Symbol

Throughout literary history, elm has featured more as a symbol than as an object knowingly perceived. As will become apparent, elm is a multiple symbol. When it features as a symbol of death and malignancy, it is convenient to refer to it as the Stygian Elm. The roots of this conceptual elm are surprisingly many. Of these, instinctive dislike is probably the least consequential as it appears to be very rare. William Cobbett had something of it, expressed, for instance, when he wished that all the Wiltshire elms were locusts (*Robinia pseudoacacia*).[171] Sir Compton Mackenzie had more of it:

I couldn't face the prospect of those melancholy elms in flat English fields in winter.[172]

The Stygian Elm has three roots in Classical literature, one Greek and two Latin. The earliest is in the *Iliad*. Homer described the grave of Eetion

περὶ δὲ πτελέας ἐφύτευσαν
νύμφαι ὀρεστιάδες, κοῦφαι Διὸς αἰγιόχοιο.

and around it the nymphs of the mountains, daughters of Zeus, the Egis-bearer, planted elms.[173]

The major Latin root arises from Virgil's vision of the Underworld, where

in medio ramos annosaque bracchia pandit
ulmus opaca, ingens, quam sedem somnia volgo
vana tenere ferunt, foliisque sub omnibus haerent.

in the midst a dark elm of vast extent spreads its branches and aged arms, around which, false dreams cling and attach themselves beneath each leaf.[174]

Erasmus Darwin supposed, incorrectly, that it was Virgil's Stygian Elm that ornamented the Portland vase, where

by fall'n columns and disjoin'd arcades
On mouldering stones, beneath deciduous shades,
Sits HUMANKIND in hieroglyphic state.[175]

Robert Bridges makes even more explicit allusion to Virgil's tree in his poem *The great elm*, the subject of which is compared with

That fabled Elm of Acheron,
Within the gates of death.[176]

The minor Latin root of the Stygian Elm comes from the augurs. Consequent on its infertility, in the sense that it bore no fruits considered serviceable to man or animals, the elm was treated as a tree of ill omen, whenever the augurs had occasion to survey the landscape for purposes of divination.[177]

The use of elm for coffin boards, described in chapter 8, provides a further root to the Stygian Elm. This theme enters English literature with Chaucer's

cofere unto carayne.[178]

It was developed most laboriously in Thomas Hood's poem *The elm tree*. This is in the nature of a

Fig. 91. Thomas Hood 'The Elm', illustrated by Gustav Doré.[179]

nineteenth-century morality culminating in the sentiment:

91

> For the very knave
> Who digs the grave,
> The man who spreads the pall,
> And he who tolls the funeral bell.
> The Elm shall have them all![179]

The literary association between elms and coffins became so generally accepted that Dickens was able to indulge Mrs Mould with a little black humour in this area in *Martin Chuzzlewit*.[180]

A further root of the Stygian Elm finds its origin in the tendency of the English Elm to drop large branches unexpectedly, sometimes with fatal effect. Richard Jefferies made much of this in *Wood magic*:

> elms are so patient, they will wait sixty or seventy years to do somebody an injury.[181]

Rudyard Kipling worked up the same sentiment into verse:

> Ellum she hateth mankind, and waiteth
> Till every gust be laid,
> To drop a limb on the head of him
> That anyway trusts her shade.[182]

The penultimate root of the Stygian Elm arises from the circumstance that the Jew's ear fungus, *Hirneola auricula-judae*, interpreted in folklore as a reembodiment of the ear of Judas, occurs not infrequently on elm, in addition to its principal host, the elder.

Last of all, the elm, having become a universal symbol of death, acted out its role by succumbing, amid widespread lamentation, to the latest cycle of Dutch elm disease.

The Stygian Elm achieves its most capacious growth in English literature. None has embedded itself as ineradicably in English consciousness as the 'rugged elms' that Thomas Gray set over his village graveyard.[183]

One is not surprised to find another Stygian Elm growing vigorously in a Hardy landscape. John South in *The woodlanders* cringes under it:

> The tree 'tis that's killing me. There he stands, threatening my life every minute that the wind do blow. He'll come down upon us, and squat us dead.

In due course the elm is felled and John Smith dies.[184]

Two Stygian Elms designated Tweedledum and Tweedledee dominate the landscape of John Whiting's play Saint's Day. To these Paul Southman and Charles Heberden finally exeunt to be hung.[185]

A Spanish Stygian Elm features in Antonio Machado's poem *A un olmo seco*.

> Al olmo viejo, hendido por el rayo
> y en su mitad podrido,
> con las lluvias de abril y el sol de mayo,
> algunas hojas verdes le han salido.
>
> *On the old elm, riven by lightning and half rotten,*
> *the April rains and May sunshine bring out a few*
> *green leaves.*[186]

The tree symbolizes the poet's dying wife for whom a springtime of recovery is a despairing hope.

Alongside the dark Stygian Elm there coexists the bright Paradisiac Elm, totally opposite to the former in symbolic role. It is a figure of the Golden Age, of Paradise and of Love. Like the Stygian Elm, the Paradisiac Elm has manifold roots, and like it also, has different roots in Greek and in Latin Classical literature.

The Greek root derives from Theocritus who issued the invitation:

> δεῦρ' ὑπὸ τὰν πτελέαν ἑσδώμεθα τῶ τε Πριήπω
> καὶ τὰν κραναᾶν κατεναντίον.
>
> *Come, let us sit under the elm, facing Priapus and*
> *the springs.*[187]

It is apparent from the verse of Walter of Châtillon that elms continued on occasion to shade loving couples during the Middle Ages.[188] Many more did so during the Renaissance and after. Two of Giambattista Marino's *Sonnetti amorosi*, *Il luogo dei suoi amori* and *A un olmo*, had their action beneath an elm.[189] The former is one of the few elm poems set to music by a major composer, Monteverdi using its text for a pleasant five-part madrigal. Dalliance beneath an elm is infrequent in English literature, presumably because of the draughts.

The Latin root of the Paradisiac Elm is the viticultural technique, referred to in chapter 8, of marrying vines to elms. This practice was mentioned by Latin authors in describing rural beatitude. It was also used as a symbol of married or free love and this application became eventually the most frequent and most widespread of all literary allusions to elm.[190] 116

The vine and elm theme entered literary use with Catullus, who, taking the vine to symbolize the bride, wrote of it:

si forte eademst ulmo coniuncta mariti,
multi illam agricolae, multi coluere vivenci.

*if it chance to be married to an elm, many farmers
and many bullocks tend it.*[191]

With Virgil and Horace, and Petronius, if the verses attributed to him are his, the vine and elm theme is principally used to describe paradisiac life in the country.[192] Its amatory significance is prominent in Ovid and Statius.[193]

The early Christians of Italy, mainly Greek-speaking, also took up the vine and elm theme. In the Shepherd of Hermas, the following passage occurs:

ἡ ἄμπελος κρεμαμένη ἐπὶ πτελέαν τὸν καρπὸν
πολὺν καὶ καλὸν δίδωσιν, ἐρριμμένη δὲ χαμαὶ
ὀλίγον καὶ σαπρὸν φέρει.

*The vine, when hanging from the elm, gives
abundant and beautiful fruit, but when lying on the
ground bears sparse and rotten fruit.*[194]

It is explained that this is a parable of the relation that should subsist between poor and rich, the former aiding the rich with their intercession and the latter providing the poor with material necessities. The vine and elm theme, so interpreted, did not persist in the mainstream of Patristic literature. It is mentioned in the *Stromata* of Clement of Alexandria and in one of the homilies of Pope Gregory the Great.[195] Occasionally it turned up in some rather odd places, as, for example, an Arian commentary on Matthew's Gospel, the *Opus imperfectum*, long attributed, but with no justification, to John Chrysostom.[196] An echo of the theme is encountered in the life of Fursey, the Irish monk whose trip to Heaven and Hell opened up the route that Dante was to follow. It was in Heaven that he met Meldan who instructed him, on his return to earth, to restrain churchmen in authority from too ready a use of excommunication

ne ulmum pro vite et alnum pro oliva plantent

*lest they plant elm instead of vine, and alder instead
of olive.*[197]

Medieval poets tended to be more interested in their own experiences than in finding neat Classical allusions, and so the vine and elm theme diminished in frequency. Alain de Lille used it to describe spring, the season

Quo vitis gemmata sinus amplexa maritos
Ulmi, de partu cogitat ipsa suo.

*in which the budded vine clasped to the connubial
bosom of the elm, reflects on its coming to birth.*[198]

It features also in the *Roman de la rose*,[199] and the English poet Alexander Neckam, who shared a wet nurse with Richard I, spares it a line:

Ulmus adest viti conjuga grata comes,

*the elm is here, married to the vine, her beloved
companion.*[200]

The Renaissance tipped the scales the other way and classical allusion became esteemed far more than observation of Nature. So, in spite of the fact that vines have seldom been married to elms outside certain tracts in Italy, they feature thus in the verse, Latin or vernacular, of every country in any way touched by the Renaissance, Pontano, in his poem *De amore*, led the way but the theme really hotted up in the *Basia* of Joannes Secundus, whose Neoera was to embrace him

Vicina quantum vitis lascivit in ulmo

as the vine wantons with its supporting elm.[201]

As would be expected, the most intricately contrived classical allusion is to be found in the Spanish verse of Luis de Góngora:

los olmos casando can las vides –
mientras coronan pámpanos a Alcides
clava empuñe Liëo

*In the marriage of elms with vines, while the vine
shoots crown Alcides [Hercules], Lyaeus [Bacchus]
grasps his club.*[202]

Distance from the region where vines were married to elms was no impediment. Later, in North America, Ralph Waldo Emerson was to moralize that

Man's the elm, and Wealth the vine.[203]

In Renaissance England, Edmund Spenser leads in the

vine-prop Elme,[204]

to be followed by Sir Philip Sidney and William Browne.[205] Sir John Davies coupled the theme with that of the cosmic dance:

What makes the Vine about the Elme to
dance
With turnings, windings, and embracements
round?[206]

A frequent development of the vine and elm theme was to contrast it, as a figure of virtuous love, with the ivy and elm theme, a figure of adulterous passion.

The two themes are so used by Adriana in Shakespeare's *Comedy of errors*.[207]

Milton placed both vine and elm in Paradise, where Adam and Eve

led the vine

To wed her elm; she spoused about him twines
Her marriageable arms, with her brings
Her dow'r th'adopted clusters, to adorn
His barren leaves.[208]

The theme continued in vigour till the nineteenth century. Wordsworth, Coleridge, Shelley and Tennyson all use it. More surprising, so also does the down-to-earth Crabbe:

The Clinging Vine prest down the branching Elm
E'en to the Earth, and in her verdant Lap
The tributary Grape, yet growing, laid.[209]

The decline of Classical education over most of Europe will presumably entail the demise of this theme. One of its last appearances was when Virginia Woolf used it in *Jacob's room* and, characteristically, got it wrong – Virgil refers to vines trained up elms, not between them.[210]

So much then, for the Classical roots of the Paradisiac Elm. Its strongest root is the widespread conviction that elmscapes directly communicate the qualities of terrestrial Paradise to their denizens, particularly children. In fact, many of the descriptive quotations earlier in this chapter have a paradisiac resonance.

Paradisiac Elms are found also in China, but again, usually in association with willows. Chu Guangxi wrote:

日 暮 閒 園 裏
團 團 蔭 榆 柳

The sun has set. We relax in the garden, gathering together in the shadows of the elms and willows.[211]

They occur too in Russian literature. Gogol issued an unforgettable invitation:

как было бы в самом деле хорошо, если бы жить этак вместе, под о дною кровлею, или под тенью какого-нибудь вяза пофилософствовать о чем-нибудь, углубиться.

How nice it would be if we were to live under one roof, or if we could, under the shade of some elm, philosophize on something, and dig to its roots.[212]

Dostoevsky's Paradisiac Elm was in the Optina Monastery which he visited with Vladimir Solovev. When he described it in the *Brothers Karamazov*, however, it had become too charged to be dubbed paradisiac as understood here.

Бывает в нощи. Видишь сии два сука? В нощи же и се Христос руце ко мне простирает и руками теми ищет меня, явно вижу и трелешу Страшно, о страшно!

It happens at night. See those two branches? At night it is Christ holding out his arms to me and searching for me with those arms. I see it clearly and tremble. It's terrible, Oh, it's terrible![213]

However, it is in English literature that the paradisiac qualities of the elm are most especially displayed. Milton's *L'Allegro* where the poet,

Sometime walking, not unseen,
By hedgerow elms on hillocks green,[214]

delineates one of the earliest and best-known poetic descriptions of a paradisiac landscape. He uncovered the manner of his poetic inspiration in a college prolusion:

Testor ipse lucos, & flumina & dilectas villarum ulmos, sub quibus aestate proxime praeterita (sic Dearum arcana eloqui liceat) summam cum Musis gratiam habuisse me jucunda memoria recolo.

I myself call to witness the groves, streams and beloved village elms under which, last summer – if I may speak of the secrets of the goddesses – I recall with delight that I received the highest favour of the Muses.[215]

Many authors dwell on the Paradisiac Elms of their childhood. Keats recalls

A laughing school-boy, without grief or care,
Riding the springy branches of an elm.[216]

Chesterton, when he brings home the seven Days of the *Man who was Thursday*, identifies it by the elms reminiscent of their childhood.[217] Robert Graves turns back to

Tarrier's Lane, which was the street
We children thought the pleasantest in the Town
Because of the old elms growing from the pavement
And the crookedness where the other streets were straight.[218]

92, 9

Edward Thomas reveals the impact of elms on an old countryman:

> Those elms had come unconsciously to be part of the real religion of men in that neighbourhood, and certainly of that old man. Their cool green voices as they swayed, their masses motionless against the evening or summer storms, created a sense of pomp and awe. They gave mystic invitations that stirred his blood if not his slowly working brain.[219]

The only relaxed moments in *Nineteen eighty-four* are where George Orwell portrays the Golden Country, where

> the boughs of the elm trees were swaying very faintly in the breeze, their leaves just stirring in dense masses like women's hair.[220]

The culmination of elm panegyric was achieved by the American author Henry David Thoreau, who concluded that men were a bunch of yahoos, brutish in comparison with the nobility of the elms above them:

> The poor human representative of his party sent out from beneath their shade will not suggest a tithe of the dignity, the true nobleness and comprehensiveness of view, the sturdiness and independence, and the serene beneficence that they do...They do not, like men, from radicals turn conservative. Their conservative part dies

Fig. 92. John Milton 'L'Allegro', illustrated by William Blake.[214]

out first; their radical and growing part survives.[221]

Though the Stygian and Paradisiac Elms dominate the symbolic elmscape, other symbolic elms occur that are assignable to neither. One such is the elm of the bizarre landscape of the Arthurian cycle. It appears with Chrétien de Troyes:

Desoz un orme en un prael
Trova une pucele douce.

He found a gentle maiden on a green beneath an elm.

It is found in Malory and again in Tennyson:

the royal crown
Sparkled, and swaying upon a restless elm
Drew the vague glance of Vivien and her
Squire.[222]

But the damsel beneath the elm or the knight whose horse is tied to it is as likely to be evil as virtuous.

Much more widespread in literary elmscapes is the elm conceived as feminine. Farrukhi pulls out the stops of the Persian poetic imagery. For him the elm is

چون دلبری اندر عقیقین و شاخ   چون لعبتی در بسّدین پیرهن
نالنده همچون من زهجران یار   لرزنده و پیچنده برخویشتن

Fig. 93. John Milton 'L'Allegro', illustrated by Birket Foster.[214]

like a heart-stealing beauty in a ruby shawl, like a
harlot in a coral skirt, moaning like me for the
absence of my beloved, trembling and twisting in on
itself.[223]

There is an abundance of examples nearer
home. Sometimes it is the individual tree that is
conceived as woman. In other cases feminine qualities
are attributed to the elmscape. Hopkins is reticent. He
records the branches of a Wych Elm

coming and going towards one another in
caressed curve and combing.[224]

George Orwell is more explicit. He celebrates his return
to England from Spain, back to

the green bosoms of the elms.[225]

Morgan Forster goes further and reveals the secret of
the Cambridge Backs:

The great elms were motionless, and seemed still
in the glory of midsummer, for the darkness hid
the yellow blotches on their leaves, and their
outlines were still rounded against the tender
sky. Those elms were Dryads – so Rickie believed
or pretended, and the line between the two is
subtler than we admit. At all events they were
lady trees, and had for generations fooled the
college statutes by their residence in the haunts
of youth.[226]

The femininity of elmscape is implied in Hardy's
description of the Vale of Blackmoor as 'languorous',
with 'heavy soils and scents'.[227] Its womb-like
character is hinted at in Alfred Williams' classic
portrayal of the Vale of the White Horse, which
confers on its denizens

a sense of vast, cumbersome wealth, of
unspeakable riches, a luxury of possession, and
imprisonment with it.[228]

It is more explicit still in H. Mallalieu's contrast
between the open landscapes painted by the Norwich
school, which disturb those used to the

womb-like security of wood and valley afforded
by other more secure counties.[229]

In the examples just quoted, femininity was, on
the whole, something desirable, whether as sexually
provocative or as maternally protective. However, as
the psychoanalysts insist, there is another side to
femininity, its destructive aspect, the assemblage of
traits characterizing the hag and female vampire.

Destructive femininity may be perceived as within
one's self as in Sylvia Plath's poem *Elm*, where the tree
represents her own tortured libidinal consciousness.[230]
More often, it is predatory on others. It would be hard
to improve on Eugene O'Neill's description of this in
the stage direction for *Desire under the elms*:

Two enormous elms are on each side of the
house. They bend their trailing branches down
over the roof – they appear to protect and at the
same time subdue; there is a sinister maternity
in their aspect, a crushing, jealous absorption.
When the wind does not keep them astir, they
develop from their intimate contact with the life
of men in the house an appalling humaneness.
They brood oppressively over the house, they
are like exhausted women resting their sagging
breasts and hands and hair on its roof, and
when it rains their tears trickle down mono-
tonously and rot on the shingles.[231]

It emerges from everything above that the
unique place of elm in literature, especially English
literature, rests on its multiplicity of levels of
significance. It is simultaneously object knowingly
perceived, manipulable sound, and symbol. The
symbol is ambivalent, simultaneously Stygian and
Paradisiac. This ambivalence is obtrusive in the
writings of Richard Jefferies. In *The story of my heart*,[232]
he describes how his nature mysticism unfolded as he
looked up through elm branches to the sky. Yet in
Wood magic it is the treachery of the elm on which he
insists.[233] In his article *In praise of elms*, M. Armfield
described the ambivalence of the elm as a confluence
of

humour and seriousness, of salt and sweet, of
the bizarre and a serenity that is almost classic
in its beauty.[234]

Arthur Machen in his Decadent novel *The hill of dreams*
expressed the ambivalence of elm in prose. It was a
symbol simultaneously of childish reassurance and of
threatening death.[235]

Christopher Hassall in *The unsafe elm*, summed
up its conflicting aspects in verse:

who could believe (having so often in
childish pursuit
Felt its rough refuge) when it was officially
pronounced
'Unsafe'? Was it not 'safe' to everlasting?[236]

I I
The artist's reaction[1]

When dealing with elm in literature, four levels of reaction by writers were found. In the first, trees were not perceived at all, or if perceived, were not regarded as of sufficient interest to mention. In the second, trees were perceived, but only as a general category, and particular kinds were not differentiated. This level of reaction could be embroidered by a writer mentioning different kinds of tree for literary effect though not perceiving the differences that his writing implied. Recognition of trees at the generic level, for example as elm or oak, occurred at level three. In the fourth level, distinctions between kinds of elm, such as between English Elm and Wych Elm were made. These four levels result from the nature of English and most other natural languages. It will be seen that a different situation presents itself in the visual sphere.

Take the no-tree level. This is usually clear cut in a literary context. Particular trees, trees in general, or words with an immediate reference to trees do not occur. In a picture, this may also be so in two cases. The first is that of the naturalistic artist whose mode of representation is such that immediate inspection indicates that there are no trees in his work. The second case is that of the fully abstract picture where objects intended as trees have no place. But in between, as for example in much Post-impressionist and some only partially abstract painting, there may be uncertainties. A painted area may be a tree, or could be meant to be a tree. One cannot say yes or no.

Consider now the second level, that of trees in general. One is certain now that an area of paint represents a tree. It is next to impossible, however, to paint a generalized tree. In most cases it is at least clear whether the representation is of a conifer or of a broad-leaved tree. Possibly Alfred Wallis gets nearest to the generalized tree. Some of his trees could with equal ease be taken as conifers or Cornish Elm.

Only seldom can a painting be even said to represent a generalized broad-leaved tree. Gainsborough and Turner approach this level of generality. It is often impossible in their works to find any feature

Fig. 94. Chichester Canal (WS), by Joseph Turner.[2]

indicating that a representation is one kind of tree rather than another. The most that can be said is liable to be negative. It is unlikely to be a beech but it could be an oak, ash or elm.

Much more often, an area of paint is clearly a broad-leaved tree and more like a particular generic category, say oak, than any other. But one cannot be sure. There is no egress from the limbo of uncertainty. This situation is by far the most common as regards pictures with trees. When trees are represented, it is often possible to say what kind of tree is probably intended. It is only seldom that a confident judgment can be made that the tree can only be an oak, ash or elm. Turner's *Chichester Canal*,[2] for example, probably represents elms, but there is no certainty. If the fact that the location is in a region where English Elm is frequent were ignored, the probability of the identification would be further reduced.

The generic level in the visual arts also differs from the corresponding level in literature. It may be possible to identify a painted tree as an oak or ash. It is, however, as difficult to paint a generalized elm as a generalized tree. If an elm is recognizable at all in a painting, it is likely to be identifiable at the fourth level, as a particular kind of elm, say English elm or Cornish Elm. Indeed, some kinds of elm are less like other elms in general appearance than various other kinds of tree.

Representations of Dutch Elm, for example, can easily be confused with those of oak, even when painted with close fidelity to perception.

So, in contrast to the situation in literature, with its four sharply defined levels, four different levels are relevant to representations of the elm in the visual arts. These are the no-tree level, the level of generalized broad-leaved tree, a level intermediate between the generalized broad-leaved tree and a particular kind of elm, and the level of precisely identifiable kind.

The presumption so far has been that pictures speak for themselves. There may be other means by which illustrations of trees can be identified. An artist may give a title which identifies the trees in his picture or he may accept a title given by someone else. One presumes, for example, that Gustave Doré intended to represent an elm in his illustration for Thomas Hood's poem *The elm tree*, mentioned in the preceding chapter. Nothing in the picture identifies the tree beyond the generalized broad-leaved category. Or the nature of the trees that an artist has depicted may emerge from his writings or conversation. There is, of course, no guarantee against his misidentifying his trees and instances have occurred, as in literature, where the artist has been mistaken.

It may also be the case that an artist uses some conventional formula to indicate a particular kind of

Fig. 95. Touches for tree foliage, by James Harding: *a*, oak; *b*, elm.[5]

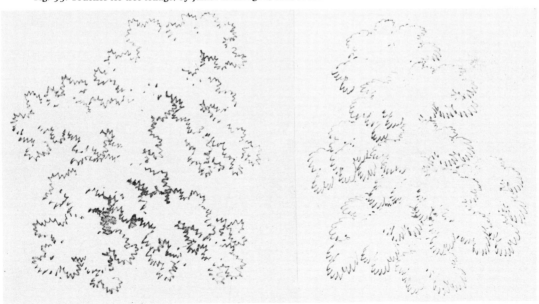

a   *b*

tree. Its application need not result in a representation that looks much like the tree intended. Provided the formula is recognized, the tree can be identified. Drawing masters, particularly in the last century, inculcated various formulae or touches for depicting the foliage masses of different trees. The touch for oak

95 was spiky. The elm touch was scalloped.[3]

Finally, if an artist has represented a particular locality with some degree of verisimilitude, it may be possible to identify his trees from a knowledge of what grows or grew in the locality. This is possible, to some extent, with John Constable, some of whose localities are very precisely known. There are snags. Constable, like other artists of similar calibre, carried no brief for topographical accuracy. The foreground ash, for instance, in *Valley Farm*, which is otherwise a somewhat drastic refurbishing of Willie Lot's cottage by Flatford Mill (Sf), may have been transferred from Hampstead (GL).[4]

It is also the case that some elms have found themselves in a painting merely because they got in

138, the way. Waltham Abbey, for instance, attracted a
139 number of artists, but conspicuously in front of the

most paintable elevation stood a large and ancient English Elm. It had to be included. So dominating was its presence that what purport to be representations of the church are really tree studies in an ecclesiastical setting. Paddington church (GL), similarly, was subservient to the elms that surround it.   141

In dealing with literature, it was shown that the process of composition often involved a recession from the real world as far as elm was concerned. It will be necessary to enquire to what extent the work of the visual artist has involved a parallel recession from reality. In neither case is it implied that recession from reality is a bad thing. That interesting problem is a topic for a quite different book.

It might be supposed that the starting point of a picture of an elm is an elm. This appears often not to have been so. Artists have always been greatly influenced by their predecessors' work. Take again the example of John Constable. The trees he painted certainly owe much to the trees he perceived. Equally they owe as much, perhaps more, to Gainsborough's trees, which in turn, derived more from the trees of such Dutch painters as Hobbema, Ruisdael and

Fig. 96. Cornfield, oil sketch, by John Constable.[8]

Waterloo, than from any trees he observed at first hand. The Dutch painters elaborated a formula for depicting the oak trees with which they were well acquainted. This became accepted as a general basic formula for any broad-leaved tree by later artists. All that was necessary, in particular cases, was to modify the formula in minor respects to represent the particular characteristics of individual broad-leaved trees of whatever kind. So it is that Constable's elms are to some extent Dutch oaks. Constable's so-called naturalism resided not in dispensing with the models provided by earlier work, but in modifying the existing formulae more drastically in the light of what he perceived for himself.

One reason for the great dependence of artists on earlier models for representing trees is the technical difficulty of depicting them,[5] consequent largely on the intricacy of the texture, whether of the tree in leaf or after leaf fall.

Reference has already been made to the existence of formulae devised by drawing masters to differentiate between different kinds of tree. The existence of such models is liable to come between the artist and what he purports to represent. In drawing a tree he is likely to reproduce the formula he has learned, modified as before, to represent the individual tree he has under observation.

Perhaps an artist is not concerned with representing a particular elm he has observed. His object may be to depict the elm mentioned in some literary work. Many artists, for example, have sought to represent the elm in Virgil's *Aeneid*, or, as indicated

Fig. 97. Cornfield, final painting, by John Constable.[7]

92, 93 in the previous chapter, in Milton's *L'Allegro*, or the elms in Gray's *Elegy*.[6] Whether what they depicted was also an elm that they had seen was of no consequence.

71 It sometimes happens that an artist has a flair for representing elms because of a quirk in his own style. The rococo line of Thomas Rowlandson, for example, which he applied so happily to a myriad of circumstances, was preadapted to representing the billowing outlines of the English Elm. Lionel Constable painted with a rather sinuous and feathery effect. This was not particularly happy for some trees. It was ideal for the Narrow-leaved Elm of East Anglia which has the sinuous and feathery qualities for which Lionel 99 Constable's style was a preadaptation.

Often, particularly in more recent times, elms have been drawn as directly as may be from nature, as for example, in open-air pencil drawings or water colours. Production may, however, be much longer. John Constable's major works, for instance, typically progressed via a pencil drawing in the field and an oil

sketch to the finished oil painting. Some went through a further stage, the engraving of a mezzotint by David Lucas, under the scrutiny, in this case, of the principal artist.

97 What happens to an elm in such a production chain? Take Constable's *Cornfield*.[7] No pencil sketch has survived but there is an oil sketch which probably 96 represents the second stage.[8] The tree in the left foreground is the feature around which the *Cornfield* appears to have been composed. It was most likely a Hybrid Elm such as still grows in the same hedge. The exact spot, on the unmetalled road from East Bergholt (Sf) to Dedham (Ex) is known from a letter of John's son Charles.[9]

The tree of the oil sketch provided the site and outline of the lower three-quarters of the front member of the row of three elms in the centre of the *Cornfield*. The two trees behind never existed. There is a field boundary meeting the lane at this point but it does not run into it at the angle shown in the painting. The angle of the back two trees to the vertical is

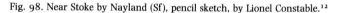

Fig. 98. Near Stoke by Nayland (Sf), pencil sketch, by Lionel Constable.[12]

impossible. In a sheltered situation as here, all trunks, like than of the front tree, would rise vertically and parallel.[10] The crown of the front tree has been raised. The branching configuration of the upper canopy, if examined carefully, will be found to be impossible. Some of the branches have a direction and thickness that preclude their arising naturally from the canopy at the intermediate level.

The oil sketch provided little by way of leaf texture. The finished painting has much more but the effect is somewhat heavy. In fact it is rather what one would expect if Constable, working on his final painting at Hampstead (GL) had refreshed his memory by looking at the elms nearby. What he would have found at Hampstead would have been the densely canopied English Elm, not the feathery, more open canopied East Anglian elms. Of course the painting is

a masterpiece. The progression, however, from oil sketch to finished painting is simultaneously a recession from reality, as represented by the Suffolk Hybrid Elm that was the starting point. Even leaf texture, which might be expected to be more naturalistic in a finished oil painting than in an oil sketch, seems not to be so in the present instance.

Lionel Constable's *Near Stoke by Nayland*, formerly attributed to his father John,[11] exists in a preliminary pencil sketch and a final oil painting.[12] In this case, the tall elm in the painting is far closer to what an East Anglian Elm in this neighbourhood looks like than the corresponding tree in the pencil sketch. However, as pointed out above, this was probably not the result of Lionel's having intended a particularly naturalistic representation. It happened that Lionel's style preadapted him to paint East Anglian elms

98

99

Fig. 99. Near *Stoke* by Nayland (Sf), final painting, by Lionel Constable.[12]

felicitously. No painting by his father recreated the Suffolk landscape as authentically as this picture by Lionel.

There are four surviving versions of George Stubbs' *Reapers*.[13] The first is a painting dated 1783, now at Upton House (Wa). Its background is one of the best English Elm hedgerows in British painting. Stubbs painted a second version of the same theme in 1785. This is the version now at the Tate Gallery. It differs in two major respects from the earlier painting. The human figures are now organized into a more studiously composed pattern. The elm hedge has disappeared and has been replaced by a botanically nondescript hedge of unidentifiable generalized broad-leaved trees. The superiority of the second version was blazoned when the appeal for its purchase by the Tate was under way. The only dissentient conceded the superior composition of the second version but insisted on its inferiority as a rendering of the English countryside.[14] This instance demonstrates again how, measured by its elms, artistic creation is liable to be progress into unreality.

The later versions of the *Reapers*, a mezzotint and an oval painting in enamel on biscuit earthenware, follow the second version.

For prints, there was a comparable production chain. The stages might be an initial water colour, a draughtsman's copy for the engraver, and finally the engraving itself. Usually artist and engraver were different persons. It was then open to the engraver, in tackling an elm, to make it more realistic than the artist had managed or intended, or less. This was largely brought about by the degree of command over leaf texture. Some engravers, like the Dalziel brothers, had a flair for this. Others did not.

Most of the preceding discussion has been about visual art with a clear representational intent. There is an important tradition in English art that partially transcends this category, that of Samuel Palmer and his followers.

Palmer's seminal experience was at the age of three, in the company of his nurse:

When less than four years old, as I was standing with her, watching the shadows on the wall from branches of elm behind which the moon had risen, she transferred and fixed the fleeting image in my memory by repeating the couplet, –

'Vain man, the vision of a moment made,
  Dream of a dream and shadow of a shade.'
I never forgot those shadows, and am often trying to paint them.[15]

It is asking too much to suppose that the infant Samuel recognized elms as such at so tender an age. Doubtless a certain amount of elucidation in hindsight elaborated the memory. Certainly he had a remarkable nurse. That elms continued to influence him in later life is confirmed by his comments on Milton mentioned in the previous chapter.

The outcome of the experience was the series of highly evocative paintings of Palmer's Shoreham Period. The continuing *leitmotiv* is the moon shining through foliage. It is no longer naturalistically drawn. From the viewpoint of dynamic psychology, however, it is clear that there is elm in all these trees. The Shoreham paintings have a brooding quality, a presage of surrealism, and perhaps some veiled sexual symbolism. They established one of the most distinctive styles in British art. In the painting reproduced in this book, the central tree is Palmer's rendering of the elm of perception. To the right the moon shines through the dark foliage put out by the primordial elm.[16]

Palmer was a major influence on two subsequent artists. Frederick Griggs was the earlier. Most of his elm engravings are in a naturalistic mode and are discussed elsewhere. Occasionally, however, as in his etching *St Ippolyts*,[17] a rendering of Ippolitts (He) churchyard, the mode is very different and immediately recalls Palmer's work. The later is Robin Tanner.[18]

Any attempt at ordered exposition breaks down over William Blake, a personality of comparable significance as writer and as artist. His treatment of elm is very different in the two arts. There is little elm in his writings. There are few overt elms in his illustrations and these tend to be recognizable only from contextual information. One accepts, for example, that the small trees at the bottom of *The Great Sun* are elms because these are the trees mentioned in Milton's *L'Allegro* for which this picture is an illustration.

Far more significant are Blake's numerous examples of the vine and tree *leitmotiv*, almost certainly derived from the vine and elm of the Classical authors. It is used in the coloured relief etchings that

Fig. 100. Reapers, 1783 version, by George Stubbs.[13]

Fig. 101. Reapers, 1785 version, by George Stubbs.[13]

illustrate such works as the *Songs of innocence, Songs of experience* and the *Marriage of Heaven and Hell*.[19]

104 Blake's symbol is in fact more complex than this. It also incorporates Virgil's elm of Hades and the Tree of Jesse as conventionally represented in Gothic art. Many writers have used the vince and elm theme perfunctorily. With Blake it is made to burgeon afresh, the culmination of its long history as an enduring symbol in European art and literature.

Media

Visual recognition that a tree in leaf is an elm depends on integrating the distinctive features of silhouette, architecture of the trunk and main branches, configuration of the foliage masses, bark texture, foliage texture, and the colour of the leaves. For a tree in winter, the foliage characters are lacking. The texture of the fine branches can, however, now be perceived.

3, 4 English Elm, Cornish Elm, Guernsey Elm and
6 Lock's Elm each have their distinctive silhouette. Other
5 kinds of elm are less recognizable in this respect. The silhouette of some can hardly be distinguished from that of other trees.

Architecture of the trunk and branches, again, depends on variety. The bole of the English Elm is 63, 84 characteristically straight and massive. In the Narrow- 2 leaved Elm it is usually less straight and more slender. In general, elm branches curve gently. The abrupt knees and bends usual in oak and ash branches are seldom found.

The foliage masses of the English Elm are dense 63 and cushion-like. In the Narrow-leaved Elm they are 2 feathery. Bark texture also depends on variety. In most kinds of elm the bark is deeply grooved. In the English 84 Elm it consists of a mesh of scaly plates.

The human eye has remarkable resolving power and this is responsible for the greatest difference experienced between looking at an elm and looking at a painting of an elm. At moderate distances, the eye picks out individual leaves and the play of light upon them. The texture thus recognized is highly distinctive. It is often sufficient in itself to identify an elm and even what kind of elm. Partly for technical reasons and

Fig. 102. Gleaning field, by Samuel Palmer.[16]

partly from intention, it is the feature most frequently lost when an elm is painted or drawn.

Leaf colour is the least diagnostic of the characters mentioned. It also varies with the kind of elm and the time of year. In spring and early summer the foliage of all kinds is a fresh, light green. In many of the Narrow-leaved Elms, this colour is retained throughout the summer. In the English Elm, the leaves change to a dark, almost blackish green. The leaves of all elms turn yellow in autumn, but the quality of the yellow varies with the kind of elm. One of the odd things about elm depiction is the avoidance by artists of the elm in full autumn colour. No painter of the first rank appears to have tackled this subject in spite of the brilliance of the spectacle. The reason why is obscure.

As the previous chapter shows, literary precedent is abundant. One is almost driven to suppose that, as soon as the temperature drops, the artists retire shivering to their firesides, averting their gaze from whatever Nature might be about outside.

The feature that above all identifies the elm in winter is the fineness of the lesser branches. Again, the remarkable resolving power of the eye enables the human observer to register this quality. But again too, and for the same reasons as with foliage texture, this salient feature is normally lost when an elm is illustrated.

Entirely different characters are diagnostic for elm twigs, leaves and flowers, observed at close distance. These are of principal interest to botanists

Fig. 103. The old road: elegy for the English Elm, I, by Robin Tanner.[18]

and are considered in the last section of this chapter.

Four levels of representation of tree were identified in the previous section. It has now to be considered how artists achieve the fourth level, that of portraying an identifiable kind of elm. Invariably, certain salient features are seized upon. Others, equally important in direct perception, are discarded. The choice as to what features are to be retained depends largely on the medium chosen and on the technique of manipulating it.

There has been some discussion as to what media are best for depicting recognizable trees. One writer concluded that monochrome representation was best, followed by water-colour drawing, followed by oil painting.[20] It is rash to generalize. Each of the more common media has yielded outstanding representations of elms. It is worth considering briefly, however, their respective scope and limitations.

Fig. 104. Vine and elm theme, by William Blake.[19]

Both oils and water colours provide colour contrasts. If the artist is mainly preoccupied with light effects or with manipulation of colour for non-representational purposes, his colours may contribute very little to the recognition of a tree as a kind of elm. Even if he has a representational intention, his use of colour may not offset the recession from naturalism that his brushwork is likely to impose. Generally speaking, the broader the brush strokes the less recognizable becomes the foliage texture, or, if a tree in winter is being represented, the fine-branch texture. The contrast between a good oil painting of an elm and a photograph emphasizes how greatly the loss of texture transforms the visual experience.

Obviously, if a naturalistic effect is intended, the more that salient features are discarded, the more important it is that those that remain should be exactly right. So, in oils and water colour, if elms are to be recognizable as such, it is imperative that such features as silhouette and branch architecture be rendered realistically to compensate for the ambiguity of the foliage or fine-branch texture.

The lack of colour contrast has proved little handicap in the potentialities of the monochrome media for representing recognizable elms. Pen and ink, pencil and even charcoal have all been used to produce outstanding elms. The same goes for wood engraving, metal-plate engraving and etching, and lithography. As with media involving colour contrast, the monochrome media may sacrifice texture, particularly in charcoal drawing and some styles of wood engraving. Effectiveness in representation of a tree such as elm then depends on controlled treatment of other features, particularly silhouette and branch architecture.

Art photography has to be approached rather differently from the other media. In the media previously discussed, the artistic quality of works involving trees has had very little to do with their recognizability. In the case of tree photography, except where special effects are being sought, there is a high probability that the photograph will reveal the identity of the tree irrespective of the quality of the photograph as a work of art. In the case of a Gothic building, say, a painting may recreate the scene just as effectively as a photograph, largely because the relevant fine texture is relatively coarse. In the case of a tree, there is likely to be, as pointed out earlier, an immense difference between painting and photograph, largely because the relevant fine texture, which the photograph is likely to preserve and the painting to lose, is much finer. In fact, if anyone in the future who has never experienced an elmscape at first hand should wish to understand the nature of its impact, he would be advised, not to pace the landscape sections of the art galleries, but to browse through the photographs in the back numbers of *Country Life*. The features that qualify an elm photograph as a work of art are not investigated here.[21]

Coloured pictures

It has already been pointed out that when British painters first paid serious attention to trees, what they portrayed were modified versions of Dutch oaks. These were not the only models. Though Dutch artists showed how individual trees could be depicted, the principles of composing a picture in which trees were conspicuous were taken far more from the much admired and much collected landscapes of Nicolas Poussin, Claude Lorrain and Salvator Rosa. Claude, in particular, was regarded as the guide in rendering Nature under its beautiful aspect, which is discussed in chapter 9. He also had a prolific output of tree drawings, also much admired, of which none but a handful are more closely identifiable.

Parallel with these influences was the tradition of English topographical painting, very much in the hands of Dutch artists in its earlier period. The landed gentry wanted a visible record of what their property looked like, faithful in detail and with composition left to look after itself. The prime concern of the topographical artists was with houses. Often the trees surrounding them were depicted according to some conventional style, quite unrecognizable as to kind. Occasionally, however, a topographical artist took a closer look at the trees in a landscape and portrayed objects with some degree of correspondence to their actual appearance. Jan Siberechts, for instance, 53 clearly depicted the trimmed elms grown along the Thames valley.

However, at the time when landscape was becoming of serious concern to British painters, it was nature under its sublime and picturesque aspects that was most esteemed. Consequently the tranquil landscapes of the south of England, exemplifying Nature under its beautiful aspect, and where elms were most in evidence, were largely ignored. So, in the landscape paintings of the eighteenth century, elms are rarely present. Trees, which independent information indicates might be elms, were liable to be painted as generalized broad-leaved trees or as thinly disguised Dutch oaks. Thomas Gainsborough, for instance, in spite of *Cornard Wood*, and in spite of the oft-quoted remark that 'Nature was his teacher and the woods of Suffolk his academy', hardly ever painted an identifiable tree. Since it is known that he used broccoli for trees when constructing models for his pictures, this is not greatly surprising.[22]

Of eighteenth century painters in oil who depicted elms as they were, George Stubbs was outstanding. It is ironic that, as mentioned earlier, he went back on his achievement. The other contemporary artist who did well with elms was Thomas Rowlandson, 71 whose medium was water colour.[23] It was suggested earlier that part of Rowlandson's success in this respect resulted from the preadaptation of his rococo line for portrayal of the English Elm. This is not the whole story. Rowlandson's draughtsmanship was so assured that when he entered terrain where other elms flourished, as Cornwall, his elms reflect the new elmscape.

Of lesser figures, Thomas Hearne, a water colour artist, deserves mention. Much of his work was topographical but with a good feel for trees.[24]

In the opening sections of this chapter some qualifications were elaborated with respect to the degree of naturalism achieved by John Constable in his portrayal of elms.

These do not minimize his influence on subsequent English landscape painting. Constable, however much influenced by previous painters, did make a remarkable advance in endowing his trees with identifiable qualities. He also established the status of the elmscapes of the southern counties as respectable subjects to paint, though they presented Nature under its beautiful rather than its sublime or picturesque aspects.[25] A whole series of his major paintings are centred on Flatford Mill and Willy Lot's cottage (Sf) where the background elms are Narrow-leaved Elms.[26] The *Cornfield*, as mentioned above, was probably based 97 on a Hybrid Elm. His painting of *Parham's Mill* (Do) has a background elmscape of English Elm.[27] In his paintings of Salisbury Cathedral he painted both English Elm and Dutch Elm.[28]

Of Constable's contemporaries over the turn of the century, the painter in oils who handled them particularly well was William Delamotte.[29] On the

Fig. 105. View near Oxford, by Peter de Wint.[30]

Fig. 106. View near Oxford, by Sir Frank Short after Peter de Wint.[30]

whole, however, it was water-colour artists who depicted elms most felicitously, especially Peter de *137* Wint[30] and Richard Bankes Harraden. Standing quite *107* apart from these is John Sell Cotman. Like most of the Norfolk painters he seldom portrayed elms. When he did, the results were outstanding, unmistakably elm, but imbued with an oriental decorative quality to which Cotman alone seems to have had access.[31]

A large number of minor artists pursued the path of naturalistic landscape portrayal opened up by John Constable. Possibly significant is the fact that those who succeeded best in realistic depiction of elms were often of less account in other respects.[32] Amongst those whose elms most recreate a real elmscape are *99* Lionel Constable, a master of Narrow-leaved Elm, *108* Birket Foster,[33] Rex Vicat Cole,[34] who captured something of the steamy quality of English Elm on *133* damp lowland soil, and William Henry Hunt, whose skill in rendering fruit and birds' nests could also be applied to bark texture.

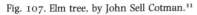

Fig. 107. Elm tree, by John Sell Cotman.[31]

Although their output of elms was very small, the pre-Raphaelite painters require separate mention. Clearly, according to pre-Raphaelite principles, any elms in their painting should be identifiable. Ford Maddox Brown demonstrated that this was so. His skyline elms are perfect and convey not only what an *109* English Elm looks like on the horizon, but what it feels like to look at one on a tranquil summer evening.[35] His other elms, the background to his major oil painting *Work*, painted in the main road, Hampstead (GL), are as assured though less evocative.[36] What Ford Maddox Brown did for the English Elm, Holman Hunt did for *127* Cornish Elm, capturing, on one occasion, the totality of all those features that make the silhouette of this kind so unmistakable.

The other pre-Raphaelite painter who achieved complete mastery of the elm at closer distance was Arthur Hughes. His picture *Home from Sea* was set in *148* Chingford (GL) churchyard. The story told by the human figures is not of concern here. The elm branches that provide the foreground have seldom been depicted more truthfully.

This is a convenient point to refer to two artists who, although not pre-Raphaelites, also achieved mastery of elm foliage at close distance. One is Francis Danby who, before switching his attention to more grandiose things, did well by the elms of the Avon and *110* Frome valleys (Av).[37] The other is Atkinson Grimshaw, who handled elm foliage of the north of England, effectively.[38] They are two of the very few painters of consequence who have achieved adequate renderings of Wych Elm foliage.

Impressionist painters were largely indifferent to elm. Their tree was poplar. Camille Pissarro did, however, paint some elmscapes during his English visits.[39] They are only sometimes successful. The English Elm in summer foliage has a note of grandeur and of solemnity. Pissarro's elms are liable to lack substance. Of English Impressionists two showed a special interest in elm, Phillip Wilson Steer and Sir George Clausen. Some of Wilson Steer's elmscapes are too generalized to be recognizable. Others capture the unique silhouette of the English Elm admirably.[40] *111* Clausen's impressionist elms are very different. Though a very long way from Victorian naturalist painting, his English Elms have a solemn, slightly *112* threatening, monumentality that immediately identifies them.[41]

Later twentieth century painting has pursued

Fig. 108. English landscape, by Rex Vicat Cole.[34]

Fig. 109. Hayfield, by Ford Maddox Brown.[35]

Fig. 110. Boy sailing a little boat, by Francis Danby.[37]

Fig. 111. Elm trees in field, by Philip Wilson Steer.[40]

such diversified paths that a systematic description of its handling of elms is hardly possible. Clearly, attention need only be directed to works with some representational intention.[42] If any twentieth century painter has to be singled out as particularly responsive to elm, this would have to be Paul Nash. His early elm studies are in monochrome or near monochrome, and are considered in the following section. During his stay

150 in Kent he was drawn to portray the Wych Elms that line the Royal Military Canal. At a later period he came across two fallen elms in 'Monster field' and these became the *leitmotiv* of the 'Monster' pictures, in which dead elms reincarnate into a new but fearsome existence.[43]

89 Mark Gertler and Gilbert Spencer recorded the
90 elmscape at Garsington, discussed in chapter 10. John
151 Piper also portrayed the elms along the Royal Military Canal. Arthur Rackham is now taken seriously. His world of elfin forest was his own, but his trees were based on real trees and, on occasion, as in his
88 illustrations for *Peter Pan in Kensington Gardens*, readers are presented with a new interpretation of a genuine elmscape.

A word is also owing to the much snubbed Rowland Hilder. As with the composer Liszt, the distance between his best work and his worst is unusually great. Both were masters of filigree. Hilder's moment of truth came, his biographer states, when he encountered the 'frost-rimmed furrows and the tall leafless elms' of Salop.[44] No artist has represented the 113 winter tracery of elms with more assurance than Hilder at the top of his form.

Monochrome pictures

Monochrome elm illustrations may be either drawings or prints. It is convenient to consider the various principal media separately. Pen-and-ink drawings of elm, not made as preparatory sketches for some other rendering, seem to be rare.[45] Monochrome washes have, however, been used. Paul Nash's remarkable series of early elm studies were monochrome or near monochrome. These demonstrate how 114 perfect control of silhouette alone can compensate for lack of colour, differentiation and no rendering of surface texture.[46]

Some of the finest pencil drawings of elms were made by John Constable. The best was his drawing of elms at Old Hall, East Berholt (Sf), which were

Fig. 112. Haymaking, by Sir George Clausen.[44]

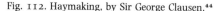

probably East Anglian Hybrid Elms.[47] In recent times some of the best pencil drawings of elm have been made by Lonsdale Ragg, whose major relaxation from being an archdeacon was to draw trees. His standard is variable but the best of his tree studies are good.[48] Some good pencil drawings have also been done by Adrian Hill.[49]

Charcoal may not seem a promising medium for trees since so little detail can be reproduced. It has been effectively used by Anthony Day in his series of drawings of pollard elms at Wicken (Ca).[50] Day has the further distinction of being one of the few artists to record the effects of Dutch elm disease.

Prints are either primary works of art in themselves or reproductions of some primary work of art, usually an oil painting or a water colour or other drawing. It is more convenient again to consider the various media separately, whether or not they are used as primary product or for reproduction of some other work.

Wood engraving, after its widespread vogue in the fifteenth and sixteenth centuries, practically died out till revived in the early nineteenth century. It was then again used extensively in book illustration. Generally speaking, as mentioned in the first section, artist and engraver would be different persons, though this was not always so. Often, the artist would draw on the wooden blocks to be engraved.

As with other media, many wood engravings of trees are completely indeterminate. Instances where elm trees are identifiable as such result from the flair of the original artist, the skill of the engraver, particularly in his rendering of texture, or a happy combination of the two. Some of the most effective of Victorian elm wood engravings were by the Dalziel brothers, already mentioned, after drawings by Birket Foster.[51] The *Illustrated London News* was illustrated almost entirely by wood engravings during the nineteenth century. One of its best elmscapes was engraved by Edmund Evans from a drawing by Samuel

Fig. 113. The garden of England, by Rowland Hilder.[44]

Read. Gustave Doré is one of the best known artists
91 who drew for wood engraving. The fact that his genius
did not extend in the direction of particularizing his
trees has already been noted.

The wood engravings that portray English
elmscapes most adequately and convey something of
the brooding solemnity of the English Elm are those
of Frederick Griggs. It has already been mentioned that
Griggs was one of the artists that produced works in
the tradition of the Shoreham period of Samuel
Palmer. This was only one side to his accomplishment.
In the many wood engravings that he made for the
Highways and Byways series, elmscapes are frequently
represented, very often miniature masterpieces that
126 recreate the visual impact and mood experienced from
scenery dominated by English Elm or Narrow-leaved
Elm.[52] Griggs is known to have been fascinated by this
tree.[53]

A number of later wood engravers have
attempted elms, as Gwen Raverat,[54] Eric Ravilious,[55]
Clare Leighton,[56] Agnes Miller Parker[57] and Charles
Tunnicliffe.[58] On the whole, they do not succeed. Joan
Hassell has done rather better, but though she has
succeeded in conveying the heavy stillness of the
English Elm, she has been less successful with its 115
overall visual effect.[59]

Line engraving on copper, or later on steel, was
nearly always a reproductive process based on an
independent painting or drawing. One of the
outstanding exceptions was Edward Calvert's *Bride*,
engraved on copper.[60] This little engraving is of 116
interest, not only as one of the outstanding prints from
Blake's circle of friends, but also as demonstrating the
continuing vitality of the vine and elm theme.

As with wood engraving, a great deal of the
effectiveness of line engravings on metal depended on

Fig. 114. The three, by Paul Nash.[46]

Fig. 115. Down the road, by Joan Hassall.[59]

Fig. 116. The bride, by Edward Calvert.[60]

the skill of the engraver in representing foliage texture.

Mezzotinting can be despatched quickly, as it was principally used for reproducing portrait paintings. There are very few mezzotinted elms. Amongst those calling for mention are David Lucas' mezzotints after paintings by John Constable, who breathed heavily down Lucas' neck the whole period of his work on them.[61] Sir Frank Short was one of the most 106 distinguished exponents of the art in recent times. His mezzotint of Peter de Wint's *Road near Oxford* conveys the massivity of the foreground elms more effectively than his original.

It might have been thought that the flowing line of the etcher would have made this method a first choice for engraving trees. This appears not to be so. Etchers in general seemed to prefer other subjects such as derelict houses and broken-up boats. There were some important exceptions. In the last century, Jacob Strutt etched an outstanding series of tree portraits, 120 including a number of elms.[62] Some competent elm 152 etchings were also made by Henry Ninham.[63]

In the present century, good etchings of elms 129 have been made by Charles Cheston and Ian Strang.[64] 87 The outstanding example is Seymour Haden's etching of the elms of Kensington Palace Gardens (GL). 103 Frederick Griggs[17] and Robin Tanner[18] were mentioned above as portraying elms in the spirit of Samuel Palmer. In both cases, etching was the medium involved.

The last of the print media that needs to be considered is lithography. Some of the nineteenth century topographical lithographers produced some 80 creditable elms, especially Charles Hullmandel.[65] Amongst later Victorian lithographers who came out well with them, George Barnard deserves special mention.[66] Of recent lithographers, one who perhaps would not have been expected to excel with elmscapes 128 was Robert Bevan. However, he spent his holidays away from the London suburban scene, which was his principal subject, in the elm country of eastern Devon. The elms made their mark.

Botanical drawings

Botanical drawings are sometimes so compe-tently composed that they qualify as specimens of fine art. Aesthetic considerations are, however, secondary in their production. Their function is to facilitate identification. This means, in the case of elm, that not only must the illustration declare itself as an elm, but it must clearly indicate too what kind of elm it is. Since identification of kinds of elm is principally based on the fine details of the leaves of dwarf shoots and of the fruits, these are the principal subjects of botanical drawings. Identification of the illustration will normally be confirmed by the attachment of the botanical name. A note of reservation is necessary. Considerable confusion has been caused to elm systematists by the attachment of wrong names to some crucial drawings.[67]

The false dawn of botanical illustration of English elms gleamed in the sixteenth century herbals. Not a single illustration is what it purports to be.

The starting point is the French edition of R. Dodoens' herbal.[68] It has a wood cut of what is presumably a Flemish Field Elm. It is galled by the aphid *Tetraneura ulmi*. This illustration will be called D, after the author. The next herbal relevant to this enquiry is that of M. de Lobel, published in 1576.[69] This has wood engravings of the shoots of two Field Elms, both again presumably Flemish. One is designated *Ulmus* and the engraving shows galls produced by the aphid *Eriosoma lanuginosum* on the leaves. The other is designated *Tilia mas* [male lime]. It is an elm and the galls on its leaves are those produced by the elm aphid *Tetraneura ulmi*. These two woodcuts will be designated L1 and L2.

The first illustrated English herbal was that of H. Lyte.[70] The text is taken from Dodoens with only minimal adaptation to the English elm flora. His elm illustration is D as in his source.

In 1581, de Lobel published a new herbal.[71] He retained illustrations L1 and L2 and added a third, L3, for an elm with broader leaves, probably a Flemish Hybrid Elm.

The German botanist J. T. Tabernaemontanus brought out a herbal in 1588–91.[72] It has four woodcuts of elm, the first galled by *Eriosoma lanuginosum* and called Rustbaum, the second two both labelled Effenbaum, and the fourth called Steinlinden but galled by the elm aphid *Tetraneura ulmi*. It is not clear whether the artist who prepared Tabernaemontanus' blocks did them from German elm material or whether he merely re-drew illustrations already published. Tabernaemontanus' set of four blocks will be designated respectively T1, T2, T3 and T4.

The next English herbal was that of J. Gerard,[73]

the first edition of which was published in 1597. The text was largely a re-hash of Dodoens' latest writings, but with rather more adaptation to the English flora. Gerard's publisher had been able to lay hands on the blocks used by Tabernaemontanus. So what Gerard calls the common elm, that is the English Elm, was illustrated by T3. His second elm, which he called *Ulmus latifolia* was given T2. He also described the spurious *Tilia mas* and illustrated it with T4. None of the wood cuts had any resemblance in detail to any English elm.

The British elms were first tidied up, as described in chapter 7, by T. Johnson who was responsible for the second edition of Gerard's herbal, published in 1633.[74] The four elms named by Johnson are well described and are identifiable with certainty. The illustrations are totally misleading. Johnson's publisher had access to all the blocks described above. English Elm was given L1. Wych Elm got L3. Of his two Narrow-leaved Elms, *Ulmus minor folio angusto scabro* received D while *Ulmus folio glabro* was allotted T2. *Tilia mas* continued its fictitious existence embellished with L2. None, again, of any of these illustrations has any detailed likeness to the elms they are supposed to illustrate.

Four years later, J. Parkinson brought out his herbal.[75] His four elms are the same as Johnson's. His illustrations are also the same except for English Elm where he uses a new but irrelevant block, probably borrowed from Flanders, showing fruits as well as leaves. So, in the first 60 years of botanical illustration in England, progress in delineating its elms was zero.

The first illustration drawn from English material was published in 1677. It is a copper engraving by Michael Burghers of the Narrow-leaved Elm described by R. Plot under the name *Ulmus folio angusto glabro*.[76] It at least shows the correct proportions of the leaves and the doubly toothed leaf margin.

Botanical illustration of the elm languished in the eighteenth century with one outstanding exception, Georg Ehret's drawing of a Narrow-leaved Elm identified for him as *Ulmus folio glabro*. This is in gouache and is one of the very few competent botanical drawings in this medium.

Thomas Bewick contributed a wood engraving of an elm shoot to R. S. Thornton's herbal. It is Bewick at his worst.[77]

The next elm illustrations of consequence are those drawn by James Sowerby for *English Botany*.[78] Sowerby's water colours were engraved and the prints coloured by hand. For the first time, a range of elms was competently illustrated from English specimens.

In the main, botanical illustrations, including those of elm, were reproduced in the nineteenth century as wood engravings. Amongst the better known are those in the illustrated edition of the *Handbook of the British Flora* of G. Bentham.[79] The elm illustrations are not distinguished.

Wood engraving ceased to be used for reproducing botanical line drawings with the introduction of line-block printing towards the end of the nineteenth century. Of the line drawings so reproduced, those of Edward Hunnybun, which illustrate the text of the never completed *Cambridge British Flora*, are outstanding.[30] Botanical drawings are very liable to appear stiff, as though drawn from dried herbarium specimens, which many of them were. Hunnybun's elm illustrations are remarkable not only for their accuracy of detail, but for the impression they give of pliability and freshness.

Florence Strudwick has some commendably clear elm drawings to her credit.[81] Those by Mary Ferguson fall down in the heavy treatment of the tertiary veins, a major challenge to the botanical artist.[82] One of the most recent sets of elm drawings has been produced by Stella Ross-Craig.[83] The detail is admirable, though the use of parallel rulings to indicate shading creates an impression of stiffness.

42 43 44, 46, 49 45 48

12

Elm in the northern counties

North is taken here in its traditional sense. It denotes those counties that, in whole or part, lie north of the crossing of the Trent by the Great North Road. These are Cheshire, Cleveland, Cumbria, Derbyshire, Durham, Greater Manchester, Humberside, Lancashire, Lincolnshire, Merseyside, Northumberland, North Yorkshire, Nottinghamshire, Salop, South Yorkshire, Staffordshire, Tyne and Wear, West Midlands and West Yorkshire. The north, so understood, is the part of England where the characteristic elm is Wych Elm. Some three-quarters of the north, however, lies in the region Z1, where elm is infrequent. Before the Elm Decline, Wych Elm appears to have been common over most of the northern counties, except on the highest ground. After the Elm Decline, it appears to have disappeared from a large part of its former area, to survive mainly among the middle reaches of the principal rivers.

Cheshire

While the present county largely corresponds to the old, its northern border has changed. Most of the Wirral peninsula has gone to Merseyside. The former north-east corner of Cheshire has been taken by Greater Manchester. A strip of land north of the Mersey has been transferred from Lancashire to Cheshire. The ancient ecclesiastical boundary between England and Wales, which has been adopted in this book, went slightly beyond the present civil boundary. It took Hawarden into England. It also included in England the district of Maelor. Warning notices surrounding this district today draw the attention of motorists to the alcoholic desert ahead.

Most of Cheshire is a flat clay plain. This is interrupted in places by often abrupt sandstone hills. Towards the east, the plain gives way to the upland moors of the Pennines. The major rivers are the Mersey and Dee.

Elm is not a conspicuous component of the Cheshire landscape. What there is appears to be mainly the product of relatively recent planting. Both Wych Elm and English Elm occur. The only name used in the county was *elm*.

Most of Cheshire lies in region Z1, where elm is infrequent. Scattered Wych Elm and English Elm occur, and very rarely Hybrid Elms and Narrow-leaved Elm. The eastern part of Cheshire, almost all in region Z1, was formerly known as Lyme Forest.[1] This name, which is thought to enter into other Cheshire names, as Lyme Hall, Lyme Handley, Audlem and Lymm, is usually derived from Celtic *lem-, meaning elm. No common British tree is less characteristic of this area than elm. It was concluded in chapter 3 that derivations of English place names from Celtic *lem-, with this meaning, are uncertain. None is less certain than this one.

Similar doubt attaches to place names such as Wychwode in Lostock Park,[2] Wychefelde in Sandbach, and Wych Moor in Lower Peover, for which derivations based on Anglo-Saxon *wice*, have been suggested. All are in the salt spring area where *wych*, derived from Anglo-Saxon *wic*, has come to mean salt working. There is no independent evidence that *wych* was ever used in Cheshire to mean elm. So again, these place names are best excluded from any consideration of the local history of elm.

One early name, Elmenwall, in Prestbury, known from the thirteenth century, is probably based on *elm*. There is some elm in the neighbourhood and the name could indicate the earlier occurrence of Wych Elm. A dug-out boat of elm excavated near Warrington has been radiocarbon-dated to around A.D. 1000.[3]

The larger of the two areas in Cheshire where Wych Elm is frequent is part of region G3,[4] which runs down the Welsh border. Most of it lies on higher ground on the Welsh side of the border. The smaller is part of the extensive region G2 which covers much of the higher ground in Staffordshire and Derbyshire.

It extends into Cheshire around Church Lawton.

The areas where Wych Elm is frequent are precisely those where its continuous existence since Boreal times would be suspected, that is on the rising ground to east and west of the central plain.

What elm now occurs in the central plain is likely to go back no further than the early eighteenth century.

Cleveland

Cleveland is a new county with Middlesbrough as its centre. It extends along the North Sea coast from Hartlepool to Loftus. The part north of the Tees was formerly in Durham. The part south was originally in the North Riding of Yorkshire.

The county is mainly lowland coastal plain. Towards the south-east, higher ground is encountered, the northern part of the Yorkshire Moors. There is one major river, the Tees, and this has one major tributary running in from the south, the Leven.

Elm is not a frequent tree except locally and in recent plantations. Wych Elm is the main kind present. It is known locally as *elm*. Scandinavian *alm* may have been also used in earlier times.

More than half the county lies in region Z1,

Fig. 117. Elm regions of the northern counties. Diagonal shading, Wych Elm; horizontal shading, English Elm; vertical shading, other Field Elms; crosses, Hybrid Elms.

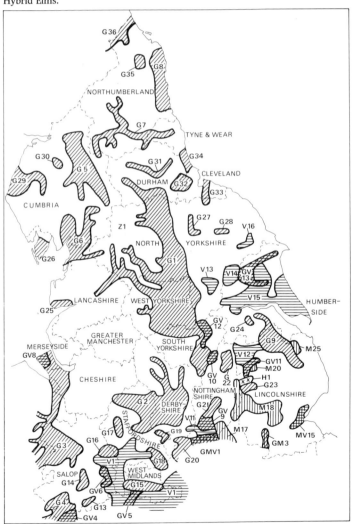

where elm is infrequent. Wych Elm occurs sporadically and English Elm rarely. The river name Leven, supposedly based on Celtic *lem-*, meaning elm, is discussed in the North Yorkshire section of this chapter. Kirk Levington is named after the river. Wych Elm is present in abundance only in the upper reaches of this river and these lie over the border.

There are two areas in which Wych Elm is commoner. One lies along the border with Durham. It is part of region G32,[5] the rest of which is in the southern part of the latter county. English Elm occurs rarely in this region. The other, G33,[6] lies south of the mouth of the Tees and extends a short distance southward into North Yorkshire. English Elm occurs rarely here too.

Whether Wych Elm is native in the county is doubtful. If it is, it would be along sheltered streams in higher ground. The occasional English Elm in Cleveland is not likely to go back further than the early eighteenth century.

Cumbria

This county was created by combining the old counties of Cumberland and Westmorland and adding the part of Lancashire formerly north of Morecambe Bay and the tongue of the West Riding of Yorkshire that extended westward to Sedburgh. The parish of Kirkandrews on Esk, the Debatable Land, is excluded as historically Scottish.

It is largely an upland area. The Lake District mountains lie at the centre. The Pennines lie along the eastern border. These two upland regions are linked by the Shap Fells. The major inland lowland areas are the Eden valley in the north and the Kent valley in the south. There is a coastal fringe of lower ground some 15 km wide in the north but very narrow in the south. The Furness peninsula is a relatively isolated area at intermediate elevations. The land immediately south of the Solway Firth is marshy and liable to inundation. In addition to the rivers already mentioned, the Derwent, flowing westward from the Lake District, and the upper reaches of the Lune and South Tyne are important landscape features.

Wych Elm is the kind usually found. It is frequent along the middle sections of the principal rivers.

Presumably *elm* was the vernacular name used by the Anglo-Saxons. It appears to have been displaced in most areas by Scandinavian *alm*. This remained in dialectal use till the present century. By the sixteenth century, *elm* had been reintroduced.

Over half the county lies in region Z1 where elm is infrequent. This covers the highest and most exposed terrain and also the coastal lowlands. Wych Elm occurs at low frequency. Some English Elm is present around St Bees. Elsewhere it is very rare. An avenue of English Elm at Arkleby Hall was destroyed by gale in 1884. The low frequency of elm in western Cumbria is confirmed by the absence of allusion to it in the early court rolls of the Leconfield estate which occupied a large part of this area.[7]

The largest region where elm is frequent is G5.[8] This covers the middle reaches of the Eden. Wych Elm is abundant and there is some English Elm at Great Salkeld. The place name Llwyfenydd, meaning elm wood, occurs in a poem attributed to the Welsh-speaking poet Taliesin. The poem is generally dated about the sixth century. It is supposed that this name survives as Lyvennet, a branch of the River Eden.[9] If this identification and etymology are correct, it would confirm the presence of Wych Elm in this part of Cumbria when Welsh was spoken.

Later elm place names are based on Scandinavian *alm* rather than on Anglo-Saxon *elm*. Almebanke in Shap is recorded in the fourteenth century.[10] Almgill in Kirkby Stephen is known from the sixteenth century.

English Elm had been introduced into Naworth Park in the seventeenth century, having been brought by sea to Newcastle upon Tyne (TW), presumably from London, and then taken to Naworth overland.[11] It is seldom found, except as a park tree, in this part of England.

Region G6,[12] also with frequent Wych Elm, lies partly in Cumbria and partly in Lancashire. The Cumbrian part covers the River Leven, Lake Windermere and the valleys of the Kent and upper Lune.

The river name Leven is usually derived from Celtic *lem-*, meaning elm.[13] There is much elm nearby but, as elsewhere, all names so derived are uncertain. Later elm place names, as before, are based on Scandinavian *alm*. Almgill, in Kendal, is recorded in the sixteenth century. In the following century, Alm Sike in Dent[14] and Almpot in Ravenstonedale are encountered.

Three of the remaining regions where Wych Elm is frequent are associated with river valleys. G29[15] extends seaward from the heart of the Lake District along the Derwent. G30[16] is a small region in the upper course of the River Caldew. G7 coincides with the main branches of the Tyne and is mainly in Northumberland. The westernmost part just extends into Cumbria at Alston.

The last region of frequent Wych Elm is not associated with a major river. It is region G26,[17] covering the southern end of the Furness peninsula.

All the regions of frequent Wych Elm are on terrain where this elm has every appearance of being native. Its persistence on these since Boreal times can hardly be doubted. The occasional English Elm stands are not likely to be older than the seventeenth century.

Derbyshire[18]

Except for a small area bordering on Sheffield, which it has lost to South Yorkshire, Derbyshire has retained its former territory. It has gained the former Cheshire panhandle along Longdendale.

The county falls into four unequal divisions. The northern three-quarters is upland, the southern end of the Pennines culminating in the High Peak. It is composed mainly of Carboniferous Limestone, Millstone Grit or Coal Measures. In the north-east of the county is a separate upland area of Magnesian Limestone. South of the first upland is a clay plain traversed from south-west to north-east by the Trent. South-east of this is a third and lesser upland, largely of sandstone. The major rivers, after the Trent, are its two northern tributaries, the Derwent and the Dove, the latter forming the boundary with Staffordshire.

Wych Elm is widespread in the northern and southern uplands. Elm was so-called in the north-east corner of the county in the eleventh century and by the seventeenth century in the south. Whether the name *wych* was also used is extremely doubtful.

The main region of infrequent elm, Z1, covers three areas of Derbyshire: much of the High Peak, the Trent and lower Dove valleys, and the eastern edge of the county, largely Coal Measures outcrop. Wych Elm occurs sporadically in all these areas. There is some English Elm in the Trent and lower Dove valleys, and very occasionally Narrow-leaved Elm or Lock's Elm. The early presence of some Wych Elm in the north-east area is attested by the place name Elmton,[19] recorded in Domesday Book. In the Trent Valley, Elm Closes are recorded at Coton in the Elms in the seventeenth century. The elm here now is English Elm. The derivation of Horwich in Fernilee in the north of the

county, recorded in the thirteenth century, from *wych* is highly doubtful. The same name occurs in Greater Manchester where also the term *wych* seems not to have been used for elm. In Church Broughton, in the valley of the lower Dove, there was an English Elm avenue of note. There was also an elm avenue at Renishaw Hall in the north-east.

Most of the northern upland area lies in region G2,[20] which also covers an extensive area in northern Staffordshire. Here Wych Elm is frequent. There is occasional English Elm in the south.

Lime Wood in Morley has been derived from Celtic **lem-*, meaning elm. Like most such derivations, this is uncertain. Neither can it be accepted that Wichefielde in Baslow, recorded in the fourteenth century, must be based on *wych*. Firm ground is reached in the sixteenth century. The manorial court rolls of Holmesfield in Dronfield refer to elm felling,[21] and some elm is noted also as present in Duffield Forest.[22] The meaning of the place name Witch Bridge in Hope parish must be left open. An Elme Barne is recorded at Wirksworth in the next century.

Notable elms in the Derbyshire part of the region G2 include the Holloway Elm that stood on the boundary of Ashover. Marriages by magistrate were once reputedly solemnized under the Marriage Elms near Ashe Hall, Sutton on the Hill.[23] There were English Elm Avenues at Etwall Hall. Derby was one of the towns that had elm water mains.[24]

A small part of northern Derbyshire, around Whitwell, comes into region G1. This is the very extensive area of frequent Wych Elm that covers the Yorkshire Dales.

Region G20 is mainly in western Leicestershire. Some of the parishes of the adjoining southern upland parishes of Derbyshire are also in it. Wych Elm is frequent and English Elm occurs occasionally. There was an English Elm avenue at Knott Hill in Ticknall parish.

One region of frequent English Elm just comes into south-east Derbyshire. This is region V11 which is mainly in southern Nottinghamshire.

The principal locations of Wych Elm, in both the northern and southern uplands are where one would expect to find this elm surviving in the wild. English Elm is not likely to have appeared on the Derbyshire scene before the close of the Middle Ages. Some of it may have been planted to supply Derby with its elm pipes.

Tyne and its larger tributaries. Elm charcoal of Roman age has been identified at Corbridge.[76] Another Embley, formerly Elmlee, this time in Whitfield, confirms the existence of Wych Elm in this region in the twelfth century. A seventeenth century timber survey at Alnwick Castle records the presence of elm around Prudhoe.[77] Next most extensive is region G8.[78] This runs down behind the coast from Beadnell Bay to Blyth, and includes the lower courses, often in ravines, of the Aln, Coquet, Wansbeck and Blyth. A little English Elm is also present.

G35[79] is a small region in the upper reaches of the River Till. There is also a little English Elm. Region G36[80] is mainly in Scotland where it covers a wide area in the Tweed valley. It extends also along the English side of the Tweed.

Wych Elm principally occurs in the middle courses of the large rivers or in the lower courses where these flow through ravines. It has presumably persisted in such sites since it entered the country. The occasional English Elm is unlikely to have been introduced much before the eighteenth century.

North Yorkshire

In the drastic reshaping of Yorkshire, North Yorkshire inherited the North Riding, the part of the East Riding that lay north and west of the Wolds, and those parts of the West Riding that lay north and west of Harrogate, and east of the industrial area. The city of York fell into its lap. A strip of land south of the lower Tees went to Cleveland. Durham got the upper Tees.

The basic geographical subdivision is between the Pennine uplands, part Millstone Grit, part limestone, to the west, the Yorkshire Moors to the west, and the intervening lowland known in its central part as the Vale of York. This pattern needs to be filled out a little. A narrow ridge of Magnesian Limestone runs from north to south on the western side of the lowland Vale of York. The Yorkshire Moors are fringed on the south and south-west by an outcrop of Corallian Limestone. Running from west to east south of the Yorkshire Moors is the flat lowland of the Vale of Pickering. Between the northern end of the Vale of York and the River Derwent which drains into the Vale of Pickering, rise the largely limestone Howardian Hills.

Most of the Yorkshire dales, Swaledale, Wensleydale, traversed by the River Ure, Nidderdale, upper Wharfedale and Airedale are in this county. These are all drained by tributaries of the Ouse. A corridor of lower ground links Airedale and Ribblesdale at Skipton. The Derwent is the major river in the east. There is much marshy terrain in the Vale of Pickering and both north and south of York in the Vale of York.

Wych Elm is abundant in the dales and on the eastern fringe of the Pennines. There are some areas where English Elm is frequent in the Vale of York. The designation used by the English was *elm*. Over wide areas in the county it was replaced by Scandinavian *alm*.

Region Z1, with infrequent elm, covers well over half the county. All the higher and more exposed parts of the Pennines and most of the Yorkshire Moors are in it. It extends also over much of the intervening lowland. Wych Elm occurs at low frequency on all but the highest ground. There is some English Elm in the lowland areas.

The presence of some elm at Drax in the twelfth century might be indicated by the place name Elmholm,[81] but since this is also referred to in early documents as Helmholm, the meaning is doubtful. In Kellington, not far from Drax, an *elmtre* is mentioned in the fourteenth century. There was an *almestubbe* in the Middle Ages at Tockwith. In all cases, the elm involved would be Wych Elm.

The city of York lies in Z1. It was however provided with elm water mains.[82] Some of the elms of its New Walk, planted in the 1730s, survived into the present century.[83] *119*

An elm of note stood on Alm Green, Thirsk. Here, after his murder at Cock Lodge, near Topcliffe, Henry, fourth Earl of Northumberland, was beheaded in the fifteenth century. It was also the place where the borough elected its Members of Parliament.[84]

Much of the largest of the regions of frequent Wych Elm, G1,[85] is in North Yorkshire. This extends up all the dales that lie in the county and also covers the rising ground at their eastern ends. This is the part of England where Wych Elm appears to grow under optimum conditions. The place name Leeming is usually derived from Celtic *lem-*, meaning elm. This is feasible, but like all derivations from *lem-*, not much confidence can be attached to it.

One of the earliest contemporary accounts of an elm is to be found in Hugh de Kirstall's narration of the foundation of Fountains Abbey in the twelfth century. The early monks, as mentioned in chapter 8, sheltered under a large Wych Elm. It was believed to have been extant in Leland's time. It is just possible that the tree pointed out to Leland was the tree called the Fountains Elm in the eighteenth century.[86]

The place names Omber, formerly Almebergh, in Markington, and Almkeld, in Middleton near Ilkley, are encountered in the thirteenth century. An Alm Croft is reported at Church Fenton in the next century and Almesford in Pannal in the century after. Aumgill, equivalent to Almgill, is recorded at Airton in the seventeenth century. The seventeenth century manorial court rolls of Spofforth record an unauthorized sale of an elm.[87] It is noteworthy that all the names are based on Scandinavian *alm*. Prince Rupert, whose career is mapped by elms, dined under one at Bolton Abbey, as recorded in chapter 9. Elm water pipes were formerly in use at Skipton.[88] There was a famous Wych Elm, the Abbot's Elm, at Easby Abbey,[84] and a Dutch Elm avenue at Snape Hall.[89]

Two small regions of frequent Wych Elm are found on the Corallian Limestone fringe to the Yorkshire Moors. G27[90] runs along the Hambleton Hills. G28[91] is centred on Kirbymoorside.

Region G33, also with frequent Wych Elm, is mainly in Cleveland but Great Ayton, which is in it too, is in North Yorkshire. This place is on the River Leven, whose name is usually derived from Celtic *lem-*, meaning elm. Wych Elm is frequent in the upper part of its valley. One is tempted to accept the derivation in this case but the doubt attached to all derivations based on *lem-* remains.

In the extreme west of the county, around Clapham, in the valley of the Wenning, a tributary of the Lune, Wych Elm is frequent. This area is in region G6,[92] which is mostly in southern Cumbria and northern Lancashire.

There are two small regions where English Elm is frequent. V13[93] is south of York, around the confluence of the Ouse and Wharfe. V16[94] is in the Vale of Pickering and on the flank of the nearby Wolds, centred on Sherburn.

It is impossible to doubt the indigenous nature of Wych Elm in the Yorkshire dales. Nowhere in England does it appear better adapted to its terrain. It is likely that it is also native in the other upland areas in which it is now frequent.

Fig. 119. New Walk, York (NY), by Nathan Drake.[83]

English Elm is only common in two small areas, each far removed from the main English Elm region V1. It is not likely to have been introduced to these much before the eighteenth century. The elms in V13 perhaps supplied water pipes to the city of York.

Nottinghamshire[95]

Save for very minor adjustments, the present county coincides with its precursor.

The overall relief pattern is one of high ground in the west and south-east and an intervening clay vale between. Proceeding westward from the lowland, the ground rises to a broad sandstone outcrop. Further west still, along the Derbyshire border, there is a narrow strip of Magnesian Limestone to the north and Coal Measures to the south. The south-east upland, the Wolds, is covered by Boulder Clay. The Vale of Belvoir, which extends into Leicestershire, lies north-east of the Wolds and is also of clay. There is marshland in the extreme north of the county. The River Trent traverses the main clay vale from end to end. The Soar, its principal southern tributary, forms the boundary between Nottinghamshire and north-west Leicestershire.

Elm is only locally common in the county. Wych Elm occurs in the north. Wych Elm, English Elm, Narrow-leaved Elm and Lock's Elm are all to be found in the south. All are known as *elm*.

Some three-quarters of the county is within region Z1 where elm is infrequent. This includes the sandstone upland, most of the Trent valley and most of the Vale of Belvoir. Wych Elm is scattered in the north of this area. In the south, English Elm, Narrow-leaved Elm and Lock's Elm are widespread but infrequent.

There was a notable elm called Langton Arbour near Blidworth. An elm associated with Byron in the grounds of Newstead Priory is described in chapter 9.

The principal area of frequent Wych Elm is region G22.[96] It occupies the rising ground to the north-east of the clay vale, between Gringley on the Hill and Grove. Wych Elm is also frequent in region G21[97] which runs south-west to north-east in a broad band from Nottingham to Oxton.

The so-called Nottingham elm, alternatively and misleadingly called the Siberian elm, was planted in Nottingham Park, a former extraparochial district now engulfed in the Nottingham conurbation. It is a Hybrid Elm supplied in the last century by the Castle

Nurseries,[98] a firm no longer extant. It is possibly of French origin.

English Elm is frequent in the small region V11.[99] It runs between West Leake and Widmerspool in the Wolds south of Nottingham. It also covers a small area in the adjoined part of Derbyshire.

There are two areas where Narrow-leaved Elm is frequent. The first, region M18, is mainly on the upland of south-west Lincolnshire. In Nottinghamshire, it covers the villages east of the Trent between Girton and South Collingham, and north of the Vale of Belvoir around Thoroton. The large elm on the green at South Collingham is mentioned in chapter 9. M17[100] occupies the south-east corner of the Nottinghamshire Wolds. Most of it is over the border in Leicestershire.

In two areas Wych Elm and English Elm are both frequent. One of these, region GV9,[101] is in the south of the county. It covers the parishes alongside the Trent to the immediate south of Nottingham. Elm buds and bud scales were found in the infilling of a Roman well at Bunny.[102] These would probably be Wych Elm. Clifton, whose grove is considered in chapter 10, is in this region. It was perhaps planted by Sir William Clifton in the seventeenth century. The other region, GV10,[103] is on limestone terrain in the north of the county around Carlton in Lindrick.

On the Wolds, south of V11, is a small region, GMV1,[104] partly in Leicestershire, where Wych Elm, English Elm and Narrow-leaved Elm are all frequent.

The status of Wych Elm presents no problems. It principally occurs on high ground, especially where calcareous. It could have persisted in such places since prehistoric times.

The history of the other elms has to be approached in the light of what is known of the history of human settlement. The county appears to have been but sparsely settled in prehistoric times. During the Roman period, the land occupied was mainly along Fosse Way and along the road between Doncaster and Ermine Street north of Lincoln.

Anglo-Saxon penetration was relatively late. The areas first taken over were the Fosse and the Trent valley above Newark. There was a large royal forest, Sherwood Forest, mainly in Z1, on the sandstone upland. Early place names indicative of wooded terrain and late settlement occur within the bounds of Sherwood Forest. They are also found in the country bordering it both to the west and east.

The sparse pre-Roman settlement makes it unlikely that any elm of human introduction would have reached the county before the Roman period. Both English Elm and Narrow-leaved Elm are on the edge of their main areas of distribution. In neither case do their present areas of frequent occurrence coincide with the settlement pattern of the Roman Period.

English Elm probably entered the county from Leicestershire, across terrain only settled late in the Anglo-Saxon period. This seems the earliest date for which its arrival in Nottinghamshire is at all likely.

Narrow-leaved Elm presumably came from Lincolnshire and Leicestershire. Its distribution does not reflect the Roman settlement pattern based on the Fosse. As with English Elm, the late Anglo-Saxon period seems the earliest feasible date for its entry.

Lock's Elm is distributed irregularly as usual. A relatively late date of introduction from the Witham valley in Lincolnshire, not earlier than the beginning of the eighteenth century, is likely.

Salop

This county has retained its former boundaries. Its border with Wales however does not correspond with the ancient ecclesiastical boundary adopted here. This puts the country around Oswestry in Wales, and the town of Montgomery and the land around Church Stoke and Hyssington in England.

The topography of Salop is complicated. The fundamental distinction is between the northern clay plain, with a projection south along the Severn valley, and the uplands in the south-west of the county. Within the plain, one area that is well demarcated is the Weald Moors, a marshy expanse north of the Telford conurbation. The south-west uplands consist of a series of roughly parallel ridges and valleys, running north-east to south-west. In the east is the platform culminating in the two Clee Hills, the Brown Clee and Titterstone Clee. To the west of this is the limestone ridge of Wenlock Edge, with Ape Dale to the west of it and Corve Dale to the east. Further west still, and proceeding from east to west are the ridge culminating in Caer Caradoc, the Church Stretton depression, the Long Mynd and the Stiperstones. The upland area of Clun Forest occupies the south-west corner of the county.

The Severn is the major river. Entering the county from Wales, it flows first east, and then, below Iron Bridge, runs south. One of its tributaries, the Teme, provides part of the south-west boundary with Wales.

Wych Elm mainly occurs in the west and south of the county. There is some English Elm in the south-running section of the Severn. Both kinds are known in the county as *elm*.

Region Z1, with infrequent elm, covers most of the northern plain. It also occupies a substantial area in the south-west upland, especially over acidic soils. Both Wych Elm and English Elm occur sporadically and there is some Lock's Elm. In the seventeenth century there was a roadside landmark elm at Cound.[105] There was an elm avenue at Burford.[106]

The largest area where Wych Elm is frequent is region G3.[107] This extends northward along the Welsh border, from where the Severn valley crosses it, to the Mersey estuary. There is some English Elm and some Lock's Elm. The surname de la Wych, current in Shrewsbury in the thirteenth century, is unlikely to refer to Wych Elm.[108] It is more likely to be cognate with the term *wych*, applied to salt works in Cheshire and in Hereford and Worcester.

Region G4,[109] also with frequent Wych Elm, covers much of the southern half of the Welsh border country. In Salop, it includes the parishes along the Teme valley. The villages around Wenlock Edge are in G14.[110] Here, Wych Elm is frequent. English Elm is found occasionally.

There are two lesser regions with frequent Wych Elm. These are G13, covering a few parishes east of Ludlow, and G16, to the east of Telford. There was an Elme Field at Sheriffhales in the latter region in the seventeenth century.[111]

The main region with frequent English Elm, V1,[112] enters the county up the Severn valley.

Both Wych Elm and English Elm are abundant in the county just south of Bridgnorth. This is region GV6.

As usual, the larger Wych Elm regions are those in which continuous occupation since Boreal times is likely.

Salop was only lightly occupied in prehistoric times. Even in Roman times there was sparse settlement except at Wroxeter and in the Severn valley immediately to the west. The Anglo-Saxon penetration of the county was late and there are no place names of early type. There was much royal forest. The fourteenth century list comprised the Forests of Lithewood, Morfe, Stretton, Shirlet and Wrekin. All or most of each of these are in region Z1 where elm is infrequent. Morfe Forest is exceptional, since this is in regions GV6 and V1. The Forest of Wyre is a further wooded area in the south-east corner of the county.

There are four areas in the county with place names suggestive of wooded conditions and late settlement. One is south-west of Shrewsbury and another south-east of the same town. A third lies between Much Wenlock and Bridgnorth and the fourth is south of Bridgnorth. These areas are either in region Z1 where elm is infrequent or in regions GV6 and V1.

The usual inverse relation between royal forest and wooded terrain, on the one hand, and the elms introduced by man, on the other, makes it unlikely that the English Elm now growing on the site of Morfe Forest was introduced there before the end of the Middle Ages.

Morfe Forest and Kinver Forest in Staffordshire between them cut right across the narrow tongue of region V1 that takes English Elm into Salop and Staffordshire. It seems likely therefore that the English Elm not only on the site of the forests but also north of them was only introduced after they had ceased to be significant formative influences on the landscape, that is after the Medieval period.

Lock's Elm is widespread and scattered, often on long-distance routes. Its possible distribution in the county by drovers is worth investigating.

South Yorkshire

In the new subdivision of Yorkshire, South Yorkshire comprises that part of the West Riding that centres on Barnsley, Doncaster, Rotherham and Sheffield. The Derbyshire environs of Sheffield have also been attached.

The high ground of the Pennines lies to the west, and the lowland plain, a southward prolongation of the Vale of York, lies to the east. There are extensive tracts of marshy ground, in particular the Hatfield Moors and Thorne Waste in the lowland plain. Magnesian Limestone outcrops as a narrow ridge to the west of the plain. Coal Measures are exposed between this and the Millstone Grit of the Pennines. The principal river is the Don.

Wych Elm is frequent on sheltered terrain at intermediate altitudes. English Elm is common around Doncaster. The vernacular designation was *elm* except where it had been replaced by Scandinavian *alm*.

Region Z1, where elm is infrequent, covers the higher Pennines and the marshland areas of the lowland plain. The place name Limb, earlier Lyme, in Dore has been derived from Celtic **lem-*, meaning elm.[113] The usual reservations have to be made about this interpretation. Much more satisfactory is the Elmtreacker [elm tree acre] recorded in Ecclesfield in the thirteenth century, presumably indicating the presence of Wych Elm.[114]

The principal area where Wych Elm is frequent is the South Yorkshire part of the region G1,[115] which covers the eastern flank of the Pennines. Some English Elm is present too. The place name Elmhirst in Darfield is on record in the fourteenth century. Another place with the same name in Cawthorne is encountered in the following century. In the seventeenth century, there was an Elm Royd in Silkstone and a locality called Elme at Wath was sited in Wath upon Dearne. Agreements for wood sales in the seventeenth century from Cawthorne and Silkstone reserve elms and various other trees.[116] A small part of region G2 crosses the southern border of the county from Derbyshire.

Wych Elm and English Elm are both frequent around Doncaster, in region GV12.[117] Almholme in Arksey is mentioned in the thirteenth century. *Le Elme Tree* at Doncaster was a local feature in the sixteenth century. Most likely all these records refer to Wych Elm.

Wych Elm in sheltered sites at moderate elevations could go back to the original Boreal forest. The English Elm in GV12 is a long way from the northern limits of the principal English Elm region, V1. As elsewhere in the north of England, it was probably not introduced much before the beginning of the eighteenth century.

Staffordshire[118]

The present county corresponds to the old, save that the urbanized regions around Birmingham and Wolverhampton have been detached and transferred to the new county of West Midlands.

The high ground of the county is at its northern and southern ends. In the north there is the southern end of the Pennines, part limestone, part Millstone Grit. In the south is a plateau, largely sandstone. Between the two uplands is a low clay plain through which runs the Trent. A northern tributary, the Derwent, divides Staffordshire from Derbyshire. The

Tame, running south to Tamworth, is the principal southern tributary.

Wych Elm is frequent in the north of the county and on the southern plateau. English Elm is common in the south-west lowland. The designation used throughout the county was *elm*.

The more exposed upland and most of the lowland plain, including the Trent valley, lies within region Z1, where elm is infrequent. English Elm and Lock's Elm are to be found occasionally. The place names Newcastle under Lyme and Burslem are both derived from Lyme Forest.[119] The forest was principally in Cheshire. In the section covering that county, the supposition that Lyme was from Celtic *lem-*, meaning elm, was dismissed as unlikely. There was certainly some Wych Elm in earlier times.[120]

An elm bow and staff were cited in a late thirteenth century plea roll concerning an assault at Essington.[121] The surname atte Elme was current in Kings Bromley in the fourteenth century.[122] A Wych Elm felled at Field, Church Leigh, in the seventeenth century may have been the largest tree ever to have been recorded in England.[123]

The principal region where Wych Elm is frequent is G2[124] which covers much of the Pennine uplands of Derbyshire and Staffordshire. A fourteenth century roll attests the presence of elm at Bramshall.[125]

There are five minor regions with frequent Wych Elm. G17[126] is just south of Stoke on Trent. The surname del Wych is recorded from Swynnerton in the fourteenth century,[127] but there is no independent evidence that *wych* ever meant elm in Staffordshire. G18 is a small region of frequent Wych Elm around Lichfield. The place name Elmhurst is known from the thirteenth century.[128] At the same time, *le Elme* near Lichfield was a local landmark.[129]

G19[130] lies between Needwood Forest and the Trent around Tutbury. There was a notable specimen Wych Elm at this town, referred to in the seventeenth century.[131]

G16 is mainly in Salop. It includes Weston under Lizard on the Staffordshire side of the border. An action for waste involving felling of elms at Beighterton, in this parish, is entered in a fourteenth century *de banco* roll.[132] The last of the five is G20, which is mainly located on the higher ground of western Leicestershire. It comes into Staffordshire around Clifton Campville.

The English Elm of Staffordshire is mainly in region V1.[133] A tongue of this region runs northward to the west of Wolverhampton and extends to the west and south of Cannock Chase. A little Wych Elm is present too.

Wych Elm occurs mainly on sheltered terrain at moderate elevations and could well have persisted on such sites since Boreal times.

The county attracted little prehistoric settlement. In Roman times the main occupied sites were along Watling Street between Wall and Water Eaton. The subsequent Anglo-Saxon penetration was twofold. One group moved up the Trent valley and along the Tame. Another, the Pecsaetan, occupied the Pennine uplands. There were two royal forests. Cannock Chase is in Z1 where elm is infrequent. Kinver Forest is in V1 where English Elm is common. A third wooded area, Needwood Forest, is also in region Z1. Place names indicative of wooded cover and late settlement are concentrated in five areas. Four are wholly or partly in Z1. Of these, one lies along the northern half of the Salop border, another lies west of Uttoxeter, the third lies between Cannock Chase and Needwood Forest, and the last is on the southern plateau. The fifth area, to the north-west of Wolverhampton is in region V1.

In the Salop section, it was argued that the English Elm of region V1 in Salop and Staffordshire was unlikely to have arrived before the close of the Middle Ages. The Staffordshire part of V1 is partly on the site of Kinver Forest and partly in the area of woodland place names north-west of Wolverhampton. It is most unlikely that English Elm was of ancient occurrence anywhere in the county.

Lock's Elm is thinly scattered, often along long-distance routes. Its distribution does not conflict with the hypothesis that it may have been introduced, ultimately from Lincolnshire, by drovers.

Tyne and Wear

The modern county of Tyne and Wear consists of those parts of the former counties of Northumberland and Durham that form the immediate hinterland of Newcastle upon Tyne and Sunderland.

Relief is relatively low though rising westward as the Pennines are approached. The major topographical features are the valleys of the Tyne, its tributary the Derwent, and the Wear.

Wych Elm occurs along the Tyne and Derwent valleys. The local designation is *elm*.

Some two-thirds of the county is in region Z1, where elm is infrequent. This covers the coastal plain and upland areas away from the Tyne and Derwent.

Region G7,[134] with frequent Wych Elm, lies along the Tyne and its major tributaries. In the county of Tyne and Wear, this entails the main river upstream from Newcastle and the Derwent.

The distribution of Wych Elm, along the middle stretches of major river valleys, could well indicate a habitat occupied continuously since Boreal times.

West Midlands[135]

This new county was formed by amalgamating those areas formerly in Staffordshire, Warwickshire and Worcestershire that had been absorbed into the conurbations of Birmingham, Coventry and Wolverhampton, together with what intervening country could be found.

Birmingham is on a sandstone plateau. Coventry is on a lesser elevation over Coal Measures. There is another Coal Measure exposure west of Birmingham. From the two areas of higher ground the land slopes down to surrounding clay lowland. There are no rivers of consequence.

Wych Elm is frequent on the Birmingham plateau. English Elm is abundant in the east of the county. The designation *elm* has been current since the eleventh century.

Region Z1, with infrequent elm, covers the north of the country. South of this is region G15,[136] on the Birmingham plateau, and extending eastward from it. Here, Wych Elm is frequent. There is some English Elm and Lock's Elm. The place name Elmdon, mentioned in Domesday Book, presumably attests the presence of Wych Elm at the time of the Norman Conquest.[137] Elms were growing in Birmingham in the fifteenth century.[138]

The eastern part of the county is in region V1,[139] the main area of English Elm.

Wych Elm could be indigenous on the Birmingham plateau. South of this it occurs infrequently till the Cotswolds. It thus appears to be at the southern limit of its main area of occurrence.

The county of West Midlands was only sparsely inhabited up to and during Roman times. It was only

Fig. 120. Tutbury Elm (St), by Jacob Strutt.[131]

occupied on any scale relatively late in the Anglo-Saxon period. Its southern half was part of the Forest of Arden.

Introduction of English Elm to the east of the county is not therefore likely to be earlier than the Anglo-Saxon penetration. Assuming that it came from north-east Warwickshire, its date of introduction would, presumably, be later than there. In chapter 13, the English Elm of north-east Warwickshire is thought to have been introduced no earlier than in the Anglo-Saxon period. The West Midlands population is likely to be later than this.

Lock's Elm has a scattered distribution as usual.

Its occurrence on the road from Birmingham to Coventry perhaps supports the theory of its distribution along long-distance routes, ultimately from Lincolnshire.

West Yorkshire

West Yorkshire is the part of the West Riding that envelops Bradford, Leeds, Halifax, Huddersfield and Wakefield.

The Pennines are on the west. The land slopes down as one proceeds eastward, traversing first Coal Measures, and then the Magnesian Limestone outcrop which runs from north to south along the eastern

Fig. 121. Tutbury Elm (St), by Thomas Peploe Wood.[131]

border of the county. There are three major rivers, the Wharfe, forming the northern boundary, the Aire and the Calder. All drain eventually into the Humber. All flow through well-marked dales.

Wych Elm is frequent throughout the county save on the exposed moorland. Its usual designation is *elm*. Occasionally it is replaced by Scandinavian *alm*.

The only part of the county where elm is infrequent is the highest ground towards the west. This area is part of region Z1.

The rest of the county is in G1,[140] the principal region of frequent Wych Elm in England. It runs down along the east flank of the Pennines and up the dales. Scattered English Elm is present towards the east.

The earliest documentary evidence for the local Wych Elm is the Domesday Book entry for (North and South) Elmshall.[141] The Wakefield manorial court rolls, which cover a number of parishes around Wakefield, are a valuable source of information for the presence of Wych Elm in the early Middle Ages. The surname del Helm was current at Sowerby in Halifax parish in the thirteenth century. In the fourteenth century, the surname ad Ulmum, or att Elme is found at Wakefield. At Sandal Magna, the surname varied between atte Helme and att Elme. A roll records the unauthorized felling of an elm in Wakefield Wood.[142]

In the fourteenth century, too, there was an Elmflatt at Methley, and, in the next century, a Brentboth Elme [burnt booth elm] at Hepworth. Alme Green in Addingham is encountered in the sixteenth century. The manorial court rolls of Dewsbury feature *le great elme*.[143] Kirkless Priory held land at Shelf, Scoles and Liversedge. Amongst the trees growing there in the sixteenth century were *elmys*.[144] Richard III's Trees at Sandal Castle are mentioned in chapter 9. There was an elm avenue at Esholt, Otley. Elm water mains were laid at Halifax.[145] There were also elm water pipes, as well as oak and deal pipes, at Huddersfield.[146]

Wych Elm has every appearance of indigenous status in the dales. The occasional English Elm would be of recent introduction. Little is likely to have reached the county before the early eighteenth century.

Fig. 122. Old elm at High Ackworth (WY).[140]

13
Elm in the southern counties

The area covered in this chapter is England south of the Trent but excluding the eastern counties. It comprises Avon, Berkshire, Buckinghamshire, Corn-wall, Devon, Dorset, Gloucestershire, Hampshire, Hereford and Worcester, Isle of Wight, Oxfordshire, Somerset, Surrey, Warwickshire, West Sussex and Wiltshire. [123]

The elm generally present is English Elm. The principal region in which this occurs in abundance, V1, is represented in every county except Cornwall. [23]

Since the English Elm is an introduced tree, it is necessary to pay attention to the history of human settlement in each county. Topography is important not so much as directly affecting elms but as determining the sites of human occupation from which elms have been disseminated.

Wych Elm is of sporadic occurrence in the southern counties, being found principally on higher ground and on calcareous soil. It seems to have survived in such localities since prehistoric times. Elsewhere Wych Elm probably disappeared around the time of the Elm Decline.

Fig. 123. Elm regions of the southern counties. Diagonal shading, Wych Elm; horizontal shading, English Elm; vertical shading, other Field Elms.

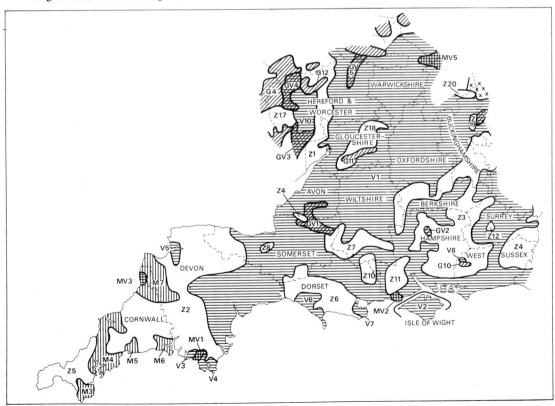

Avon[1]

This new county is composed of the southern part of the former county of Gloucestershire and the northern part of former Somerset. Bristol and Bath are the urban nuclei.

The southern end of the limestone range of the Cotswolds is in Avon. The clay Vale of Berkeley lies between it and the Severn estuary. Between Bristol and Bath and the Somerset border is a geologically complex terrain of hilly relief. There are marshy levels by the Severn north and south of Aust and south of Clevedon. The Bristol Avon is the principal river.

English Elm is everywhere present. Wych Elm occurs in addition in the southern hill country. The name *elm* was anciently in use in the north of the county. In the south *wych* appears to have been current. Wych Elm was differentiated as *wych hazel*.

Most of Avon is in the main English Elm region VI.[2] It covers the entire county except the southern hills. Sporadic Wych Elm is present in some places.

The historical evidence for the occurrence of elm starts with a tenth century Anglo-Saxon charter relating to Weston super Mare. This is one of the very few in which elms occur as boundary markers. The most pertinent passage runs:

to ðam twam wycan standað on gerewe eal swa ðaet gemere gaeð swa up to ðam wice stynt.

to the two elms standing in the row [of trees] as [along] the balk which runs up to where the elm stands.[3]

In the fourteenth century, the place name Elmeleghe is recorded in Henbury[4] and the surnames atte Elme at Churchill and atte Wych at Norton.[5] Le Nelmestrete at Thornbury and the Patch Elm at Rangeworthy make their appearance in the next century.

In Bristol, elm board was being utilized in the fifteenth century.[6] The place name Elmley in Frampton Cotterell is on record in the next century. Sales of elm timber are on record at this time at Corston, South Stoke and Walcot.[7] Bristol's water supply ran through elm pipes in the eighteenth century.[8]

The Watch Elm at Stoke Gifford and the Corston Elm are both mentioned in chapter 9. Notable trees included the Battle Elm at Oldbury on Severn, and the Sheperdine Elm, formerly the Conyberry Elm in the same parish. There were elm avenues at Henbury House and Blaise Castle.[9]

In the central part of the hilly country south of Bristol, English Elm and Wych Elm are both frequent. This is region GVI,[10] which extends into northern Somerset. Wych Elm is recorded under the designation *wych hazel* in this and neighbouring parts of region VI in the eighteenth century.[11]

Wych Elm is likely to be indigenous in the part of region GVI covering the Mendips just over the border in Somerset. Its presence in the southern hills of Avon might also go back to prehistoric times. Alternatively, it could have been taken later by man from the Mendips.

Prehistoric settlement in Avon was early, especially in the southern hill country. This region was also densely occupied in Roman times. There was also much settlement along Fosse Way.

The Anglo-Saxon entry was late. The River Avon became a long enduring tribal boundary between the Hwicce to the north and the West Saxons to the south. There was a royal forest at Kingswood near Bristol. It had only one elm in the seventeenth century.[12] The marshy tracts by the Severn and the clay lowlands behind them were presumably tackled in the later stages of Anglo-Saxon expansion. Place names suggest that the country to the north-east of Kingswood Forest was also wooded and settled late.

The date of introduction of English Elm has to precede that of the Weston charter. The late advent of Anglo-Saxon invaders makes it highly unlikely that they were responsible for its first appearance in Avon though they doubtless took it into the previously unsettled clay lowlands. If it is accepted, as argued in chapter 5, that the general pattern of distribution of the English Elm had been established by the time the Anglo-Saxon invasions took place, it is extremely unlikely that it would not have been present in Avon, which is so near to the putative centre of origin of this elm in England. It fits in with what is known elsewhere, if it is supposed that the West Saxons applied the term *wice* to the English Elm they found, while the Hwicce called it *elm*.

If English Elm was already present in the southern hills in Roman times, then the question has to be asked how it got there. The theory was advanced in chapter 3 that this elm was first introduced into Salisbury Plain in Wiltshire. The period of the Roman occupation, with good communication between the towns on the Fosse and the river valleys of northern Wiltshire, seems the most likely date for such an introduction. The lack of elm amongst the wooden artefacts of the Somerset lake villages suggests that English Elm had not become common in the country immediately south of Avon in the later Iron Age.

Berkshire[13]

The modern county corresponds to its predecessor minus the Vale of the White Horse, taken by Oxfordshire, and plus Slough, a consolation prize from Buckinghamshire.

The chalk downs run from west to east across the western half of the county and there is a small chalk plateau in its north-east corner. A small area of chalk upland also outcrops at the south-west corner. South of the chalk are clays and sands of Tertiary age, the sands predominating in the south-east of the county.

The Thames forms the northern boundary to the eastern half of the county. Its major tributaries are the Kennet, entering it from the west, and the Loddon, coming in from the south. All three rivers flow in well-marked valleys.

English Elm is frequent along the Kennet and Thames valleys. Its earlier designation was probably *wych*. It has been called *elm* since the sixteenth century.

Most of the Berkshire chalk downland, most of the Tertiary hills north of the Kennet and the Tertiary sands in the east fall into region Z3 with little elm. The Thames valley at the Goring gap and between Cookham and Bray also comes into Z3. English Elm occurs at low frequency. There is some Wych Elm on the chalk downs and very occasionally, as at Frilsham, some Narrow-leaved Elm. The existence of some elm, perhaps Wych Elm, in the twelfth century, may be indicated by the place name Wikenholt recorded in that century in Lambourn.[14] The fourteenth century surname atte Wychmount in Bucklebury might refer to elm.[15] The Elm Landes recorded here in the next century leave no doubt. English Elm is perhaps more likely.

82 The rest of the county is in the main region of frequent English Elm, V1.[16] The principal locations are the Thames valley between Bisham and Tilehurst, the north-east corner of the county around Windsor, the Kennet valley, and parts of the eastern end of the Berkshire downs. The elm board used at Windsor in

the fourteenth century could have been local.[17] The existence of English Elm in the Thames and Kennet valleys is definitely established by the place name Elm Grove at Reading, an action for waste involving elms at Ufton Nervet in the fifteenth century[18] and the frequent mention of elm in the sixteenth century Thatcham churchwardens' accounts.[19] There were Elm Landes at Sonning in the seventeenth century.

Reading received its water via elm pipes.[20] Notable trees included the elm on Forbury Hill, Reading, and the Procession Elm in Thatcham.[21] The elmscapes to the south of Reading achieve their classic description in the writings of Mary Mitford.

The place name Wychmere in Winkfield may indicate the presence of elm in the part of the region V1 lying in the north-east corner of the county in the fourteenth century. Later references to elm at Eton are assembled in chapter 10. A Bowyer's Elm was standing there in the fifteenth century.[22] The most celebrated elm avenue in the county was the Long Walk in Windsor Park, planted in the 1680s.[23]   *124, 125*

The relatively infrequent Wych Elm of the county is mainly on the chalk downs. Possibly it survived here from prehistoric times.

Early human settlement was in two areas. One was on the chalk downs, especially towards the west. The other was along the river terraces of the Thames around its confluence with the Kennet and downstream around Maidenhead. The land occupied in Roman times was much the same, but with extension into the middle course of the Kennet.

The earliest Anglo-Saxon settlement was mainly in the vicinity of the Roman sites along the Thames. In the later stages of Anglo-Saxon expansion, Berkshire was disputed land between the Middle Angles and the West Saxons. East Berkshire perhaps pertained with Surrey to the transient domain of the Middle Saxons.

The whole of eastern Berkshire, much of it on barren Tertiary soils, lay in the medieval royal forest of Windsor. Place names indicative of wooded conditions and late settlement occur not only within the early bounds of the forest but up to 15 km further west. Most of this wooded area is in region Z3 where elm is infrequent.

The historical evidence confirms the presence of English Elm in the Thames and Kennet valleys in medieval times. Its date of introduction is not likely to be later in the Berkshire section of the Thames valley than around London, if the assumption is justified that the Thames was the route along which it passed from its west of England source to the environs of London. If the argument in chapter 5 is correct, English Elm would have reached the London area before the Anglo-Saxon invasion. Its introduction to the Thames terraces in Berkshire is therefore likely to be no later than Roman times. Since the Kennet was not thickly settled then, it is more likely that it did not receive its elms till the later stages of Anglo-Saxon expansion. One can only speculate how long before the end of the Roman occupation English Elm had been introduced. It is tempting to suppose that the expansion of English Elm was one consequence of the improved internal communication brought about by the Romans. At all events, it is likely to have reached Berkshire some time between its introduction into Oxfordshire and into the London region.

The origin of the rare Narrow-leaved Elm of the eastern chalk downs is an enigma as yet unsolved.

Buckinghamshire[24]

Modern Buckinghamshire only differs from the older county in having lost the environs of Slough to Berkshire.

The main high ground is the chalk upland of the Chilterns which cross the southern half of the county from north-east to south-west. North of the Chilterns is a clay plain interrupted in the southern half of the county by a narrow ridge of Corallian Limestone. The ground rises again on the north border of the county where older limestone outcrops. On the eastern border to the immediate south of Bletchley, there are Greensand hills. South of the Chilterns, the underlying rocks are Tertiary deposits, with soils liable to be sandy. The Thames is the southern boundary. The Colne, one of its northern tributaries, is part of the county boundary in the south-east corner. The other major river is the Great Ouse which pursues a sinuous course in the north of the county.

English Elm is abundant in the lowlands north

Fig. 124. A short ride in the Long Walk, Windsor (Br), by William Heath.[23]

A SHORT RIDE in the LONG WALK or the PONIES POSED

and south of the Chilterns. It seems that elm was known as *wych* in the Middle Ages. The term *elm* had come into use by the sixteenth century.

There are three areas where elm is infrequent. The most extensive is over the Chilterns. These lie in region Z3. This region also covers part of the Tertiary exposures further south. Wych Elm and English Elm are present but infrequent. The field name Wichcroft, recorded for Wendover in the thirteenth century, perhaps indicates the presence of elm.[25] Wych Elm would be the more likely kind at this time. Elm boards were being purchased at Chesham in the seventeenth century.[26] King Stephen's Tree at Chequers, mentioned in chapter 9 was a notable specimen elm, even though its attribution is improbable.

The other two areas with little elm are smaller. Z19 covers the Greensand hills south of Bletchley. Z20 is in the northern tip of the county and adjoining parts of Northamptonshire. There is some English Elm in both regions. Narrow-leaved Elm is present too in Z20.[27]

The rest of the county, north and south of Z3 is in V1,[28] the main region of frequent English Elm. Wych Elm occurs rarely. Narrow-leaved Elm is widespread, though never abundant, in the north of the county.

The presence of elm in the south of the county is attested by entries in the fourteenth century manorial court rolls of Iver.[29] Notably specimen trees *126* included Milton's Elm at Chalfont St Giles, mentioned in chapter 9, and the large elm on the green at Horton. The 'rugged elms' of Gray's *Elegy* were in Stoke Poges churchyard.[30]

North of the Chilterns, elm is recorded at Aston Abbots in the sixteenth century.[31] A seventeenth century account book surviving amongst the Stowe Manuscripts records the sale of elms growing at Barton Hartshorn, Buckingham, Padbury, Radclive, Stowe,

Fig. 125. The return from Ascot (Br), by Sir Alfred Munnings.[23]

Tingewick and Water Stratford.[32] The elms in the common field at Calverton were the subject of a seventeenth century manorial court order.[33] Elm was also present at Chicheley,[34] Great Kimble[35] and Whaddon.[36]

Two notebooks of late eighteenth century date, dealing with timber around Chicheley, distinguish between *elm* and much less common *wich elm*.[37] It is not clear what the latter means. Possibly it denotes Wych Elm as defined in this book. It might mean a Hybrid Elm. Both these kinds occur just over the borders with Bedfordshire and Northamptonshire. Narrow-leaved Elm is also present at low frequency in northern Buckinghamshire. It cannot as yet be ruled out that this was the elm intended.

Wesley's Tree, where Wesley preached to the inhabitants of Stony Stratfield in 1777, is mentioned in chapter 9. It survived into the present century. There were many fine specimens of English Elm on village greens in this part of Buckinghamshire, as at Aston Abbots, East Clandon, Newton Blossomville and Stone. There were elm avenues at Chicheley Hall, Hanslope Park, Tyringham and Wotton Underwood.[38] The idyllic elmscape of Milton's *L'Allegro*, discussed in chapter 10, was probably the part of region VI at the foot of the Buckinghamshire Chilterns.[30] Aylesbury formerly obtained its water through elm mains.[39]

The Wych Elm in the county is mostly in the Chilterns. It has probably persisted there since Boreal times.

Prehistoric human settlement was in two areas. One was along the river terraces of the Thames. The other was along the Icknield Way on the scarp face of the Chilterns. More intensive occupation of the Thames valley area occurred in Roman times. The Chiltern sites continued. Settlement also became important on the Corallian outcrop near Aylesbury. There were also occupied sites on both sides of the Buckinghamshire stretch of Watling Street.

Anglo-Saxon settlement was initially on or near Romano–British sites, though not along Watling Street. A great deal of the county appears not to have been penetrated by the Anglo-Saxons till relatively late in the pre-Conquest period. Latterly it was Middle Anglian territory with the Cilternsaetan along the Chilterns.

There were three royal forests in or partly in the county. Salcey and Whittlewood Forests are partly in Buckinghamshire and partly in Northamptonshire. Region Z20, with infrequent elm, corresponds largely with Salcey Forest. Bernwood Forest lay along the border with Oxfordshire, continuous with Shotover and Stowood Forests over the border. Early place names indicate that much of the northern third of the county was wooded in early times and therefore likely to have been settled late. Wooded terrain is also indicated by place names around Beaconsfield and in the south-west corner of the county.

The historical evidence for English Elm indicates its occurrence in the county in the fourteenth century. It is argued in the next chapter that English Elm had

Fig. 126. Boveney church (Bu), by Frederick Griggs.[28]

reached the London area by Roman times. If this be so, it is likely to have been present in the Thames valley, between London and its presumed introduction site somewhere in the western counties. Most of the Buckinghamshire part of region VI is on lowland terrain north and south of the Chilterns, areas settled relatively late in the Anglo-Saxon Period. It seems, therefore, that the introduction of English Elm to most of Buckinghamshire should be dated to the later stages of Anglo-Saxon penetration of the county.

Narrow-leaved Elm is mainly in the north of the county and presumably entered from north-east Northamptonshire. This was forested and sparsely settled country till late in the Anglo-Saxon period, and the Narrow-leaved Elm now there is not likely to be older than this. It is, however, not likely to have been introduced long after English Elm had been established.

Cornwall[40]

The boundaries of the present county differ little from those of the old. The tongue of Devon that formerly ran into Cornwall north-west of Launceston has been absorbed.

There are upland moors over granite in the Land's End peninsula to the north-east of Helston, and to the north of St Austell. The most extensive of the granite uplands is Bodmin Moor. Higher ground, largely over serpentine, forms the bulk of the Lizard peninsula. The north-east of the county lies over Culm Measures. The Tamar, dividing Cornwall from Devon, is the major river. The sheltered estuaries at Falmouth and of the Camel and Fowey rivers contrast strongly with the exposed land behind them.

In distinction from all the other counties treated in this chapter, English Elm is rare. Cornish Elm, instead, is abundant on sheltered terrain.

Elm was reputedly known as *elaw* or *ula* by late Cornish speakers. The English immigrants appear to have used the terms *elm* for Cornish Elm and *wych hazel* for Wych Elm.

There are two areas where little elm occurs. Z2 covers an extensive area of western Devon and also most of Cornwall east of the Camel and Fowey estuaries. Bodmin Moor is wholly included. Cornish Elm is thinly scattered on lower ground. A Hybrid Elm, probably a cross between Wych Elm and Cornish Elm,

occurs at Pillaton. The place name Bodelva in St Blazey, first recorded in the thirteenth century, possibly incorporates the compound *elwyth*, also meaning elm. More likely, the second element is a personal name.

In the sixteenth century, Leland noted that the entry to Tintagel was by 'long Cornwall elmetrees layde for a brigge'.[41] This is perhaps the earliest recognition that the elm of Cornwall was a distinct kind. Already by the sixteenth century, St Keyne's well, mentioned in chapter 9, had an elm amongst the four different trees that shaded it. A manorial court roll of Pelynt reserves 'oak, ash, elm and witchhals [wych hazel]' to the lord of the manor.[42] This perhaps confirms the existence of some Wych Elm or Hybrid Elm in eastern Cornwall.

Z5, the other area with infrequent elm, covers the exposed western end of Cornwall, including Land's End but not the Lizard. Cornish Elm occurs here at low frequency in more sheltered localities. If the reports of the presence of elm in the Newlyn submarine forest bed are correct, they presumably indicate the presence of Wych Elm in Neolithic times.[43]

Three place names, Crellow in Stithians, Trevella in Cranstock, and Trevelloe in Paul, possibly incorporate the Cornish name for elm, but there are etymological difficulties about each. The most likely elm place name, first recorded in the thirteenth century, is Eglosallow, perhaps meaning church of the elms, the alternative designation of Illogan, but even this is uncertain.[44]

The borough of St Ives was purchasing elm board in the sixteenth century.[45] John Wesley, as mentioned in chapter 9, preached under elms near Camborne.

In the Land's End peninsula, much Dutch Elm was planted in the eighteenth century, in combination with sycamore and elder, for shelter belts.[46]

There are five areas where Cornish Elm is frequent. M3[47] covers the Lizard peninsula. M4,[48] the *127* largest region, occupies the corridor of lower ground connecting the Falmouth and Camel estuaries. The place name Trevella perhaps indicates the presence of Cornish Elm in St Erme in the fourteenth century. It was abundant at St Breock in the sixteenth century.[49] Estate surveys of the seventeenth century record elm at Bodmin and St Enoder.[50] Of later plantations, the elm avenues radiating around the

Falmouth pyramid, later re-sited, were especially commented upon by visitors.[51]

The Fowey estuary is in region M5. Estate records of the seventeenth century attest the existence of elm at Golant, west of the estuary, and at Lansallos, on the east of it.[50]

St Germans is at the centre of the small region M6. There was a field called Elm Park in Landrake in the sixteenth century. The church wardens' accounts of St Stephen by Saltash mention elm in the century following.[52] The last region, M7,[53] covers the north-east corner of Cornwall and extends eastward into the adjoining part of Devon. The table board that graced the vicarage of Poundstock in the seventeenth century was of elm, quite likely of local provenance.[54]

Enclosed on all sides by M7 is a small region, MV3, around Bude, where English Elm and Cornish Elm are both frequent. The sixteenth century churchwardens' accounts for Stratton include payment

for sawing elm.[55] Cornish Elm, at this date, is the more likely kind.

Fossil pollen profiles seem to indicate that Wych Elm never occurred in quantity in Cornwall and that it died out over most of Cornwall at the Elm Decline. A few stands seem to have survived in eastern Cornwall.

The history of the introduction of Cornish Elm is reconstructed in chapter 5. It is argued there that this elm was brought from Brittany by man after the Roman period but before Cornwall was occupied by the Anglo-Saxons. The eighteenth century church terriers show how widespread in Cornwall it was by that time.[53] Only for region M3 is evidence lacking. The present pattern of distribution, reflecting very much the occurrence of sheltered terrain in an otherwise exposed landscape, could well go back to the start of the present millennium. Present evidence is not sufficient to locate the site of introduction.

Fig. 127. Helston (Co), by Holman Hunt.[47]

The English Elm in MV3 presumably came from the Devonshire part of region V1. Its date of introduction could be no earlier than the eighteenth century.

Devon[56]

This large county has preserved its former boundaries save for the cession of a small piece of land on its north-west border to Cornwall.

The two major elevations are the granite upland of Dartmoor and the Old Red Sandstone upland of Exmoor. The south of the county is also on Old Red Sandstone though at lower elevation. Between the two sandstone outcrops is a wide band of Culm Measures which give a heavy clay soil. On the longitude of Exeter is a band of New Red Sandstone weathering to a fertile red soil. A narrow tongue of this extends westwards over the Culm Measures to Hatherleigh. The east of the county is a geologically complex mass of younger rocks whose outcrops result in much broken relief. The River Tamar divides Devon and Cornwall. The other principal rivers are the Torridge and Taw draining northward, and the Dart, Teign, Exe, Otter and Axe, which drain southward.

English Elm is frequent in the south-east quarter of the county. It appears to have been known originally as *wych*. The term *elm* had come in by the seventeenth century. Wych Elm was differentiated as *wych hazel*.

Dartmoor, Exmoor, much of the land over the Culm Measures of north Devon and the land between the Dart and the Kingsbridge estuary, lie in region Z2 where elm is infrequent. This region also covers eastern Cornwall and western Somerset. English Elm[57] and Wych Elm both occur occasionally. The place name Witchells in Shirwell, recorded as Wichehole in the fifteenth century, perhaps indicates the presence of some elm, probably Wych Elm, north of the Taw estuary.[58]

The main area with frequent English Elm is the Devon part of region V1.[59] It corresponds roughly with the New Red Sandstone outcrop and the part of Devon *128* east of this. Wych Elm occurs sporadically.

Two river names, the Leman, a branch of the Teign, whence Lemonford in Ashburton parish, and the Loman, a branch of the Exe, whence Craze Lowman and Uplowman, have been derived from Celtic **lem-*, meaning elm. No confidence is possible

Fig. 128. The smithy, Luppitt (Dv), by Robert Bevan.[59]

over these derivations, though there is a little Wych Elm in the vicinity of both rivers which could conceivably have given its name to them. An eleventh century Anglo-Saxon charter refers to a *wice cum* in Ashburton. This would probably denote Wych Elm. In the seventeenth century a locality called Elm existed at Colyton and there was a place called Down Elms in Upottery. The Tiverton churchwardens' accounts of the same century refer to setting and pruning elms.[60] Exeter Cathedral close was planted with elms in the seventeenth century and again in the 1770s.[61] The city was provided with elm water mains in the eighteenth century.

Two specimen elms call for special mention. One was the Cross Tree at South Tawnton.[62] The other was the Cross Tree at Moretonhampstead, known latterly as the Dancing Tree.[63] Both trees shaded the village cross. The second was used for dancing as described in chapter 9.

There are three smaller regions with frequent English Elm. V5[64] is at the mouth of the Torridge river. It is doubtful whether the surname atte Wyche, recorded at Westleigh in the fourteenth century refers to English Elm at this date. Just possibly it refers to Wych Elm, which occurs infrequently in the neighbourhood. More likely *wych* here is a variant of *wick*, usually spelled *week* in Devon, and most often meaning a dairy farm.

The other two, regions V3 and V4,[65] are on the south coast between Plymouth and the Kingsbridge estuary. They flank a small region, MV1, where both English Elm and Cornish Elm are common.

The main area with Cornish Elm is in north-west Devon, part of region M7[66] which extends into north-east Cornwall. A little English Elm is present too. The *elming* plank mentioned in the seventeenth century Hartland churchwardens' accounts would probably be Cornish Elm.[67]

The general pattern of distribution of elm today is reflected by the mentions of elm in the 1727 and 1745 terriers of parishes within the county.[68] These show, in particular, much elm between the Exe and Dart estuaries in region V1, and much also in regions V4 and V5. The wedge of Z2 between V1 and V4 is also revealed. The principal difference is that there was much less elm to the immediate east of Dartmoor and in the far east of the county. The absence of elm at Axminster in the sixteenth century is suggested by a customal that forbade tenants to cut oak, ash or crab, but made no mention at all of elm.[69]

Wych Elm is abundant nowhere in the county but it tends to be found in situations, in particular by upland sheltered streams, where it could have persisted since prehistoric times. Its presence under the designation *wych hazel* is recorded at Tiverton in the early eighteenth century.[70]

Devon was very sparsely inhabited from the Later Bronze Age to the end of the Roman occupation. The Anglo-Saxon penetration by the West Saxons was also late. Dartmoor, in region Z2, became a royal forest. Early place names indicate that much of the land in central Devon between Exmoor and Dartmoor was well wooded and likely to be settled late. A similar indication of wooded terrain is provided by place names towards the Dorset border.

English Elm was certainly present at the longitude of Exeter at the end of the Middle Ages. It can hardly have entered till after the West Saxon invasion. There is comparatively little English Elm in western Dorset, nor was there formerly in eastern Devon, so it is likely that English Elm was brought from western Somerset. A closer dating of its arrival would be difficult.

The English Elm in outlying regions V3, V4, V5 and MV1 presumably came from V1 after it had reached its present westward limit.

The two areas in which Cornish Elm is present may represent an eastern fringe to Celtic Cornwall. There are 'Celtic' dedications such as St Bridget (Virginstow), St Germanus (Germansweek) and St Nectan (Hartland) in M7. There are none in MV1 but there are two, St Budoc (St Budeaux) and St Winwaloe (Portlemouth) along the stretch of coast in which it lies. It is possible that Cornish Elm was introduced into these regions from Cornwall proper before they were anglicized.

Dorset

Dorset has grown slightly beyond its original configuration. Its eastern boundary has been advanced to the Avon and a little beyond it to include Christchurch, and it has gained a few parishes north of Sherborne from Somerset.

The dominant feature of the Dorset landscape is the chalk downland. The outcrop is V-shaped, the point of the V lying to the west. Within the V are

Tertiary beds, largely sands. The peninsula of the Isle of Purbeck constitutes the south-east corner of the county. It is an upland area, much of it limestone. There are Greensand hills north of the chalk, around Shaftesbury. The northern part of the county is largely lowland clay, the Vale of Blackmoor, but to the north-west of the Vale, the land rises to limestone hills. South of the chalk downland is broken hilly country, with limestone outcrops as at Portland and clays, as in the Vale of Marshwood.

The principal rivers are the Stour and the Frome, the former entering the sea near Bournemouth, and the latter draining into Poole harbour. The Avon forms part of the eastern boundary.

English Elm is frequent in the Vale of Blackmoor, in parts of eastern Dorset, in the Isle of Purbeck and south of the chalk downland between Bridport and Portland. The older designation was *wych*. It was being replaced by elm in the fifteenth century.

Most of the chalk downs and the Tertiary sands are in region Z6. This extends a short distance over the border with Somerset. Elm in general is infrequent though English Elm is widely but thinly distributed. Its presence in earlier times may be indicated by the place name Wyccheworth in Wareham parish, on record in the fourteenth century, and by Boneliston Elme at Stinsford, recorded in the following century.[71]

Dorchester is in Z6. Elms were planted along several of the roads out of it in the late eighteenth century. These feature in the novels and verse of Thomas Hardy.[72]

There are two lesser regions where elm is infrequent. Z10 centres on Cranborne. The surname atte Wych, recorded for Cranborne and Alderholt in the fourteenth century, suggests the presence of some elm.[73] In the extreme north of the county is region Z7. This region covers a largish area of southern Wiltshire. There is some English Elm[74] and Wych Elm.

The main English Elm region, V1,[75] covers the north of the county, including the Vale of Blackmoor, and much of the eastern chalk downland. It meets the sea north and east of Poole harbour. The surname atte Wych was current in five places in the north in the fourteenth century: Over Compton, Sixpenny Handley, Tarrant Hinton, Tarrant Kingston,[73] and Trent.[5] In the seventeenth century, a *wich tree* is mentioned as a boundary marker of Gillingham.[76] Elm board was being used at Sherborne School,[77] and elm pulleys were in operation at Wimborne Minster during bell casting operations.[78]

The extreme west of the county is also in V1.

Fig. 129. Studland church (Do), by Charles Cheston.[82]

In his will of 1417, Thomas Brook of Holditch in Thorncombe instructed his executors to bury him in cloth, not in 'wheche or leede' [wych or lead].[79] Burial in elm was clearly local practice.

As already remarked, English Elm is frequent behind the coast between Bridport and Portland. This is the location of V6.[80] There is little elm in Portland itself. Its local scarcity moved Leland to record that it was at least present around the church.[81] There was a field in Portland called Wych Croft in the seventeenth century.

The remaining area of high frequency of English Elm is the northern side of the Isle of Purbeck, region V7.[82]

Wych Elm seems to have largely disappeared from the county after the Elm Decline. What appear to be Wych Elm leaves of Neolithic age have been preserved in a calcareous tufa at Blashenwell in the Isle of Purbeck.[83]

Prehistoric settlement was largely on the chalk downland in region Z6 where elm is rare. Other sites that attracted prehistoric man were the Isle of Purbeck and Hengistbury Head at the mouth of the Avon. The Roman settlement pattern was much the same.

Anglo-Saxon penetration was relatively late over most of the county. The invaders were West Saxons. The principal royal forests, Gillingham Forest and Blackmoor Forest, were in the north of the county, the former largely in region Z6. Cranborne Chase lay along the Wiltshire border to the east of Gillingham Forest, much of it in region Z10.

The Vale of Blackmoor, the principal focus of English Elm in Dorset, was settled late in the period of Anglo-Saxon expansion. This was presumably when it was introduced. The source would probably be the adjoining parts of Somerset. Wiltshire is an unlikely immediate source in view of the heavily forested country that lay along the Dorset–Wiltshire border.

The date of origin of English Elm in regions V6 and V7 is quite uncertain on present evidence. It presumably came from V1, but from which particular part of it is not apparent. A late date, say early eighteenth century is feasible. There is however nothing on present evidence to contradict an origin in medieval times.

Gloucestershire[84]

The older county had to hand its southern end to Avon. Otherwise only minor border rectifications were made.

The main high ground is the limestone upland of the Cotswolds. This runs across the county from north-east to south-west. There is a steep descent to the clay lowlands of the Severn valley. There is a gentle descent to the upper Thames valley in the east of the county. West of the Severn, the Forest of Dean is on high ground. The land to the north of the Forest is undulating country over Old Red Sandstone. The Severn is the major river.

English Elm is abundant in all the lowland areas. Wych Elm is frequent in the central Cotswolds. Both are called *elm*. There are two areas with little elm in the county. One is part of region Z1. This runs down the western border of Gloucestershire and covers the whole of the Forest of Dean. Both English Elm and Wych Elm occur occasionally. Some elm was present at Kempley in the sixteenth century in this region.[85] The Raven's Elm at Dymock and Cowley's Elm at Westbury on Severn were specimen trees of note.

The other area with little elm is region Z18. This covers most of the higher Cotswolds save the central part. Wych Elm is of occasional occurrence. The place name Elmynhaye in Nympfield is on record in the sixteenth century[86] and Elmeley in Alderton in the seventeenth century. A Patent Roll mentions that elm grew at Temple Guiting in the sixteenth century.[87] The elm involved in all cases was probably Wych Elm. Dabb's Elm, on the road from Stow on the Wold to Tewkesbury, reputedly marked the place where a sheep stealer so named was hung.[88]

Most of the central lowland of the county and the lower slopes of the eastern Cotswolds, are in the region V1,[89] the main region of English Elm in the south of England. Some Wych Elm and, much rarer, some Narrow-leaved Elm also occur. The evidence for the earlier presence of elm is copious and covers the whole of the Severn valley. An elm pile has been identified from under the Roman wall at Gloucester.[90] The species is not known. At so early a date Wych Elm from the Cotswolds is more likely. The place name Lemington, near Moreton in Marsh, has been derived from Celtic *lem-*, meaning elm. English Elm is abundant in this vicinity today. If the derivation were correct, it could indicate the existence of English Elm here when the Anglo-Saxons arrived. However, like all names derived from *lem-*, this one cannot be accepted without reservation.

Secure evidence, presumably applying in general to English Elm, is available from the twelfth century. This is when the place name Elmore is first on record. In the thirteenth century there was an Elm Bridge at Hucclecote and an Elmefelde in Newent. The surname de Ulmo was current in Churchdown and atte Nelme, a common Gloucestershire variant of atte Elme, at Newent. There was also a de Ulmo family in Chipping Campden.[91] In the environs of Gloucester, a de Ulmo family appears in the thirteenth century to reappear in the vernacular as atte Nelme in the century following.[92] Elsewhere, in the fourteenth century, families flourish under the names atte Elme at Dursley,[93] atte Nelme at Cam[94] and atte Nelme at Sandhurst.

In the fifteenth century, the place name Elmesplace in Southam is encountered. A Cold Elm grew at Norton, presumably in an exposed situation. Further place names derived from *elm* appear in the next century, as Elmefelde in Walton Cardiff, Elmead in Stoke Orchard and Elm Ridge at Ham near Berkeley. A different Cold Elm grew at Alkington, also near Berkeley. The Rye Nelme flourished in Breadstone and Wykes Elm stood in Hamfallow in Berkeley. A manorial customal of this date from Standish specifies that the bodies of timber trees, oak, ash and elm, belong to the lord of the manor. The toppings could be taken by the tenant.[95]

A number of individual trees are recorded in the seventeenth century, as the Long Elme at Weston Birt, the Tall Elms of Sherborne, the Haydon Elm at Boddington, the Three Mile and Four Mile Elms in Hardwicke to the south-west of Gloucester, and, the Little Elm at Cam and Conigers Elmes at Ham in Berkeley. There was a conspicuous roadside elm near Quedgeley.[96] The Pauntley manorial customal forbade the copyholders to fell their elms without licence.[97]

The great gale of 1703 brought down many elms in Gloucestershire, including one at Slimbridge, locally reputed to be the largest in England.[98]

More notable specimen elms were to be found in the Gloucestershire section of region VI than anywhere else in England. Some have been noted already. There was a third Cold Elm at Forthampton. The Three Shire Elm stood where the former counties of Gloucestershire, Warwickshire and Worcestershire met. The Barrow Elm at Hatherop is mentioned in chapter 9. Maud's Elm at Cheltenham was the subject of a long, pathetic and improbable story. The elm grew from a stake thrust through the heart of a damsel whose death was treated, unjustly, as suicide.[99] Piff's Elm in Elmstone Hardwicke, which gave its name to the hamlet so-called, was also reputed, in its day, to be the largest elm in England.[100] Other elms of note were Chester's Elm at Gotherington, Collard's Elm at Bulley, Isobel's Elm at Ashchurch, the Picked Elm at Kings Stanley, Stocks Elm at Saul, the Tirley Elm at Tirley, and the Willis Elm at Ham near Berkeley. The Bangrove Elm near Little Washbourne and the Hope Elm near Gloucester must bring this list to a close.[101] Among elm avenues, those to Arle Court in Cheltenham and in Cirencester Park[102] were well known. Gloucester had elm water mains beneath its streets.[103]

Cheltenham Spa, mentioned in chapter 9, was embowered in elm. The main Well Walk was a double avenue of elms between the River Chelt and the Well.[104] Sezincote House, whose Indian-style architecture inspired the Brighton Pavilion (ES), was planted around with elm, as was the latter.[105]

The central Cotswolds are the site of region G11.[106] Wych Elm is frequent.

That Wych Elm is indigenous in the limestone country of the higher Cotswolds is beyond doubt. It is certain too that Wych Elm is native in suitable sheltered sites in the west part of the county. There is much indigenous Wych Elm in the nearby parts of Wales.

Prehistoric settlement was principally along the relatively open terrain of the Cotswolds. The settlement pattern in Roman times was very much the same as before the Roman occupation. Fosse Way attracted some new development. Anglo-Saxon penetration was relatively late, Gloucestershire becoming part of the territory of the Hwicce. The clay lowlands were occupied relatively late. Most of the county west of the Severn became royal forest, the Forest of Dean, part of region Z1 with little elm. Early place names indicative of wooded terrain and late settlement are found north of Newent, in the Severn valley south of Tewkesbury, and beneath the scarp of the Cotswolds around Wootton under Edge.

Though the Severn vale was densely planted with elm by the thirteenth century, this was terrain in which human settlement was late. It seems necessary therefore to date the introduction of English

Elm here late in the Anglo-Saxon period. The Hwicce presumably found English Elm already growing in some parts of the territory they acquired, if only because Gloucestershire must be near to the point of introduction of English Elm, whether this is Wiltshire, as argued in chapter 3, or elsewhere in the west of England. Also the fact that the Hwicce apparently used the term *elm*, while their neighbours to the east and south used *wice* tells against any later introduction of elm from further east. The only area which was heavily settled in Roman times and where English Elm is now frequent is the line of the Fosse which runs along the eastern slope of the Cotswolds. Perhaps it is not a coincidence that the place name Lemington, rather grudgingly considered above, is just off the Fosse and very near the small Roman town at Dorn. Certainly, in the Roman period, there was good communication between the Fosse and the Wiltshire river valleys of Salisbury Plain. It seems a very likely time for English Elm to be taken to parts of England outside the introduction site.

Hampshire[107]

Hampshire is what it was, but shorn of the Isle of Wight and a strip along its former border with Dorset.

More than half the county is chalk downland. The chalk dips under Tertiary beds, clays and sands, in the north and in the south. There is a narrow outlier of chalk, Portsdown Hill, behind the south coast. The older rocks surrounding the Weald comprise a narrow band of Gault clay next the chalk, then an upland Greensand outcrop of which the Devil's Punchbowl is the best known feature.

The major rivers are the Test and Itchen, draining into Southampton Water. The Avon divides Hampshire from Dorset. The Loddon, a tributary of the Thames, flows northward.

Elm is not frequent in Hampshire as a whole. What there is is mainly English Elm. It is mostly to be found behind the south coast and in the valleys of the Test and Itchen. The earlier designation of elm was probably *wice*. Wych Elm was differentiated as *wych hazel*.

A large part of the county is in region Z3, where elm is infrequent. This covers much of the chalk, parts of the northern and southern Tertiary outcrops and the beds surrounding the Weald. Wych Elm and English Elm are scattered throughout the area and

hybrids between the two appear also to occur. The place name Wishanger (earlier Wichanger) in Headley, first encountered in the twelfth century, may indicate the presence of some kind of elm here.[108] Elm was within reach of Stoke Charity in the fifteenth century as evidenced by the utilization of elm plank by the churchwardens.[109] It was bought for the bells at Crondall in the seventeenth century, but it is not clear how close to hand was the source of supply.[110] There was a notable elm avenue at Stratfield Saye and another in Bramshill Park.[111]

There are two smaller regions with little elm. The environs of the three-corner point of Hampshire, Dorset and Wiltshire lie in region Z10, which is mainly in the latter two counties. A large part of south-west Hampshire, principally over Tertiary sands, is in region Z11. There is a little English Elm in these areas, though some elm was present in Romsey in the sixteenth century.[112]

Most of the rest of the county is in the main English Elm region VI.[113] The principal locations are the valleys of the Avon, Test, Itchen and Loddon, and the coastal plain. Wych Elm occurs occasionally, especially in the north of the county. The presence of English Elm along the south coast in the fifteenth century can be inferred from the steward's Books at Southampton. Entries refer to elm obtained from Beaulieu, Exbury, Fawley, Hythe and Marchwood.[114] In the Itchen valley, the place name Wychurst in Tichborne is recorded in the thirteenth century. There were elms in Winchester Cathedral close in the seventeenth century.[115] The place name Whitsbury, in the Avon valley, first recorded in the twelfth century, appears to be based on *wice*. In the Loddon valley in the fifteenth century, elm is mentioned in the Basingstoke manorial court rolls.[116] The place name Wychcroft in Highclere, in the north-west of the county, may go back to the seventeenth century.

Elm water mains were laid in Gosport,[117] Portsmouth[118] and Southampton[119] and presumably drew on local supplies. The Domum Tree, around which the pupils of Winchester College formerly commenced their singing of *Dulce Domum*, was one of the more celebrated Hampshire elms.[120] One that drew the curious was the Groaning Tree near Beaulieu. This, as its name implies, creaked prophetically in the wind.[121] Faith, Hope and Charity were a conspicuous trio of elms on Southampton Common. They are mentioned in chapter 9. John Wesley, as also mentioned in chapter 9, preached under an elm near Portsmouth.

The Guernsey Elm was largely introduced into the English landscape by two Hampshire nurseries, Rogers of Red Lodge Nursery at Eastleigh and Hillier of Winchester. It was planted extensively round Southampton.

Along the south-west coast is a region, MV2,[122] where English Elm and Narrow-leaved Elm are both frequent. The place name Lymington is usually derived from Celtic *lem-*, meaning elm. Both English and Narrow-leaved elms are present now. Unfortunately, like all names of this type, its proposed etymology cannot be accepted with complete confidence. The presence of Narrow-leaved Elm between Christchurch and Lymington in the seventeenth century is known from the second edition of Gerard's *Herball*.[123]

The small region GV2 is located at the head of the River Test. English and Wych Elm are both frequent.

Wych Elm occurs widely scattered over the chalk downland and is locally frequent. Its survival here from prehistoric times is likely.

The county was not densely settled in early times. Iron Age sites were principally around the Itchen valley in the neighbourhood of Winchester, along the north-east shore of Southampton Water and in the upper Loddon valley. In Roman times, in addition to those sites, the upper Test was occupied and the coastal plain eastward from Porchester. It is generally supposed that Anglo-Saxon penetration was mainly from the north from the Thames Valley, with perhaps a minor influx from the south coast. Latterly, Hampshire was West Saxon territory. In medieval times, a high proportion of the county was royal forest. The New Forest occupied the Tertiary sands of the south-west, mostly in region Z11. The Forest of Bere, partly in region Z3, lay behind the coast further east. Buckholt and Chute Forests lay along the Wiltshire border, the latter mainly over it. It was partly in region Z3. Pamber Forest was situated on the Berkshire border. Bagshot Forest was sited around the three-corner point of Hampshire, Berkshire and Surrey and was wholly in region Z3. Alice Holt and Woolmer Forests lay around the three-corner point of Hampshire, Surrey and West Sussex. They were likewise in region Z3.

A large number of early place names indicate wooded terrain and presumably late settlement. They occur at maximum density in and around the various royal forests and over much of region Z3.

English Elm was certainly present in the county at the end of the Middle Ages. It is doubtful whether it had arrived by Roman times. Its overall infrequency and the wide extent of wooded terrain suggest relatively late introduction. There is no elm in such identifiable timber as has survived in Romano–British archeological sites. The foundation piles under Winchester Cathedral are beech, not elm. The existence of Narrow-leaved Elm along the south-west coast is more easily explicable if there were no English Elm near at hand. It seems most likely then that the English Elm of Hampshire was mainly introduced in the latter stages of Anglo-Saxon penetration. A multiple origin is quite feasible. The northern elms could have come from the Thames valley via the Loddon, those of the Test could have come from nearby Wiltshire, and the coastal population could have come from the Sussex coastal plain.

The history of the Narrow-leaved Elm of region MV2 is treated in chapter 3. An Iron Age introduction was considered feasible. The area was not densely settled either in Roman or early Anglo-Saxon times.

Hereford and Worcester[124]

As its name declares, this county embraces the former separate counties of Herefordshire and Worcestershire. Part of northern Worcestershire went into the new county of West Midlands. There have been minor rectifications along the border with Gloucestershire. The old ecclesiastical boundary between England and Wales did not coincide with the present civil boundary. Norton, Presteigne and New and Old Radnor were English. The Black Mountain parishes were Welsh. It is the older boundary that is utilized here.

The county is of highly diversified relief. Running from north to south, dividing the former counties of Herefordshire and Worcestershire, is the narrow range of the Malvern Hills. The limestone upland of the Cotswolds and their outlier, Bredon Hill, are at the south-east corner of the county. To the north-east, the land rises to the sandstone plateau on which Birmingham is situated. There is an upland of old rocks south-east of Hereford. There are more extensive uplands of old rocks along the north-western boundary. The lower country between the various uplands is floored with red soils, largely derived from

Triassic rocks to the east of the Malvern Hills and from Old Red Sandstone to the west of them.

The major rivers are the Severn, with its tributaries the Avon on the east and the Teme on the west, and the Wye, with its tributary, the Lugg.

Elm is frequent in most of the county. English Elm is very generally distributed. Wych Elm is mainly in the west. Both are locally known as *elm*. The term *elvin*, also current in Kent and East Sussex, was also in former use.

The principal region with infrequent elm is part of region Z1. It runs down as a corridor between the former constituent counties. Both English Elm and Wych Elm occur occasionally.[125] Some elm was present at Ledbury in the sixteenth century.[126] There was a conspicuous roadside elm near Burford in the seventeenth century.[127]

The other area with little elm is Z17. This covers the Golden Valley north-west of the Black Mountains, and some of the country adjoining. English Elm is present here at low frequency on lower ground. Some Wych Elm is present higher up. V1,[128] the principal English Elm Region, extends over the lower ground drained by the Severn and Avon. Wych Elm occurs sporadically. The historical record goes back as early as the eighth century when the place name Elmley, later Elmley Castle, is first recorded.[129] Another Elmley, later Elmley Lovatt, is mentioned in the next century. It has been suggested that Wichenford, mentioned in Domesday Book, is derived from *wice*.[130] It is however doubtful if *wych* was ever a current name for elm in the county. Anglo-Saxon *wic*, which most usually meant dairy farm, became *wick* in most parts of England in the Medieval period. In Hereford and Worcester, as in Cheshire, it became *wych*, and was the word used to denote salt workings. Also the Hwicce, the tribe who occupied the terrain later demarcated by the medieval diocese of Worcester, also lent their designation to various place names. It is consequently rather unlikely that Wichenford indicates the presence of elm. The surname atte Wych, current in Severnstoke in the fourteenth century, had also better be discarded as evidence for the presence of elm.[131]

To return to names based on *elm*. In St Johns, the surname de la Helme is mentioned in the twelfth century. It was replaced by atten Elmes in the next century. The place name Helme is recorded for Hallow in the thirteenth century. The surname atte Nelme was current both in Rous Lench and Church Lench in the fourteenth century. The site of the Battle of Evesham was marked by a group of elms around Battle Well in the late Middle Ages.[132] In the seventeenth century the Haseler Elm of Charlton is first mentioned. A roadside elm by Martin Hussingtree was a traveller's landmark.[133]

The Friar's Elm, also known as the Bread Elm, stood at Barnards Green. It was an English Elm beneath which the Rector of Guarlford distributed bread to the poor on Good Friday.[134] The Lyre Elm stood at Ombersley Court. The Auborne Tree was one of the surviving Revolution Elms of Kempsey mentioned in chapter 9. The Rye Elm, where one of the courts leet of Oswaldstow hundred was held, is also mentioned in that chapter. The Sansame Walk was an elm avenue on the outskirts of Worcester. Another avenue led to Hanbury Hall.[135]

Region V10[136] covers the lower ground drained by the Wye and Lugg. English Elm is frequent and there is some Wych Elm. Elm pumps were sunk at Holme Lacy in the seventeenth century.[137] The Rotherwas Elm was a notable specimen tree.

Along the Welsh border and extending a considerable distance into Wales, largely along the

Fig. 130. Friar's Elm, Barnards Green (HW), by Thomas Armstrong.[134]

valleys of the Wye and Usk, is region G4.[138] This is one of the principal regions of frequent Wych Elm in the British Isles, though only a minor part of it is on the English side of the border. The place name Almeley, Elmelie in Domesday Book, confirms the local presence of Wych Elm in the eleventh century. The other area with abundant Wych Elm, G12,[139] lies over the higher ground east of Leominster.

There are three areas where Wych Elm and English Elm are both common. Region GV3[140] occupies the southernmost corner of the county and extends over the Welsh border into Gwent. The unconventional behaviour of the English Elm in the churchyard of Ross on Wye is described in chapter 9. There were elm avenues at Hill Court and Kentchurch Court.

Region GV4[141] is a transition zone between English Elm region V10 and the mainly Welsh Wych Elm region G4. The Prophet Elm of Credenhill, mentioned in chapter 9, was one of the more notable specimen trees. A fine elm avenue led to Moor Court. The third area, region GV5, covers the county between Droitwich and the Clent Hills. Elmbridge is entered in Domesday Book. The kind of elm alluded to is more likely to be Wych Elm. There was an elm avenue at Mere Hall.[142] The indigenous status of Wych Elm in much of the county is beyond question. It is the main elm in the west and on higher ground elsewhere. G4, as already stated, is one of the major Wych Elm regions in the British Isles.

The county was not extensively occupied in prehistoric times. Settlement tended to be on higher, presumably more open, ground, as the Malvern Hills and Breedon Hill, and the flanks of the Wye Valley. During the Roman period, there were important new settlements in the lower Avon Valley, between Worcester and Droitwich, and in the Wye valley between Hereford and Kenchester.

Anglo-Saxon penetration was late. The lower Avon was taken over first. The tribe who eventually occupied the Severn valley was the Hwicce, mentioned above. The Wye valley and Welsh border was settled by the Magonsaete.

Much land became royal forest. Feckenham Forest occupied a wide area in the angle between the Severn and Avon. Kinver Forest, mainly in Staffordshire, extended into the northern tip of the county. There was a small royal forest, Haywood Forest, south of Hereford. Other forest areas were Wyre Forest, along the Salop border, and Malvern Chase along the

Malvern Hills, both largely in region Z1 where elm is infrequent.

Early place names confirm the wide extent of wooded terrain in the county and the likelihood of late settlement. They are particularly frequent between Worcester and Kidderminster, in the Severn valley east of the Malvern Hills, and in the country north-west of Hereford.

The very early *elm* place names in the part of region V1 in the present county strongly suggests that English Elm had arrived before the end of the Roman occupation. The dearth of earlier settlement on most of the terrain now occupied by English Elm makes it unlikely that this elm had arrived before the Roman period.

The nearest likely source of English Elm would be the Fosse Way settlements in Gloucestershire, already suggested as the main location of English Elm in that county in Roman times. It seems therefore that the planting of English Elm in the Severn and Avon Valleys took place in the Anglo-Saxon period. The English Elm in Feckenham Forest and the country to the north and west of it is likely to be later still.

Region V10, centred on Hereford, would seem to require a separate centre of origin. There is far less evidence of an early presence of English Elm here. Perhaps the Magonsaete obtained English Elm from the neighbouring Hwicce at some later stage in the Anglo-Saxon period.

Isle of Wight

The island, now a county in itself, was formerly in Hampshire. Tertiary deposits at low elevation outcrop in the northern half of the island. A chalk ridge bisects the island from east to west. South of this is lower terrain over Greensand. Further south still is a second outcrop of chalk down.

Elm occurs throughout the island. Evidence has not been obtained on the early name for elm. It was known as *elm* in later times.

The whole island constitutes region V2[143] with English Elm frequent throughout. It is, however, distinctly more abundant in the southern half of the island. Among elms of note were the Seven Sisters in St Lawrence. The old churchyard of Bonchurch was surrounded by large elms that feature conspicuously in the several prints of this picturesque spot.[144]

In Whippingham, Guernsey Elm has been extensively planted. In chapter 5 it was surmised that its presence might owe something to the Prince Consort.

The island was first settled in the south. The concentration of population in this half of the island remained the settlement pattern through Iron Age, Roman and early Anglo-Saxon (Jutish) times. There was a royal forest, Carisbrooke Forest, in the northern half of the island.

The greater abundance of English Elm to the south of the island and the sparse early settlement along the opposite Hampshire coast cast doubt on any introduction of this elm from Hampshire. It is more likely that it was introduced from the densely settled coastal plain of West Sussex where English Elm is abundant. If this were the source, it could have been introduced in Roman times. An introduction in early Anglo-Saxon times is possible but less likely. At least, the Jutes who are said to have settled the island after the Roman withdrawal are unlikely to have brought it from Kent, where the local elm in the southern parts is Narrow-leaved Elm, not English Elm.

Oxfordshire[145]

The present county consists of its predecessor aggrandized by the addition of the Vale of the White Horse, taken from Berkshire.

The major relief features are the parallel calcareous uplands that traverse the county. The northernmost is the limestone range of the Cotswolds.

The southern is the chalk downland of the Chilterns. A lowland clay vale separates the two. A minor interrupted ridge of Corallian limestone runs down the middle of the clay vale parallel with the major uplands. The south-west part of the clay vale, between the Corallian ridge and the chalk is the Vale of the White Horse.

The Thames flows in a winding channel from west to east across the county. Its major tributaries are the Cherwell, Evenlode and Windrush, entering from the north, and the Ock, draining the Vale of the White Horse.

English Elm is abundant throughout the county except in the Chilterns. The earliest designation was probably *wych*. The term *elm* had come into use by the thirteenth century.

The only part of Oxfordshire where elm is infrequent is the Chiltern upland. This is part of region Z3. Wych Elm and English Elm occur occasionally. The place name Wichelo, earlier Wycheleyes, in Ipsden,[146] first encountered in the thirteenth century, probably indicated the presence of Wych Elm.

The rest of the county is in region V1,[147] the main area of English Elm. Within the county, English Elm is rather more common in the south than in the north and west. Wych Elm occurs sporadically, mainly on higher ground in the west. Narrow-leaved Elm is present in a few places in the north-east. Lock's Elm also occurs in a few localities in the north.

53, 89, 9

Fig. 131. Bonchurch (IW), by A. Willmore after W. I. Walton.[144]

There is abundant evidence for the occurrence of English Elm from the thirteenth century onwards. There were five academic hostels called Elm Hall in Oxford. At least one of these went back to this century.[148] At the same time, an elm is recorded as growing at Torald's School in Oxford and there was a family in the town called de Ulmo.[149] Elsewhere in the county in the same century, the surname de Ulmo is on record at Black Bourton, Filkins and Weald, and the variant ad Ulmum at Kelmscott.[150]

In the fourteenth century, the court rolls of Cuxham[151] and Hook Norton[152] reveal the local presence of elm. An elm at Adderbury was blown over by the wind,[153] and there was a high court action for waste involving felling of elms at Middle Aston.[154] An outhouse at Harwell of elm cruck construction has been dated to the fourteenth century.[155] There was much elm at Abingdon, referred to from the fifteenth century onwards in the accounts of the abbey and entered in the survey made when it was dissolved.[156] In the latter century, John Pape bequeathed elms at Cote and Yelford to the Abbess of Elstow to repair Clanfield rectory.[157]

Other elm sites appear in the sixteenth century. The manorial court rolls of North Moreton record the felling of elms.[158] Those of Spilsbury have a memorandum on planting them.[159] Possibly the place name Wygenolds in Ashbury parish, recorded in the same century, is equivalent to Wikenholt which occurs in Berkshire and is thought to be a derivative of *wych*. An exercise in phoney etymology by the antiquarian Leland attests the presence of elm at Ewelme even though the place name is not, as Leland thought, derived from *elm*.[160]

R. Plot, the Oxford antiquary, writing in the seventeenth century, mentioned the presence of English Elm at Binsey and Middle Aston. He described the Great Hollow Elm at Bletchingdon in which the inevitable poor woman brought a child to birth. He [42] also describes a Narrow-leaved Elm at Hanwell which occurred both in avenues and wild in the park.[161] The indefatigable Michael Woodward, Warden of New College, in his visits to college property, gave detailed instructions concerning elm at Upper Heyford, Kingham and Stanton St John.[162] In the seventeenth century too, those going in the Rogationtide procession around Brightwell are found refreshing themselves with cider beneath an elm.[163] Elm was also present at Chawley[164] and Finmere,[32] and Eaton supplied elms to St John's College.[165] The Roke Elms of Benson and Coleman's Elm at Hook Norton also enter historical record at this time.

Oxford was receiving its water supply through elm pipes by the late seventeenth century.[166] One of the best known of specimen trees was Jo. Pullen's Elm on Headington Hill, which was burned down in 1909.[167] An elm at Henley on Thames, mentioned in chapter 9, had a lugubrious association with Prince [74] Rupert. Roadside landmark elms were sited at Filkins and Benson.[168] Elms of lesser repute included the Stocks Tree at Knighton in Compton Beauchamp and Thumper's Elm at Leafield. The most famous Oxfordshire elm avenue was in Blenheim Park, Woodstock. The Fairmile was a notable avenue on the outskirts of Henley on Thames, planted by Sir William Stapleton.[169] There were also elm avenues to Witney Grammar School and at Chislehampton House[170] and Shirburn Castle near Wallingford.[171]

Of the many plantings of elms at Oxford, the principal were the academic groves of Durham (refounded as Trinity), Magdalen, Merton and St [83, 105] John's Colleges, the Gravel Walk outside Magdalen, the [106] Terrace Walk outside Balliol, and the Non Ultra Walk on the northern outskirts of the old town, and above all the Broad Walk in Christ Church Meadow. These [84] are described further in chapter 10. The elms of Rewley Abbey and the triple elm around which St John's College was founded are described in chapter 9.

There were two notable Hybrid Elms in the county. One was in Magdalen College Grove, which was mostly English Elm. The Hybrid Elm was famous for its great size. It blew down in 1911.[172] The other was the Tubney Tree. This was the original of Matthew Arnold's Fyfield Elm around which maidens danced on May Day.[173] The tree was genuine, the maidens were not. *Murray's Handbook for Berkshire* appears to have the local tradition. It is witches who dance around the elm and the time is midnight.[174]

The impact on English literature of the elmscapes of Oxford and Garsington is described in chapter 10. Matthew Arnold's Elysian elmscape was the country between the Thames and the Ock. The elms at Kelmscott are associated with William Morris.[175] The elmscape south of the Ock received its classic description in Alfred Williams' *Villages of the White Horse*.[176]

Wych Elm is not frequent in the county. It occurs mostly on higher ground in the Chilterns and

Cotswolds. In both places, it probably has indigenous status.

Prehistoric settlement was mainly along the Thames above the Goring gap, and along the lower reaches of the Ock. The settlement pattern in Roman times was similar but with expansion up the Cherwell and Evenlode valleys. Anglo-Saxon entry was very early, mostly perhaps from East Anglia via the Icknield Way, but with some groups probably coming up the Thames. In later Anglo-Saxon times the county was debatable land between the West Saxons and Middle Angles. There was much royal forest, both north-west of Oxford, Wychwood Forest, and south-east of it, Shotover and Stowood Forests. Early place names indicative of wooden terrain and late settlement occur in and around the royal forests, north-west of Bicester, south-west and south-east of Oxford and around Henley on Thames.

The historical evidence is conclusive that English Elm had become abundant in the Oxfordshire part of V1 by the thirteenth century. If the argument in chapter 3 that English Elm was first introduced to Wiltshire in the Later Bronze Age is accepted, and similarly the argument in chapter 14 that English Elm had reached the London area by Roman times, it can hardly be supposed that the Oxfordshire Thames valley, en route from one to the other, received its English Elm any later than Roman times. It is conceivable that it even got it earlier. A specimen of double-looped palstave was found at Garsington. This is the artefact whose Spanish source was used in chapter 3 to support a Late Bronze Age date for the introduction of English Elm from Spain. Clearly there were trading connections in the Later Bronze Age between the ports handling goods from Spain and the Thames valley. A possible route would be via the Wiltshire Avon and the Berkshire Ridgeway. Contacts along this route would presumably persist into the Iron Age. The introduction of English Elm into the valleys of the lower Cherwell and Ock in Roman times is also quite feasible.

The scattered Narrow-leaved Elm in Oxfordshire north of the Thames valley is an outlier of the Narrow-leaved Elm population of northern Northamptonshire. In view of the former wooded nature of this part of the county, it seems likely that it got its elms, whether English Elm or Narrow-leaved Elm, considerably later than the Thames valley sites.

Lock's Elm is very scattered and likely, as in most places elsewhere, to have been present for no more than a few centuries.

Somerset

Apart from the loss of the land north of the Mendips to the modern county of Avon, and a smaller loss of land east of Yeovil to Dorset, Somerset retains its ancient boundaries.

It is greatly diversified topographically. There are five upland areas. The highest are the limestone Mendips in the north, and Exmoor and the Brendon Hills in the west. To the east of the Brendon Hills is the detached upland of the Quantocks. The Polden Hills are a narrow ridge running north-west to south-east across the Levels. Limestone hills are present in the south-east of the county. The rest of Somerset is lowland. Furthest inland this is clay, but east of Bridgwater there is extensive peaty marshland, penetrating many miles inland along the valley of the Parret and its major tributaries. This is the region of the Levels. The principal river, after Parret is the Brue on which Glastonbury stands.

English Elm is abundant throughout the lowland. The original name for elm seems to have been *wych* in most of the county and *elm* in the north. From no later than the fourteenth century, *elm* progressively replaced *wych*. This was one of the counties in which Wych Elm was differentiated as *wych hazel*.

The largest of the areas with little elm is region Z2. It covers Exmoor and the Brendon Hills and extends through Devon into eastern Cornwall. A little Wych Elm and less English Elm is present. Two areas with infrequent elm are wholly within the county. These are regions Z9, in the central Mendips, and Z8, which is centred on the Quantocks. Both Wych Elm and English Elm occur at low frequency in the latter. The remaining two areas with little elm lie in regions mostly in neighbouring counties. Region Z7 is mainly in southern Wiltshire and covers the hilly country east of Bruton. Z6 is mainly in Dorset but extends over the Somerset hills south of Ilminster. Narrow-leaved Elm occurs rarely at Winsham. The surname atte Wyche was current in Winsham in the fourteenth century.[5]

Most of the rest of the county is in the main English Elm region V1.[177] Occasional Wych Elm is present, principally on higher ground. The place name Limington, near Ilminster, has been derived from Celtic *lem-*, meaning elm.[178] There is much English Elm thereabouts today, but like all *lem-* etymologies, this one has to be treated with reserve. Rather surer ground is reached in a ninth century Anglo-Saxon charter defining bounds in the Taunton area. At one point the boundary goes *on wice hryig* [on to the elm ridge].[179]

The parish name Elm, including the present villages of Great Elm and Little Elm, is recorded in Domesday Book. The elm-slab chest in North Curry church has been dated as before A.D. 1200.[180] The surname de Ulmo, or later atte Elme, appears in Wells in the thirteenth century.[181] Elms were also growing at Street at this time.[182]

Families named atte Elme flourished at Compton, Preston and Nettlecombe, in the fourteenth century. The surname atte Wyche, presumably derived from *wych*, is commoner. It was current in the fourteenth century at Bishops Lydeard, Chard, Ditcheat, Henstridge, Kingsbury Episcopi, Kingstone and Somerton.[183] The manorial court rolls of Winchcombe also mention this surname in the same century.[184]

The purchase of elm for building purposes is on record at Bridgwater in the fourteenth century.[185] Elms were being felled at Milborne Port[186] and Tintinhull[187] in the next century. It is in the latter century that the bond tenants at Wellington were stated, in a manorial customal, to be entitled to fell and sell the elms on their tenements without licence. The implication of this unusual privilege is that elm was abundant.[188] A *pyked weeche* [pointed wych] grew at Kingstone in the fifteenth century.[189]

The Tudor antiquary Leland noted particularly the abundance of elm in Somerset during his travels, especially between Wells and Evercreech, between Ilchester and South Cadbury, and between Crewkerne and Hinton St George.[190] There is contemporary evidence for the presence of elm at Kilmersdon,[191] Wedmore,[192] and Drayton, East Brent, Luccombe and Muchelney.[193]

In the seventeenth century, elms are encountered in some places not yet mentioned. The tenants at Castle Cary were forbidden to top maiden elms,[194] and there was a boundary elm of the manor of Lytescary in Charlton Mackrell parish.[195]

Notable elms in region VI included Marshall's Elm, mentioned in the seventeenth century.[196] It gave its name to the modern hamlet. There was a Cross Tree at North Perrott. The Hundred Elm on the boundary between Carhampton and Dunster is mentioned in chapter 9. Magistrate's sittings were sometimes held inside a large hollow elm at Northover in the early years of the last century. This tree was blown down in 1833.[197] There was a fine elm avenue at Ven, near Milborne Port.[198] There was another, called the Apostles, in North Curry churchyard.

In the north of the county and partly in Avon is region GVI,[199] covering the lower parts of the Mendips. Both English and Wych Elm are frequent. North Elme, in Chew Magna, is recorded in the seventeenth century.[200]

Wych Elm certainly has indigenous status in the Mendips. Probably it is native also around Exmoor.

The abundant English Elm of the county appears to have reached its present limits by the early Middle Ages at the latest. The historical evidence suggests that it reached some parts of the county well before the Norman Conquest. It seems likely that it had not appeared when the lake settlements at Glastonbury[201] and Meare[202] were flourishing since no elm at all has been identified in the extensive inventory of wooden artefacts recovered. Even the wheel hubs were of oak or ash.

There was extensive marine inundation of the Levels in Roman times and no English Elm can have reached this part of the county till the waters had receded.

Prehistoric settlement was principally in the Mendips and the southern hills. In Iron Age times there were lake villages at Glastonbury and Meare. The Roman settlement pattern was somewhat different. The Mendips remained important. The southern hills settlement area expanded considerably and its centre shifted northward to Ilchester. The Levels were largely deserted. Anglo-Saxon penetration was relatively late.

There was one large royal forest, Exmoor, lying in region Z2 with little elm. The Mendips were also royal forest, Z9 covering the inner part. Selwood Forest, mainly in Wiltshire crosses the border into eastern Somerset. It is encompassed by region Z7. The lesser forests, those of North Petherton and Neroche, were too small to affect the elm distribution pattern. Place names indicative of wooded terrain and late settlement occur in the southern Quantocks and east and west of Frome.

Taking into account the nearness of Somerset to the presumed introduction site of English Elm in the west of England and the historical evidence of its occurrence in Anglo-Saxon and Medieval times, it seems necessary to date its introduction not later than the Roman period. The absence of elm artefacts in the lake villages suggests that it should not however be dated earlier than this. The dense Roman settlement round Ilchester, where English Elm is abundant today, would seem a likely focus for the Somerset population. A direct introduction of English Elm from Wiltshire, assuming that this was its centre of introduction, is rather unlikely. The area between the two counties was densely forested. An indirect introduction via Gloucestershire along Fosse Way, which runs through Ilchester, is more probable.

Surrey[203]

The modern county corresponds to Surrey as it was except that its north-east urbanized corner has gone to Greater London and the south-west corner of the defunct county of Middlesex has been added to it.

The chalk upland of the North Downs crosses the county from east to west. South of the North Downs and roughly parallel to them is a range of Greensand hills. North of the Downs are low-lying tertiary deposits, largely London clay. There is a clay vale, a continuation of the Holmdale of Kent, between the Downs and the Greensand hills, and an extensive clay plain, the Low Weald, south of the Greensand. A small part of the sandstone hills of the High Weald, extends into the south-east of the county. The Thames crosses its north-west corner. Its two principal tributaries on the Surrey side are the Mole and Wey.

English Elm is frequent over most of the county except the western border and the south-east corner. It appears to have been called *wych* in early times. The term *elm* was in use by the fourteenth century.

There are three zones where elm is infrequent. The western border of the county, largely over Tertiary sands, is in region Z3, which covers an extensive area in the adjoining counties to the west. Scattered English Elm is present throughout.[204] Its earlier occurrence is attested by le Elmelond in Thorpe and the surname atte Elmes in Chertsey. Both are recorded in the fourteenth century.[205] A Wychbedde is mentioned at Egham in the next century but this is referred to elsewhere as a withy bed and may therefore be a mistake.[206] There were elm avenues latterly at Littleton Park.

The south-east corner of Surrey is in region Z4. This region covers most of the Kent and Sussex Weald and part of the North Downs of Kent. The small region

Z12 lies immediately east of Guildford. It lies over chalk to the north, and over Greensand to the south.

The rest of the county is in the principal English elm region VI.[207] The place name Limpsfield has been derived from Celtic *lem-* meaning elm.[208] This is doubtful. Limpsfield is in the vale beneath the chalk scarp and is on the periphery of this particular part of region VI. There is little elm further east in this vale till the Medway cuts through it. The vale itself appears not to have been settled throughout in Roman times, though there was an isolated Romano–British villa at Titsey, near Limpsfield, by one of the roads that crossed the Weald. It seems unlikely therefore that English Elm was here in the Roman period nor is it a locality where Wych Elm would be expected.

A very early, seventh century, Anglo-Saxon charter relating to Chobham describes a boundary which runs 'to ðare greten wich' [to the great wych].[209] If this does mean elm, it is the earliest documentary evidence known.

In the thirteenth century the place name Wych Stumbel, possibly the antecedent of West Humble, is mentioned in Mickleham. At the same time a family named de la Wyche flourished at Woking. In the following century, the place name Ellsham, earlier Elmesham, in Banstead, is first recorded. The surname atte Wych was current at Bramley and at Arlington in Godalming, and there was a family called atte Wychegh' at Woking.[210] Wychmoor, in Elsted, thought to be derived from *wych*, is found in the fifteenth century. The parliamentary survey of Halliford Manor, in Shepperton and Sunbury parishes, during the Commonwealth enters elms.[211] Timber accounts in the same century attest the presence of elm at Guildford, Godalming and Bramley.[212] Nonsuch Palace, the royal retreat erected in the defunct parish of Caddington, had been demolished in the time of John Evelyn. His diary records the elms that survived it.[213] Elm water mains were laid in Guildford.[214]

There was a much frequented spa at Epsom. As so often, it was planted around with elm. Lime was also used.[215]

Most of the Wych Elm now growing in Surrey appears to have originated from amenity planting from the eighteenth century onwards. It may be native in shaded ravines in the Greensand hills, especially where the soil is calcareous.

Prehistoric settlement was in three main areas:

Fig. 132. Elms, St Annes Hill (Sr), by Henry Dawson.[204]

Fig. 133. Elm and pound, Hambledon (Sr). by William Henry Hunt.[207]

on the North Downs near Epsom, on the North Downs near Farnham, and in the Thames valley around Staines. The settlement pattern in Roman times was similar but with considerable extension around the same three foci.

The first land occupied by the Anglo-Saxons seems to have been the North Downs south-west of Croydon. It has been supposed that Surrey may have been the southern domain of the elusive Middle Saxons. It was eventually incorporated in the West Saxon kingdom.

The Surrey Weald seems to have known no agricultural settlement until late in the Anglo-Saxon period. Place names indicative of woodland and late settlement are mainly concentrated in the Weald, along the border with Kent, in the south-west corner of the county, and to the east of Woking and Guildford.

Reasons are assembled in the next chapter for supposing that English Elm must have reached the London area by Roman times. Assuming also that this elm originated in the west of England, it is likely that it reached London along the Thames valley. Probably, therefore, the Romano–British settlements in the parts of Surrey near the Thames would have received this elm concurrently. It is likely that the English Elm on the North Downs of Surrey was derived from the London area. The fact that Stane Street from London

Fig. 134. Lane near Dorking (Sr), by Dalziel Bros after Birket Foster.[207]

to Chichester runs through the North Downs settlements and the likelihood that it was the route whereby the English Elm reached the coastal plain of Sussex in Roman times rather suggests that English Elm was also introduced into the North Downs settlements at this time. The introduction of English Elm into most of the clay lowlands both north and south of the North Downs would have to await the expansion of Anglo-Saxon settlers into these areas in late pre-Conquest times.

Warwickshire[216]

The new Warwickshire corresponds with the old save for the loss of the north-west conurbation to West Midlands.

The principal upland is the limestone belt in the south. There is a minor elevation over the Coal Measures outcropping in the north-east part of the county. Most of the county is low lying, over red Triassic clays towards the north-west, and over Lias clay towards the south-east. Dunsmore Heath, in the east, is an area of glacial sands and gravels. The major rivers are the Avon and its southern tributary, the Leam. The two streams run respectively north and south of Dunsmore Heath.

English Elm is frequent throughout the county except its northern tip. It appears always to have been designated *elm*. The sporadic Wych Elm in the north was known as *wych hazel* in the seventeenth century. One suspects that this designation had come in from outside the county.[217]

The only area where elm is infrequent is in the north where region Z1 extends into the county. Some English Elm is present. There were elm avenues at Middleton Hall.

Practically the whole of the rest of the county is in region V1,[218] the main English Elm region. There is occasional Wych Elm in the north. Narrow-leaved Elm occurs sporadically along the eastern border. Lock's Elm is occasional in the north.

The River Leam, whence Royal Leamington Spa and Leamington Hastings, is usually derived from Celtic *lem-*, meaning elm.[219] English Elm is abundant today in the basin of the Leam. Narrow-leaved Elm also occurs and there is a little Wych Elm not far to the north. There was a Roman settlement round the Leam, though the county, in general, was sparsely inhabited then. It is conceivable that the river was named after the elm but the general uncertainty about *lem-* place

names prevents much weight being attached to the Warwickshire examples of them.

The place name Ilmington, first on record in the tenth century, has been derived from Anglo-Saxon *ilme* or *ylme*, believed to be variants of *elm*. Since there is no independent evidence that such words existed, no reliance can be put on this etymology. The place name Whichford, entered in Domesday Book, has been derived either from *wice*, meaning elm, or from the tribal name *Hwicce*. The early spellings have initial *w-*. Forms commencing *wh-* only appear in the thirteenth century. It would have been expected, had this name been derived from the Hwicce, that spellings with *wh-* would have been earlier. On the other hand, it is unusual, but not unknown, for words in *w-* to become aspirated to *wh-*. Moreover, there is no independent evidence that elm was ever known as *wych* in Warwickshire. With such uncertainty, this name too must be discarded from this discussion.

Firm ground is reached in the fourteenth century. The surname de Ulmo is encountered at Charlcote and atte Nelme at Copped Bevington in Salford Priors.[220] In the fifteenth century, the surname atte Elme was current in Stratford upon Avon.[221]

A sixteenth century survey of monastic estates makes incidental allusion to elms at Draycote and Kings Newnham.[222] The manorial court rolls of Wootton Wawen also mention it at this time.[223] The court rolls of Ladbroke refer to the great elm at the churchyard gate.[224] Very large numbers of elms are known to have been standing around Stratford upon Avon in this century.[225]

Negative evidence has always to be treated with caution. One can sometimes, however, make inferences from complete timber inventories. One such was made for the estate of John Archer in Tanworth *c.* 1500.[226] This mentioned alder, ash, aspen, oak and withy but no elm. Sometimes trees were mentioned but not identified further and so elm could have been covered by the general term, but it would be stretching coincidence to suppose that it was always entered as an unidentified tree. The survey suggests therefore that elm was not present in this part of the Forest of Arden at this time.

Important positive and negative evidence is provided by a complete inventory of trees at Chilvers Coton in the seventeenth century. There was no English Elm and only three trees of Wych Elm.[217] It would appear that V1 had not then reached its present north-east limit in the county.

There was a Parson's Elm at Hampton Lacy in the seventeenth century.[227] Specimen elms included Wightwick's Elm at Stivichall, felled in 1783,[228] and the Gospel Elm at Stratford upon Avon.[229] Warwick was provided with elm water mains. The elms at Rugby School are considered in chapter 10. Elm avenues existed at Charlecote Park,[230] Compton Verney, Coughton Court, Harvington Hall, Maxstoke Castle,[231] and along the road across Dunsmore Heath.

The remaining region, MV5, in which English Elm and Narrow-leaved Elm are both frequent, occupies a small part of eastern Warwickshire around Churchover. MV5 principally occurs in northern Northamptonshire and southern Leicestershire.

Wych Elm has a scattered distribution in Warwickshire today. Most is probably the result of eighteenth century and later planting. Its indigenous status over the county in general is doubtful.

The county was little settled in prehistoric times. Roman settlement was mainly along Fosse Way and between it and the River Avon. Anglo-Saxon penetration seems to have been down the Avon and Leam rivers. The headwaters of these communicate respectively with those of the Welland and Nene and so the Wash which was a major entry point for the invaders. Latterly Anglo-Saxon Warwickshire was divided between the Arosaetna in the south-west and Mercians to the north. The former were associated with the Hwicce of Hereford and Worcester. This division was preserved in the boundaries of the medieval dioceses of Worcester and Lichfield which partitioned Warwickshire in the way described.

North-west Warwickshire was settled only late in Anglo-Saxon times. This area was the Forest of Arden. Though it was never a royal forest, it was formerly well wooded and the infrequency of elm in the area even at the end of the Middle Ages has already been noted. In addition to the Forest of Arden, early place names suggest that there was wooded terrain and late settlement west of Nuneaton.

English Elm must certainly have been present in the south of the county by the fourteenth century. Reasons were advanced in the section on Hereford and Worcester for the supposition that English Elm had been introduced there in Roman times via the Fosse settlements in Gloucestershire. If this were so, it would be likely that English Elm were introduced into southern Warwickshire also, for, as already noted, the Fosse traverses this county.

North-east Warwickshire was lightly settled in Roman times. It seems necessary, however, to postulate that English Elm had been taken north along the Fosse before the end of the Roman occupation in order to explain its present distribution pattern in Leicestershire. Most of the elm in the Forest of Arden probably arrived in the post-Medieval period.

The Narrow-leaved Elm of eastern Warwickshire is clearly derived from further east. It is not likely to be earlier than the late Anglo-Saxon period. The scattered Lock's Elm would not be expected to have been established in Warwickshire for more than a few centuries.

West Sussex[232]

This county is the western half of the old county of Sussex. The chalk upland of the South Downs runs west to east across the south of the county. The other upland areas are Greensand hills to the north-west and the western end of the sandstone hills of the High Weald in the east. North of the Downs is the extensive clay plain of the Low Weald. South of the Downs is a coastal plain which widens from east to west. The major rivers are the Arun and Adur.

English Elm is frequent in the coastal plain, at the northern foot of the Downs and in the western end of the Low Weald. It appears to have been known earlier as *wych*, though the term *elm* had come into use by the sixteenth century. This was one of the counties where Wych Elm was distinguished as *wych hazel*.

There are two regions where elm is infrequent. One, Z3, covers the western part of the South Downs and the Greensand hills in the north-west corner of the county. Both English and Wych Elm occur at low frequency. The presence of some elm at Chilgrove in the fourteenth century is suggested by the surname atte Wych recorded at this time.[233] A claim by the copyholders of Duncton in the early eighth century that they had the right to the elms on their land seems to have been made in imitation of an earlier claim at Sutton, the adjoining parish to the south. There is not much elm at Duncton now.[234]

The other low frequency region, Z4, covers the High Weald and the adjoining parts of the Low Weald. Here also there is occasional English Elm and Wych Elm.[235] There was some elm formerly at Bolney, for elm pipes were being made there in the sixteenth century.[236] There was an elm avenue in Sedgwick Park, Horsham.[237]

Most of the rest of West Sussex lies in V1,[238] the main English Elm region. This covers the coastal plain, parts of the Downs and the low ground immediately to the north of them, and a corridor across the Low Weald clay linking the south Sussex population with that of Surrey. What may be the earliest evidence for the occurrence of elm is the place name Wiggonholt, earlier Wikeholt, first recorded in the twelfth century[239] and perhaps based on *wice*. The surname atte Wych, recorded at Oving in the coastal plain in the thirteenth century, and nearby at Rumboldswyke in Chichester in the next century,[233] probably indicates the presence of English Elm in the coastal plain.

In the early sixteenth century, indentures concerning land at Chichester specified that neither oak nor elm was to be cut.[240] A claim by the copyholders of Sutton that they had the right to the elms on their lands was put forward in the late seventeenth century.[234] Though the Lord of the Manor contested this, the claim probably indicates that elm was frequent. Elm was present at Oldberry in Boxgrove[241] and at Selsey[242] during the seventeenth century.

Among the more widely known elms of West Sussex were those planted behind the city walls of Chichester in the seventeenth and eighteenth centuries.[243]

Within region Z3 are two small enclaves with frequent elm. V8[244] where English Elm is common, is at the head of the Rother valley. This river is a tributary of the Arun. G10, with Wych Elm frequent, lies along the north flank of the Downs between Elstead and Bepton.

It seems that Wych Elm has persisted in the county since prehistoric times in favoured places, mainly the northern side of the South Downs.

Prehistoric settlement was mainly on the Downs and in the coastal plain. The land occupied in Roman times was much the same but good communication was established between London and the south coast by the construction of Stane Street. The south Saxons, in their turn, settled in the same area.

The Weald, mostly in region Z4, was penetrated relatively late, and not fully occupied till the Middle Ages. Place names indicative of woodland occur throughout it.

English Elm had certainly reached West Sussex by the Middle Ages. Its present area of distribution closely reflects the prehistoric settlement patterns that lasted till the Weald was finally occupied. There is little likelihood that the South Saxons were responsible for introducing it. Sussex, in the earlier part of the Anglo-Saxon period, seems to have been very much cut off from the rest of England. It is far more likely that the English Elm in the south of the county was introduced in the Roman period. The existence of a corridor of elm-rich country along Stane Street is consonant with this supposition. The relative infrequency of English Elm in Hampshire, and the forested nature of south-east Hampshire, tell against an introduction from that county. One is left with an introduction from the Thames valley via Stane Street.

Wiltshire[245]

Wiltshire has escaped all but very minor reshaping. The greater part of it is occupied by chalk upland, the principal part of which is the plateau of Salisbury Plain. This is separated from the Marlborough Downs by the clay lowland of the Vale of Pewsey. The Vale of Wardour, another clay depression, separates Salisbury Plain from the chalk downs that run into Dorset.

The limestone upland of the Cotswolds rises along the north-west border of the county. Between the outcrops of chalk and limestone is a clay vale of which the Vale of Pewsey is a branch. A minor ridge of Corallian Limestone, roughly parallel in its course to the chalk scarp, interrupts the lowland clay. There are discontinuous sandstone outcrops west of the chalk. Tertiary beds overlie the chalk to the south-east of Salisbury.

The Wiltshire Avon and its tributaries from east to west, Bourne, upper Avon, Wylye, Nadder and Ebble have cut deep valleys in the chalk. The Thames rises in the north-east corner of the county. The Bristol Avon drains much of the clay lowland.

English Elm is common throughout most of the county. Its earlier name seems to have been *wych*. Wych Elm was differentiated as *wych hazel*. The term *elm* had appeared by the thirteenth century in the north-west of the county.

There are four areas where elm is infrequent. The largest is region Z7. It occupies the part of Salisbury Plain to the immediate north and north-east of Salisbury, the lower Wylye valley, most of the Vale of Wardour, and the sandstone outcrop in the south-west. Both English Elm and Wych Elm occur occasionally. The place name Great Wishford, Wicheford in Domesday Book, appears to indicate the presence of some elm at the Conquest.[246] There was a large elm in Wylye village where the inhabitants foregathered in the seventeenth century.[247] Capability Brown spared the elm avenue at Longleat.

All the other areas with sparse elm are peripheral. Z3, which covers a wide area in Berkshire and Hampshire, extends a short distance over the eastern border. English Elm is occasional. The earlier presence of some elm is attested by the place name Witcha, formerly Wycheley, in Ramsbury parish, on record in the fourteenth century, and Elmdown, in the same parish, mentioned in the sixteenth century.[248] Capability Brown did not spare the elms that once lined Savernake Broad Walk. Z10 is centred around Cranborne in Dorset. It includes some of the villages in the upper course of the Ebble. Z11 covers the New Forest of Hampshire but extends a short distance over the part of the border nearest the forest.

The rest of the county, including the villages of the chalk upland and of the clay lowland, is in V1,[249] the principal English Elm region. Wych Elm also occurs rather infrequently but is widespread. The downland away from the villages of Salisbury Plain is treeless, but the villages themselves are sunk in the valleys of the rivers that thread the plateau, and along these English Elm is abundant. The contrast is described by Morgan Forster:

This is Nature's joke in Wiltshire – her one joke. You toil on windy slopes, and feel very primeval. You are miles from your fellows, and lo! a little valley full of elms and cottages.[250]

The archeological evidence for elm in this area goes back to the Bronze Age. Some of the Bronze Age burials were in hollowed-out elm trunks or on an elm plank.[251]

The presence of elm in the north of the county is attested by the surname atte Elme, current at Broad Hinton in the thirteenth century,[252] and by the place name Elmescrofte in Bradford on Avon. The skippet used for keeping documents at Devizes was turned from elm wood. It is dated to the fourteenth century and quite likely was of local material.[253] Elm board was certainly in local use at the time for building.[254] The place name Wycchesole in Monkton Farleigh, recorded in the fifteenth century, is thought to be based on *wych*.

In the sixteenth century, the name Witch Hill in Aldbourne may also be a derivative of *wych*. Elm is mentioned as present at Bushton[255] and North Bradley,[256] and elm board was being utilized at Salisbury.[257]

Michael Woodward, the Warden of New College, Oxford, visited the college property in Wiltshire in the seventeenth century and gave instructions about elm at Alton Barnes and Stert.[258] John Aubrey provides some valuable details. He noted that Wych Elm occurred wild in the hundred of Malmesbury and he related that a large Wych Elm at Donhead St Mary was used by bowyers in his youth. He also recorded that the storm which supposedly raged when Oliver Cromwell died blew down the large elm at Urchfont.[259] There was a London Elme at Swallowcliffe in this century. Other localities in which elm is recorded at this time are Brixton Deverill, Grittleton, Netherstreet and Sutton Benger.[260]

One of the most celebrated local elms was the Wych Elm called the Larmer Tree which stood in the Dorset border near Tollard Royal. It is further discussed in chapter 9. The Stock Tree was an elm which shaded the stocks at Rushall.

An early elm avenue at Longford Castle is mentioned in chapter 9. There were fine elm avenues at Corsham Court.[261] Francis Kilvert rhapsodized over the Wiltshire elmscapes, particularly those at Langley Burrell.[262] More recently, an equal enthusiasm for Wiltshire elms, differently expressed, has emanated from Heather Tanner.[263]

The persistence of Wych Elm in the county since Boreal times seems beyond doubt. It probably provided the elm wood for the Bronze Age burials. It is likely to have diminished in frequency however and it is now much less common than English Elm.

Wiltshire holds a big place in any discussion on the history of the English Elm. In chapter 3 it was argued that English Elm was introduced from Spain to England in the Later Bronze Age. It was also thought likely that Salisbury Plain was the place where it was first grown. At the present time, English Elm is most frequent on clay lowlands, terrain not settled in prehistoric times. The one landscape category where English Elm flourishes and that prehistoric man favoured is that of easily worked soils alongside rivers such as the Thames terraces and the banks of the rivers traversing Salisbury Plain. On this view, the valleys of the upper Avon and Bourne would have been the most likely original sites of English Elm in the British Isles.

The prehistoric settlements in the county were largely confined to the chalk uplands and this situation persisted until Roman times. There was some extension of settlement in Roman times along the roads that were constructed to link the Marlborough Downs area with Bath and with Cirencester.

Anglo-Saxon penetration was from the direction of the upper Thames Valley. Again, it was the chalk upland that was first occupied. The clay lowlands were only entered in the later stages of Anglo-Saxon expansion. Latterly, all of Wiltshire was the territory of the West Saxons.

There was much royal forest. The larger forests were all in zones where elm is infrequent. Selwood Forest, along the Somerset border, was a major barrier between Wiltshire and the settled country to the west. It was known to Celtic speakers as Coit Maur [Coed Mawr], the great wood. It was in region Z7, as was Clarendon Forest in the southern half of the county. Savernake and Chute Forests, along the borders with Berkshire and Hampshire, were largely in region Z3. The Wiltshire part of Z10 pertained to Cranborne Chase. The Wiltshire part of Z11 was a fringe of the New Forest of Hampshire. Only Braden Forest in the north and the small Pewsham, Chippenham and Melksham Forests in the west of the county had no regions with infrequent elm corresponding to them. Early place names indicate the presence of wooded terrain and presumably late settlement north and south of Bradford on Avon.

If the earlier presence of English Elm is accepted, some expansion of its area in Roman times would be expected. It certainly appears necessary to conclude, for reasons considered in earlier sections of this chapter, that it was introduced into the counties along Fosse Way and into the upper Thames valley before the end of the Roman period. However, the elmscape of the lowlands of north-west Wiltshire can hardly be other than a creation of Anglo-Saxon farmers in the later part of the pre-Conquest period.

14
Elm in the eastern counties

For the purpose of this chapter, eastern England comprises the counties of Bedfordshire, Cambridgeshire, East Sussex, Essex, Greater London, Hertfordshire, Kent, Leicestershire, Norfolk, Northamptonshire and Suffolk. The elm flora of this part of England is of far greater complexity than of those considered in the preceding two chapters. Narrow-leaved Elm is the major component of the elmscape over the greater part of this area. In contrast to English Elm, which appears to be descended from a single introduction and which exhibits only minor variability, Narrow-leaved Elm has been introduced from a number of different Continental sources and shows much variation.

As in the previous chapter, it is necessary to investigate the history of human settlement in each county since both the Narrow-leaved Elms and what English Elm is present owe their presence to human activity. Topography, again, is mainly important as having determined the sites of human occupation, rather than as directly affecting the elms.

Wych Elm occurs only in relatively few localities, usually on higher ground and where the soil is calcareous. It is likely to have survived in such localities since Boreal times. Elsewhere it probably disappeared at the time of the Elm Decline. In Essex and in Northamptonshire and the adjacent parts of Bedfordshire, there are extensive stands of Hybrid Elms, presumably derived from crossing between indigenous Wych Elm and introduced Narrow-leaved Elm.

Bedfordshire[1]

Except for minor rectifications, Bedfordshire has retained its former boundaries.

The principal upland is the chalk outcrop of the Chilterns, which runs from north-east to south-west across the southern part of the county. Greensand hills occupy the centre.

In the north is a clay plateau, largely Boulder Clay. There is a clay vale between the Chilterns and the Greensand. The Great Ouse follows a tortuous course in the northern part of the county and has cut down to the underlying limestone. Its tributary the Ivel threads the Greensand hills.

English Elm is frequent in the south of the county. Narrow-leaved Elm is common in the north. The original name for both in early times was probably *wych*. A variant of this, *wick*, may also have been current. By the sixteenth century, the term *elm* had come into use.

There are few areas with little elm. Z3 covers the Bedfordshire Chilterns. It extends also into several other counties in south-east England. Wych Elm and English Elm occur occasionally. It is probably the

Fig. 135. Elm regions of the eastern counties. Diagonal shading, Wych Elm; horizontal shading, English Elm; vertical shading, other Field Elms; crosses, Hybrid Elms.

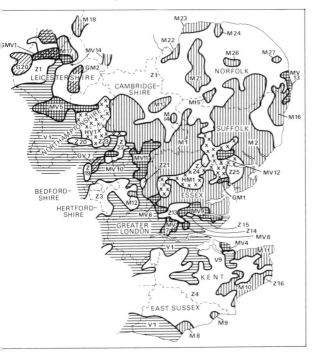

former that is referred to in the sixteenth century manorial court rolls of Caddington.

Z22 is a region with infrequent elm in the centre of the county, mainly over the Greensand. English, Narrow-leaved and Wych Elms all occur sporadically. The last is likely to be the kind of elm referred to in a fourteenth century *de banco* roll concerning an action for waste involving the felling of six thousand elms in Sheerhatch Wood in Willington. Elm is also recorded in Ampthill Parks and at Blunham in the sixteenth century.

In the seventeenth century, elm board was being sold at Sandy and *wich elme* was growing in Northill churchyard. This latter designation may mean Wych Elm as defined in this book. There was an avenue of elms at Avenue House, Ampthill. Several of the recent plantations of elm in this region, in particular at Roxton and Tempsford, are known to have been made from material supplied by Wood and Ingram's nursery at Brampton (Ca).

The other two regions with little elm are Z19, on the Greensand hills along the Buckinghamshire border, and Z23, at the centre of the Boulder Clay plateau in the north.

V1,[2] the principal region of English Elm, sends two tongues into Bedfordshire. One comes up the clay vale between the Chilterns and the Greensand hills. The other follows the valley of the Great Ouse from northern Buckinghamshire. Wych Elm and Narrow-leaved Elm occur occasionally in the Bedfordshire part of this region.

The earlier presence of elm in the first tongue is probably attested by the place name Wychhull at Luton.[3] This might be either Wych Elm or English Elm. It can only be the latter that is concerned in the churchwardens' purchase of elm board at Shillington in the sixteenth century.[4] There is a particularly full and interesting survey of timber standing in Barton in the Clay in the early years of the next century. Almost every tree mentioned is elm. English Elm is abundant around the village today.

The presence of elm, probably English Elm, in the second tongue of region V1, is first attested by the elm piles that supported old Bedford Bridge. This is believed to have been built in late Anglo-Saxon or early post-Conquest times. The surname atte Wiche current at Stagsden in the fourteenth century[5] could also denote English Elm. The sixteenth century manorial court rolls at Biddenham included an order to tenants

to plant 'vi ashes, okes, elmes, willowes or wyches'. Obviously two kinds of elm were being distinguished. The term *elm* must refer to English Elm. It is probable that *wych* here denotes the Hybrid Elm of region HV1 described below.

Elm was said to grow naturally in woods about Wootton in the seventeenth century.[6] This would probably be Wych Elm. In the same century extensive purchases of elm board were made in the building of Hawbury Hall, Renhold. The abundance of elm at this time around Bedford is illustrated by the report of the gale of 1607,[7] which brought down 60 elms in its environs.

The water mains at Bedford were formerly of elm. Avenues of elms led to the Grange and to the Bury in Kempston, and stood near Elstow church. They were a major feature of the formal layout of Wrest Park, Silsoe.[8]

Narrow-leaved Elm is common in the extreme north of the county. This is in region M1,[9] the large area of high frequency of this elm, centred on Cambridgeshire.

English Elm and Wych Elm coexist at high frequency in region GV7[10] which lies over the Greensand hills around Woburn. Some Narrow-leaved Elm is present too. Land in Husborne Crawley is described in a seventeenth century rental as well wooded with elm and ash.[11] Perhaps Wych Elm is implied. There was an elm avenue in Woburn Park.

English Elm and Wych Elm are both frequent in MV10,[12] a small region around Wilshamstead. There is some Narrow-leaved Elm here also. An elm avenue led to Houghton House in Houghton Conquist.

The north-west of Bedfordshire lies in region HV1 which is mostly in northern Northamptonshire. English Elm and Hybrid Elm are frequent. Elm was felled for building purposes at Harrold in the seventeenth century but the kind cannot be determined. There was an elm avenue at Bramham Park.

Wych Elm appears to be native in the Chilterns, on various sites on the Greensand, and on the limestone exposures in the north. It may well have persisted in all these sites since Boreal times.

Prehistoric settlement was mainly along the scarp face of the Chilterns and on the river gravels of the Ouse around Bedford. In Roman times there was some extension of settlement downstream along the Ouse. The first Anglo-Saxon incomers occupied much the same ground as their predecessors. Bedfordshire is one of the few counties where several of the Middle Anglian tribes listed in the Tribal Hidage can be located with confidence. The Cilternsaetan took over the Chilterns south-west of Hitchin. The Gifle went into the Ivel valley.

There are two areas where early place names indicate wooded terrain and late settlement. One is in the northern Boulder Clay region and corresponds in part to region Z23. The other runs down the border with Buckinghamshire.

The English Elm in the tongue of V1 that runs along the foot of the Chilterns had certainly arrived by the end of the Middle Ages. It is largely on the lowland clay that was only occupied late in the Anglo-Saxon period. This would seem to be the earliest date at which its presence would be likely. The elm would have been presumably derived from similar terrain further along the foot of the Chilterns and ultimately from the Thames valley. English Elm reaches its present limit of continuous distribution just over the Cambridgeshire border.

The English Elm around Bedford poses a problem. What was presumably English Elm had arrived at Bedford in time to provide piles for Bedford Bridge. In view of the apparent dearth of early settlement west of the town, it seems unlikely that northern Buckinghamshire would have been the source. It seems more likely that Bedford was a separate centre of introduction, with English Elm brought from a distance in late Anglo-Saxon times. The Thames valley, again, would be a likely source.

The contact zone between English Elm and Narrow-leaved Elm runs down the west side of the border between Bedfordshire and Cambridgeshire. Contact between the two kinds of elms in this area of late settlement could hardly have been made before late Anglo-Saxon times. It is unlikely to have been made long after the Norman Conquest since elm, as already noted, was present at Bedford about this time and Bedford is not far from the contact zone. The Hybrid Elms of region HV1 are discussed in the Northamptonshire section.

Cambridgeshire[13]

The new Cambridgeshire is an agglomeration of four earlier units: the former counties of Cambridge-shire, Huntingdonshire, the Isle of Ely and the Soke of Peterborough.

The principal topographical distinction is be-tween the Fenland in the northern half of the county and what to the fenlander are the highlands to the south and west. These culminate in the chalk outcrop in the south-east of the county. They also include a great deal of clay at lesser elevation, especially in the west and south-west. Limestone outcrops in the north-west and there is a little Greensand exposed along the Bedfordshire border. A small part of the Breckland sands lie along the Suffolk border.

There are a number of islands of higher ground in the Fenland. The largest is the Island of Ely (to be distinguished from the former administrative area termed the Isle of Ely). The principal minor islands are those on which Chatteris, Doddington and March, Thorney, and Whittlesey are situated. The Fenland includes the Black Fen over peat, in the south, and the Brown Fen, over silt, in the north. All the main rivers drain into the Wash. Their courses have varied over the centuries. They are, from east to west, the Great Ouse, with its tributary the Cam, the Nene, and along the north-west boundary, the Welland.

Narrow-leaved Elm is found in most parts of the county outside the Fenland. English Elm is frequent in the south-west. The name for elm over most of the county was *wych*. It had begun to be replaced by *elm* in the sixteenth century.

The Fenland, the minor Fen islands, the lime-stone area in the north-west, much of the chalk scarp face and the Cambridgeshire Breckland are in region Z1, where elm is infrequent. What elm occurs is mainly Narrow-leaved Elm.[14] A little Wych Elm is present on the north-west limestone.

The place name Elm occurs in this area, first recorded in the tenth century. It is not likely to refer to the tree. Today there is little elm at Elm other than obviously recent plantings, nor anywhere near. Some of the early forms of the place name are anomalous, and one variant, Eolum, might be a tribal designation.[15] The next place name to require consideration is Elmeney in Waterbeach parish, recorded in the twelfth century. There is a little elm here now, and the local designation in this part of the county at the end of the Middle Ages was *wych*, not *elm*. It is safer to ignore this name too.

One can be confident that the *wych* in fifteenth century Peterborough accounts was elm.[16] In the sixteenth century there was elm in the 'grovettes' of Denny Abbey,[17] and piles of *witch* were used as boundary markers in Whittlesey Mere.[18] In Yaxley, on the edge of the Fenland, *wyches* were recorded as present in the following century.

The references just cited concerned elms in or near the Fenland. On higher ground, unauthorized felling of *wiches* is entered in the fourteenth century manorial court rolls of Elton. A *wych* was a landmark in Bottisham in the next century.

Most of the higher ground in the county is in region MI[19] which extends also into Hertfordshire, Essex and Suffolk. Narrow-leaved Elm is frequent, and there is some English Elm in the south. Two kinds of Narrow-leaved Elm need to be distinguished. West of Cambridge is an elm with dark green small-toothed leaves. East of Cambridge is an elm with light green, large-toothed leaves. There are intermediates along the contact zone.

It has been suggested that the place name Lymage in Great Staughton parish is derived from Celtic **lem-*, meaning elm. There is much elm at Great Staughton village but not much at Lymage today. So, as often elsewhere, this etymology is suspect. Olmstead, in Castle Camps, of which Elmstede is an early variant, is recorded from the thirteenth century. Possibly it is based on *elm* or a variant of it. It is near the Suffolk border over which *elm* was in early currency.

The felling of *wyches* at Landbeach is recorded in the fifteenth century.[17] Elm was certainly growing at Cambridge in the sixteenth century[20] and at Brampton in the seventh century. Particular interest attaches to a note dated 1637 in the parish register of Wendy. It states that

> the timber trees with other witch and elm on ye outside of ye church yard which lys south and west belong to the vicar.[21]

The implication is that two kinds of elms were being distinguished. The more common Narrow-leaved Elm would be *wych* and the English Elm present at low frequency in Wendy would be *elm*.

The elmscape around the town of Cambridge was famous. Literary allusions to it are noted in chapter 10. Elm was a conspicuous component of the academic groves of Jesus College and Queens' College and of the Wilderness of St John's College, designed by Capability Brown. It was the sole component of St Catharine's College grove, felled in 1922. The dominating feature of the Cambridge Backs was the long elm grove that stretched from the backs of St John's College to those of King's. Many of the college walks and academic walks that ran out into the western fields were of elm. The Three Sisters of St John's College were well-known specimen elms. Land belonging to the corporation was also planted with elms, as, for example, Christ's Pieces and Parker's Piece. Various kinds of elms were planted. There was Narrow-leaved Elm on Christ's Pieces, English Elm along the Backs, and Huntingdon Elm around Parker's Piece. The former Brooklands Avenue was one of the earliest to be made of Huntingdon Elm.

Outside the town of Cambridge there were other specimen elms of note. The Bird Tree at Fulbourn[22] and the Cruel Tree in Buckden had, respectively, amatory and sinister associations. Great Gransden was distinguished by several 'corner elms'.

Capability Brown destroyed the north avenue of Wimpole Hall but left the imposing south avenue, perhaps the finest English example of an avenue of Narrow-leaved Elm.[23] There were lesser elm avenues at Horseheath Hall and at Shingay. The holy well in Holywell churchyard was under a large Narrow-leaved Elm.

Much of the later amenity planting in the county was from material supplied by the nursery of Wood and Ingram (now Laxton Bros.) at Brampton. This was the nursery responsible for the introduction of the Huntingdon Elm.

Region MI3[24] corresponds to the Island of Ely. The elms in the western part are the form of Narrow-leaved Elm with small-toothed leaves. Those in the eastern part have leaves with large teeth. The early presence of elm is well attested by two place names based on *wice*, Witcham, first recorded in the tenth century, and Witchford, recorded in the eleventh.

The small region MI4[25] comprises Soham and Wicken. The latter name, recorded from the thirteenth century, is usually derived from *wic*, meaning dairy farm. It could also come from the dative plural *wicum* of *wice*, meaning elm, which would become *wicken* in Middle English. The place name Ashen in Essex has been so derived. Wicken is quite close to Witcham and Witchford, mentioned above.

VI, the main English Elm region just reaches the county in the south-western parish of Gamlingay.

Most of south-west Cambridgeshire is in region MVII,[26] which extends some distance south into Hertfordshire. Narrow-leaved Elm and English Elm are

85, 86

136

52

both frequent. The presence of *wych* in the fourteenth century is attested by the manorial court rolls of both Sawston and Whittlesford. There was a famous

137   English Elm outside Melbourn church, traditionally associated with the gathering of the villagers to protest against the levying of ship money.[27] An elm avenue led to Pickering Manor at Whaddon. The spring that gave its name to Orwell originally gushed from a group of elms near the church. It ceased to flow as a result of coprolite digging.[28]

Region HVI, which is mainly in Bedfordshire and Northamptonshire, crosses the county border at Keyston and in the neighbourhood of St Neots. Hybrid

Elm and English Elm are both frequent. Elm of some kind is known to have been present at Keyston in the sixteenth century.[29]

A special feature of southern Cambridgeshire is the presence of village groves, mostly or entirely of elm. They are usually called *grove*, occasionally *spinney*. They are found in or near the centre of the parish in marked contrast to the indigenous oak-ash woods that normally lie on the perimeter. They occur in region MI, as at Ball's Grove, Grantchester, and the various groves at Boxworth, in MI3, as in the Great Spinney, Sutton, and at Mepal, and in MVII, as at Shepreth. They are discussed in chapter 5.

Fig. 136. Wimpole avenue (Ca).[23]

Wych Elm is rarely seen in the county except in recent plantings. Possibly some survived from earlier times on the limestone terrain in the north-west.

Prehistoric settlement in the county was at three principal sites: the Cam valley and that of its tributary the Snail, the valley of the Great Ouse around St Ives, and the valley of the Nene east of Peterborough.

Settlement extended greatly in Roman times. There was denser occupation of the Cam valley. The Great Ouse was settled along most of its course in the county outside the Fenland, and so was the Nene. Land was occupied along Ermine Street between Godmanchester and Water Newton. There was much settlement in the Fenland itself, particularly over the silt and along the Car Dyke Roman canal.

The Fenland sites suffered marine inundation at the end of the Roman period, and if elm had been introduced, it can hardly have survived.

Early Anglo-Saxon settlement was along the Cam and its branches, the Great Ouse valley between St Ives and St Neots, and the Nene valley around Peterborough. The Island of Ely was also entered. There was only one small royal forest, Sapley Forest, just north of Huntingdon. There are early place names indicative of woodland cover and presumably of late settlement west of Cambridge, south-east of Newmarket, along the Bedfordshire border south of St Neots, along the western margin of the Fenland, and west of Huntingdon. The last area was called Bromswold, a name perpetuated in Leighton Bromswold.

The historical evidence certainly suggests that the existing pattern of elm distribution had become established by the Norman Conquest. The two forms of Narrow-leaved Elm are centred, respectively, in the Cam valley but extending far eastward into East Anglia, and in the Great Ouse valley, though reaching

Fig. 137. Melbourn Elm (Ca), by Richard Bankes Harraden.[27]

almost to the town of Cambridge. Reasons were presented in chapter 3 for supposing that these two populations represent introductions of Later Bronze Age date. The existence of different forms of Narrow-leaved Elm in the Cam and Great Ouse valleys suggests introduction of elm via the Wash at a time when local communication was poor, that is, before the Roman occupation.

The expansion of the two elm populations into previously unoccupied wooded terrain would have had to await the Anglo-Saxon penetration of these areas in the later Anglo-Saxon period. It is noteworthy that the wooded land west of Cambridge has elms of the Great Ouse valley form. Presumably penetration of this part of Cambridgeshire was from the west, not from the east.

The Island of Ely has a mixed elm population, the Great Ouse valley form in the west, the Cam valley form in the east. This suggests that some time before the Norman Conquest, elms were introduced from two separate pre-existing sources, a western source for the Great Ouse valley form and an eastern source for the Cam valley form.

English Elm had arrived in south-west Cambridgeshire by the end of the Middle Ages, presumably from the land at the foot of the chalk scarp face in Hertfordshire. The fact that the contact zone between English Elm and Narrow-leaved Elm is so close to the putative centres of introduction of the latter suggests that contact must have been established early, no later at least than Roman times.

East Sussex[30]

This county is the eastern half of the old county of Sussex. It reproduces many of the topographical features of West Sussex. The chalk upland of the South Downs runs from west to east across the county, meeting the sea at Beachy Head. The sandstone upland of the High Weald, to the north of the Downs, meets the sea between Hastings and Rye. The clay plain of the Low Weald separates the Downs from the High Weald. There are two coastal marshes, the Pevensey Levels and the Rother Levels, an extension of Romney Marsh. The main rivers are the Sussex Ouse and the Rother, which provides part of the boundary with Kent.

English Elm is frequent around the villages on and at the foot of the South Downs. Narrow-leaved Elm is frequent on the Pevensey Levels and to the immediate west of Rye. Elm seems to have been known earlier as *wych*. The designation *elm* had come in by the sixteenth century. It was sometimes replaced by the term *elvin*, which is discussed in the Kent section.

The High Weald and adjacent parts of the Low Weald are in region Z4 where elm is infrequent. English Elm occurs rarely, Wych Elm also rarely. It is just possible that the latter elm is responsible for the place name Holywych in Hartfield,[31] recorded in the thirteenth century. Hastings used elm for its water mains.[32] The elm timber purchased for the repairs at Ashburnham in the seventeenth century was probably English Elm[33] and may have come from region V1 to the south.

English Elm is common, as stated above, in the villages in the South Downs and along the foot of the scarp. This area is the south-east limit of region V1.[34] References to felling elms are entered in the sixteenth century manorial court rolls of Preston[35] and Jevington.[36] Similar entries occur in the court rolls of Eastbourne Burton and Folkington in the next century.[37] The felling of *eleven* [*elvin*] trees at Lewes is recorded in the seventeenth century.[38]

In the landscaping of the immediate surroundings of the Brighton Pavilion, an elm Promenade Grove had an important role,[39] recalling the association of Indian architecture and elm at Sezincote (Gl).

M8[40] is a region with frequent Narrow-leaved Elm covering the Pevensey Levels. A Commonwealth parliamentary survey attests its presence at Pevensey in the seventeenth century.[41] Narrow-leaved Elm is also frequent west of Rye in region M9. It may have supplied the elm timber for the water pipes laid down in Rye.[42]

Wych Elm is a rare tree in East Sussex. If it did survive anywhere from earlier times, sheltered ravines in the High Weald would be the most likely sites.

Prehistoric settlement was on the chalk downs and by the sea at Pevensey and at Hastings. The settlement pattern in Roman times was largely the same. When the South Saxons first replaced the earlier inhabitants, they too occupied the same land.

The Weald, in region Z4 where elm is rare, lay outside the zone of agricultural settlement till the Middle Ages. Many early place names confirm its wooded state.

The existing pattern of elm distribution was certainly established by the end of the Middle Ages. It reflects very exactly the prehistoric settlement pattern that persisted till the Weald was entered.

It was suggested in the previous chapter that English Elm had been introduced into the West Sussex coastal plain via Stane Street in Roman times. From there it would then be taken eastward along the downs. It failed to pass Eastbourne. This suggests that another elm was already present on the Pevensey Levels. This would be the Narrow-leaved Elm there now, which is of a type still to be found in northern France. The boundaries of M8 correspond with those of the Iron Age coastal site at this location. It was suggested in chapter 3 that the Iron Age was the most likely date of introduction of elm here.

Essex[43]

The only recent change in the boundaries of Essex has been the detachment of its urbanized south-west part for the benefit of Greater London.

The county lacks striking diversity in relief or soil. It is low lying, much of it clay, Boulder Clay in the north, London Clay to the south. Chalk outcrops in the north-west and in a strip just north of the Thames. There is extensive marshland along the Thames and around the North Sea creeks. The Thames is the southern boundary into which run the Roding and the Lea. The Lea, and further north its tributary the Stort, provide part of the western boundary. The Stour is the boundary with Suffolk. Other rivers running into the North Sea are the Colne, Blackwater and Crouch. The Blackwater, known in its higher course as the Pant, has two major tributaries, the Chelmer and Brain.

Essex is holy ground for the elm systematist. Nowhere else in England, nowhere else in Europe, is so complex an assemblage of elms to be found. It has provided the main key for unravelling the intricate connections between elm variation and the history of human settlement. English Elm is mainly found in the south of the county. Narrow-leaved Elm of various sorts and Hybrid Elms occur in the centre and north.

The designation *elm* is of ancient occurrence in the east. Elsewhere the usual term seems to have been *wych*. As usual, this was progressively displaced by *elm* from at least the fifteenth century. Field names such as Wick Hazel in Belchamp Otten suggest that there may have been a variant *wick* for *wych*.[44] If this were so, some of the numerous field names with *wick* may refer to elm.

Though elm is so abundant throughout Essex, there are some areas where it is rare. Z13 is one such region in south-west Essex, centred on Chipping Ongar. English Elm and Narrow-leaved Elm occur at low frequency. There were sufficient elms to overhang the highway at Widford in the sixteenth century.[45] The seventeenth century manorial court rolls of Dodding-hurst include an injunction against felling an elm windbreak, and one of the seventeenth century boundary elms of Waltham Forest was within this region.[46]

Z14 is a small region with little elm between Billericay and Wickford. Z15 is another such region on the higher ground north of Rayleigh. There is some English Elm in the former and both English Elm and Narrow-leaved Elm in the latter.

Two of the areas with little elm are in the north. Z24 lies within the bend of the River Stour west of Sudbury. Z25 occupies the western half of the Tendring peninsula between the mouths of the Stour and Colne. In each of these there is occasional Narrow-leaved Elm and Hybrid Elms. Some elm was presumably present at Elmstead in early times, this name being recorded in Domesday Book.[41] Elm board was being bought at Wivinghoe in the sixteenth century.[48]

V1,[49] the main English Elm region, extends into Essex in a band stretching across the south of the county from the Greater London border to Foulness. It also enters the west of the county around Waltham Abbey. A little Narrow-leaved Elm is present in the eastern part of the Essex section of V1. The place name Wickford is first recorded in the tenth century. It is completely indeterminate whether it is derived from *wic*, with some such meaning as dairy farm, or from *wice*, meaning elm. The spelling Wichford is found in the thirteenth century. There are elm bridge piles at Waltham Abbey of fourteenth century age,[50] and elm piles were driven at this time also under the king's chamber in Hadleigh Castle.[51] Elm is known to have been present at Billericay, Prittlewell and Shopland in the sixteenth century and at Belhouse Park, Aveley, in the seventeenth century.[52]

Adam's Elm, otherwise known as Allen's Elm or Ellen's Elm, was a landmark near Leigh.[53] Another, much illustrated, tree was the large English Elm in the churchyard of Waltham Abbey,[54] mentioned further in chapter 11. The importance of elm in the landscape of southern Essex is brought out in the description of

138, 139

Fig. 138. Waltham Abbey (Ex), by John Varley.[54]

Fig. 139. Waltham Abbey (Ex), by J. Rogers after George Campion.[54]

elm branch patterns by Glyn Morgan, quoted in chapter 10.

The larger of the two areas in which Narrow-leaved Elm is frequent is M2.[55] This occupies a wide area in Suffolk. In Essex it extends southward in a broad band behind the North Sea coast to the Crouch estuary. It also runs far inland behind the Blackwater estuary. Several sorts of Narrow-leaved Elm are distinguishable within this area.

An early place name in this area is Tey ad Ulmos, the earlier form of Marks Tey. This is on record in the thirteenth century. In the next century a number of references to elm occur. There was an Elmefelde at one or other of the Braxteds. The manorial court rolls of Great Waltham enter the felling of elms,[56] as do those of East Hall manor in East Mersea.[57] There was a Wichefelde at Feering, and the surname atte Elme was current at Great Clacton. In the fifteenth century, a wichhegs [wych hedge] features in the manorial court rolls of Great Canfield.[58]

The celebrated Dovercourt beetles, made from the elms growing around Dovercourt, are described in Holinshed's Chronicles, published in 1577.[59] A later writer describes the elms as 'strong, knurly and knotted and crooked'.[60]

Amongst the localities not yet mentioned, Coggeshall had elm growing locally in the sixteenth century,[61] and there was an Elmefield in Witham. In the seventeenth century a perambulation of the parish of White Roding mentions a number of elms on the boundary.[62] They also grew at Stanford Rivers.

The Marsh Tree was an elm serving as a nautical landmark at Jay Wick in Great Clacton.[63] It was blown down in 1935. There was an elm avenue flanking a canal in front of Boreham House,[64] while diagonally planted elm avenues led to Langleys in Great Waltham.[65] Elm pipes were in former use at Chelmsford.[66]

Region M1,[67] which occupies an extensive area in Cambridgeshire, Hertfordshire and Suffolk, is the smaller of two areas where Narrow-leaved Elm is frequent. It covers north-west Essex. There is a little English Elm along the Hertfordshire border. The place name Elmdon is in Domesday Book, but there is some doubt about it. The usual word for elm in this area was wych, and the variant Edelmedon, recorded in the thirteenth century suggests a possible derivation from the personal name Aeðelmaer. Elmdon is best discounted as evidence for elm.

Less uncertainty attaches to a wegetrou [wych tree] growing at Hempstead in the fourteenth century. In the following century, at Clavering, both a Wychecroft and an Elmestret are recorded. Though elm had clearly entered local speech, wych continued in use. There was a Wycheffelde, for instance, at Stansted Mountfichet in the sixteenth century. Amongst specimen elms was the High Tree at Hempstead.[68] Audley End, near Saffron Walden, once had its elm avenue.

Wych Elm is infrequent today over most of Essex. It is frequent, together with Narrow-leaved Elm, in the small region GM1,[69] south of Colchester.

Along the contact zones between region V1 and the areas where Narrow-leaved Elm is frequent, are four regions where English Elm and Narrow-leaved Elm are both frequent.

MV7,[70] in the south-west of the county, is mainly in Greater London. It is discussed in the section covering that county. MV8[71] is a small region covering Laindon and Langdon Hills. There is some Wych Elm too, in Langdon Hills. MV9[72] is a larger region on either side of the Crouch. The thirteenth century elm coffin at Danbury is described in chapter 8. Elm of some kind was present in Danbury Park in the sixteenth century.[73]

Graces, also in Danbury, had an elm avenue. MV6[74] is on the north side of Epping Forest. One of the seventeenth century boundary elms of the forest was in this region.[46] The Golden Elm was a Narrow-leaved Elm with yellow leaves at Little Parndon.[75]

Running across Essex from the Hertfordshire border between Bishops Stortford and Harlow to the Suffolk border along the lower Stour, is a broad zone of mixed Narrow-leaved Elm and Hybrid Elms.[76] This is region HM1. The earliest evidence of elm in this area is provided by the timber used in the Romano–British pile dwelling at Skitt's Hill, Braintree.[77] This was mostly of oak, but there was some elm. Wych Elm, Narrow-leaved Elm or, less likely, Hybrid Elm, are possible sources. Elm grew at Colchester in the twelfth–thirteenth centuries as evidenced by the place name Elmsholt and some cleftelmtrees functioning as boundary markers.[78] A Tudor survey of timber on chantry lands indicates the presence of elm at Sheering and Stebbing. About this time, too, there was a Wychefielde in Hatfield Broad Oak. In the following century, elm timber was being sold at Cressing.[79]

Colchester had water mains of elm.[80] An elm

67

avenue led to Great Saling church and hall, but this place is more renowned for one of the largest elms in the county, a Hybrid Elm on Great Saling Green.[81]

Though there is little Wych Elm in Essex now, the large zone with frequent Hybrid Elm, region HM1, suggests that Wych Elm was once frequent but has been largely hybridized out of existence.

In view of the very complex pattern of elm variation and distribution, the history of human settlement is of special importance. The county was widely but thinly settled in early times. In the Iron Age, most of the settlement sites were either on the North Sea coast or along rivers draining into the North Sea. They included the Tendring peninsula, the middle reaches of the Stour, the area around Colchester, the Colne valley, the valleys of the Brain, Blackwater and Chelmer, draining into the Blackwater estuary, the Crouch estuary, the land behind Shoeburyness, the Thames plain between Dagenham and Grays, the middle Lea valley and the country nearby around Epping, and the Stort valley. All these settlements would be the result of entry from the approach to the Thames estuary. An early settlement site of different antecedents was that along the Cam valley in north-west Essex. This would be an outlier of settlement from the Wash. Salt production in the coastal marshes was actively pursued both before and after the Roman conquest. Its sites are marked by the so-called red hills.

The Roman settlement plan was similar to the foregoing but with some extension. The Dengie peninsula was entered, and there were extensive settlements on the Thames plain around Mucking.

With the coming of the Anglo-Saxons the picture changed dramatically. It has long been remarked how sparse are the early Anglo-Saxon sites in the Thames estuary. It seems that the invaders gave London and its surroundings a wide berth. The late persistence of the Romano–Britons in the area is feasible. When the Anglo-Saxons did settle, the earliest sites were mainly alongside the Thames, as at Shoeburyness, Mucking, and the stretch between Barking and Tilbury. The other early major settlement site was the upper Cam.

When Essex emerges into the light of modern history, it is as the kingdom of the East Saxons. By this time the Middle Saxons had disappeared. The recent county of Middlesex presumably represented part of their domain, but it is a comparatively late admini-

strative unit. It is possible that the medieval arch-deaconry of Middlesex may represent a more ancient arrangement. It comprised not only the county of Middlesex, excluding London, but also Hertfordshire east of Ermine Street and a broad band of country in Essex running south-west to north-east till it reached the Suffolk border between Steeple Bumpstead and Bures. The rest of Essex was partitioned between the archdeaconries of Essex and Colchester. The former comprised the country south of the archdeaconry of Middlesex to the west, and south of the Blackwater estuary to the east. The rest of the county was in the archdeaconry of Colchester. It was in two detached portions, on either side of the Archdeaconry of Middlesex, an arrangement which the *Victoria County History of Essex* describes as singular, but makes no attempt to explain.[82]

Much of interior Essex was royal forest. In the late Middle Ages this had shrunk to the large Waltham Forest, now Epping Forest, in the south-west, and the smaller forests of Hatfield, Writtle and Kingswood. The first three might represent remnants of a former continuous forest area. Early place names confirm the wooded nature of the Waltham Forest area. They also indicate the presence of wooded terrain at the head of the Pant and Brain rivers, in the interior of the Tendring peninsula, and south of the Crouch estuary. These areas were presumably settled late. Several correspond to regions of low frequency of elm. Waltham Forest is partly in Z13. The interior of the Tendring peninsula is in Z25. The area south of the Crouch estuary is in Z15.

The historical evidence assembled above suggests that the present distribution of the various kinds of elm had become established by the Norman Conquest. It was argued in chapter 3 that the pattern of distribution of Narrow-leaved Elm in Essex, a series of small populations mostly on maritime or riverain sites, and strongly differentiated from one another, mirrors a period when the settlement pattern was of a number of sites located so, and between which communication was poor. The source of these elms has been traced to northern France. The latest period that answers this specification is the Iron Age which was in contact with northern France. It seems most likely therefore that the Narrow-leaved Elm in the Essex part of region M2 originated from a series of introductions made in Iron Age times. It is also suggested in chapter 3 that there may have been a connection between elm introduction

140

and salt production. It seems that there may have been as many as five separate sites of introduction, namely the Tendring peninsula, the neighbourhood of Colchester, the Dengie peninsula, the Roding valley and the Lea valley.

Much of the Narrow-leaved Elm in region M1 of north-west Essex is of a different sort from that in the areas just considered. Its affinities are with the elms with large-toothed leaves in western Cambridgeshire and Suffolk. This population is believed to be of Later Bronze Age date and to have been introduced into the Cam valley via the Wash from central Europe. Its movement up the Cam into Essex would have been very early.

If this general picture be correct, there would have been five small populations of Narrow-leaved Elm of French origin and one in the north of the county of central European origin at the beginning of the Roman occupation. It is suggested in the section on Greater London that the Romans were responsible for bringing English Elm from the west of England to London. Probably it was at this time too that it was introduced into the Thames plain of south Essex. The absence of Narrow-leaved Elm from all but localized

areas in this part of Essex suggests that another elm had been introduced comparatively early.

The interior of Essex was penetrated relatively late in the Anglo-Saxon period. Much of central Essex is occupied today by region HM1 with its mixture of Narrow-leaved Elm and hybrids between it and Wych Elm. It is possible that hybridization took place when settlers brought Narrow-leaved Elm from the perimeter of the county into the interior, where it hybridized with native Wych Elm. It is curious how closely region HM1 *140* corresponds with the Essex part of the archdeaconry of Middlesex. If this is not mere coincidence, it suggests that the interior was entered by the Middle Saxons, taking Narrow-leaved Elm from the Lea valley. Supporting this conjecture is the fact that the elm place names in this area are mainly based on *wych* in contrast to coastal Essex, where *elm* names are more usual. It would have to be supposed that the hypothetical immigrant Middle Saxons spoke of *wych*, while the East Saxons proper, as located by the medieval archdeaconry of Essex, spoke of *elm*. The north-west of the county has its own form of Narrow-leaved Elm. Its history seems different from that of the rest of Essex. In medieval times, it was in the western, detached part of the archdeaconry of Colchester, perhaps indicating some recollection that it was peopled by a different stock from the inhabitants of region HM1 with its numerous Hybrid Elms.

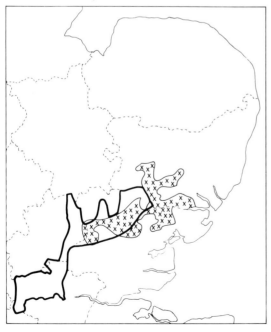

Fig. 140. The Essex hybridization zone. Crosses, Hybrid Elms; heavy line, boundary of the medieval Archdeaconry of Middlesex.

Greater London[83]

Greater London comprises almost the whole of the former county of Middlesex, and such parts of Essex, Hertfordshire, Kent and Surrey as have contributed to the London conurbation.

It is mostly a lowland area, mainly on London Clay, but extending southward to the dip slope on the North Downs. The Thames is the principal topographical feature, bisecting the county from west to east. The major tributaries are all on the north, namely the Roding, Lea and Colne, the last providing part of the western boundary.

English Elm is abundant everywhere. Narrow-leaved Elm is common around Dagenham. The designation *elm* was in use in the immediate vicinity of London from early times. In the part of the county formerly in Essex, *wych* was current and perhaps the variant *wick*. The expression *wych elm* was used for Narrow-leaved Elm.[84] The place name Wickelme Ponds in West Ham, recorded in the seventeenth century,[85] is one reason for suspecting a word *wick* for elm.

One small area in the south-east has little elm. It is in region Z4 which covers much of Kent and Sussex. English Elm is occasional, and at Keston is a little Narrow-leaved Elm.

The rest of the county, excepting only the area around Dagenham, lies in V1,[86] the principal English Elm region.

Negative evidence suggests that there was but little mature elm near London in Roman times. The extensive excavated remains of Roman pilework and wharfage along the London waterfront is mainly of oak but with other species also. No elm has been found. Such timber water pipes as have survived from Roman London are perforated oak blocks. This complete absence of elm in artefacts for which it was later a preferred timber suggests that if present, it was in insufficient supply to be a substantial timber source.

The twelfth century pilework under London Bridge, however, was of elm.[87] In this century too the first references are found to the Middlesex county executions at Tyburn, described in Latin as *ad Ulmos*

109,
114

or *ad Ulmetum*.[88] The surname atte Elmes, borne by a purchaser of property at Fulham, had come into use in the following century.[89]

The city of London, as appears from fourteenth century records, also sent felons *as humeaus* [to the elms] for execution. The elms in this case were at Smithfield.[90] It was in this century that the elm pilework recently uncovered at Trigg Lane was driven, also the elm piles under the Jewel Tower at Westminster.[91]

In the villages north of the Thames, fourteenth century manorial court rolls record the presence of elm at Walham Green in Fulham and at Tottenham.[92] South of the river, the place name Elmstead in Bromley also attests its presence.[93] Elm board was used for building at this time at Eltham and at Lesnes Abbey, Erith.[94]

In the fifteenth century further mentions of elm around London are found. There was an Elmys Tavern near the Charterhouse, and there were elms on the present site of Dean's Yard at Westminster Abbey.[95] Elms were lopped and pollarded at Chiswick.[96] On the other side of the river, elms were felled for piles for

Fig. 141. Paddington (GL) church.[86]

141 London Bridge at Beckenham.[97] They were felled without due authorization at Tooting Bec.[98]

The picture of London embosomed in elm is confirmed by Tudor documents. John Stow, in his *Survey of London*, described the elm hedgerows in Whitechapel and those that had existed shortly before in Ratcliff. He also tells the story of the eccentric curate of St Katharine Cree church who preached from a high elm in the churchyard.[99] There was an Elm Close on the site of Long Acre, and an elm in Chancery Lane served as a parish boundary mark.[100] 'High Helmes' flourished at Shoreditch. Francis Bacon planted elms at Gray's Inn but these only replaced older ones gone to decay.[101] Abbot Feckenham's elms at Westminster are mentioned in chapter 9. They remained as a welcome amenity in the abbey grounds till 1779. The elms at Fulham Palace featured in the Marprelate Tracts and are considered in chapter 8. The Queen's Tree at Chelsea, an elm whose site is marked today by a public house, is first mentioned in 1586. An unlikely legend brings Elizabeth I and Lord Burleigh beneath the tree to shelter from a downpour.[102]

Elms were planted at Hampton Court in the reign of Henry VIII. The curiously shaped elm there called King Charles' Swing could have been even earlier.[103]

There are many references to elm in other outlying localities in the sixteenth century. They feature in the manorial court rolls of Hanwell and Tooting Graveney.[104] The Barn Elms of Barnes make their first appearance, and Whett's Elm, a landmark at Eltham, receives its first mention.[105] Elms also provided landmarks for navigating the Thames. The western half of Woolwich Reach was formerly called Podds Elms Reach, named after Podds Elms, a row alongside the river on the south side. They were faced by another row, Saunders Elms, on the north side.[106]

By the seventeenth century the evidence for the overall distribution of elm is well nigh complete. There were still elms in St Paul's Cathedral churchyard, though the storm of 1611 had brought down the tallest.[107] The Lincoln's Inn elms were sufficiently well known to provide the rendezvous with the Devil in Ben Jonson's *The divell is an asse*.[108] There was an Elme Court in the Middle Temple and elms in the garden of

Fig. 142. Rosamond's Pond (GL).[111]

the Inner Temple.[109] The 'royal walk of elms' in St James's Park[110] survived till the formal landscaping of the park in the time of Charles II. The new layout comprised both elm and lime avenues, the northern avenue being the Mall, with both elm and lime rows at first. It is described more fully in chapter 9. Rosamond's Pond, in the south-west corner of the park, was the resort of those who sought relief from unrequited love by drowning.[111] It was girt about by elms. The Mall became a fashion parade and elm malls were planted elsewhere in imitation. They were to be found at Moorfield, Chiswick and Hammersmith. Richmond Green, outside another royal residence, was also planted with elm.[112]

The specimen elm that most impressed the seventeenth century was the Great Hollow Tree at Hampstead,[113] mentioned in chapter 9. The Seven Sisters, an underground station on the Victoria Line, can be traced back to one of the Three Wonders of Tottenham, a 'tuft' of elms enclosing a walnut tree. The elms outlived the walnut.[114] The Nine Elms of Battersea are also first heard of in this century.

In the seventeenth century too, elms grew in St Martin in the Fields and Marylebone, and further out at Hornsey, Twickenham and Hillingdon.[115] A survey of timber on the estates of All Souls College, Oxford, in Middlesex, enters elms at Edgware, Kingsbury and Willesden.[116] The place name Wickelme Ponds in West

Fig. 143. The Great Hollow Tree, Hampstead (GL), by Wenceslaus Hollar.[113]

1. The Bottom aboue ground in Compaſs is —— 28.foote
2. The Breadth of the doore is —— 2.foote
3. The Compaſs of the Turret on the Top is —— 34 foote
4. The Doore in Height to goe in is—6 foot. 2 Inches
8. The Height to the Turret is —— 33.foote
11. The Lights into the Tree is —— 16
18. The Stepps to goe vp is —— 42
19. The Seat aboue the Stepps,
Six may Sitt on, and round about roome
for foureteene more
All the way you goe vp within
the Hollow Tree

Ham, mentioned above, suggests that Narrow-leaved Elm may once have grown there. Elm was also present at Loughton.[117] In what was formerly Kent, Evelyn records the presence of old elms on his Deptford estate.[118] The abbey ruins at Lesnes were, by the seventeenth century, overgrown with elms and other trees.[119]

The New River was cut to bring piped water to London in the early seventeenth century. The street mains were elm pipes, vast quantities of elms being grown for this purpose. Pipe manufacture was a major incentive for growing elm in the Thames basin.

The principal amenity planting of elm was in Kensington Palace Gardens. It is considered further in chapters 9 and 10. The major London parks, Hyde Park, Green Park and St James's Park, already mentioned, had much elm timber. The elms in Hyde Park featured prominently in the public outcry over the sitting of the Great Exhibition of 1851. This affair

62

87, 88

75, 76

is described in chapter 10. Greenwich Park was landscaped in the formal French style in the seventeenth century. Extensive use was made of elm.[120] Regent's Park was a later creation. It is but one element of a grandiose landscape design devised by John Nash. Starting from the Prince Regent's residence at Carlton House, a royal route proceeded along Regent's Street, through Oxford Circus into Portland Place, reaching its grand finale in the Broad Walk of English Elms in Regent's Park. This is in alignment with Portland Place.[121] The garden at Hampton Court had one rare feature, a cradle walk of elm, later called Queen Mary's Bower, after William III's consort, Mary II.

The London squares were all planted with elm when Per Kalm, the Swedish traveller, visited England in the eighteenth century.[122] The substitution of elm by plane in the London squares was one of the major landscape changes in London during the nineteenth

Fig. 144. The Seven Sisters (GL).[114]

THE SEVEN SISTERS, TOTTENHAM.

Fig. 145. The Seven Sisters (GL), by Hans Unger.[114]

century. Extensive use of elm was made in the various pleasure gardens such as those of Vauxhall, Cremorne and Marylebone.[123]

After the demise of the Great Hollow Tree at Hampstead, other Hampstead elms entered the limelight.[124] Irving's Elm, mentioned in chapter 9, was an outdoor preaching station of Edward Irving. The Nine Elms, also mentioned in chapter 9, were frequented by Lord Mansfield and his friend Alexander Pope. Judges' Walk, in legend a retreat for the courts during the Great Plague, seems originally to have been elm, though replanted in lime. The Gibbet Tree for felons that stood by the roadside on the heath is mentioned, like the elms above, in chapter 9.

There were other notable specimen elms to the north of London. One marked the site of the Battle of Barnet.[125] Latimer's Elm, which stood near Monken Hadley,[126] was felled in 1935. Both are considered in chapter 9. Chingford church was surrounded by elms which appealed not only to topographical artists but also to the pre-Raphaelite painter Arthur Hughes considered in chapter 11.[127] The elms at Harrow School feature in chapters 8, 9 and 10. The Nine Elms

in Battersea had become a row before the house of William Watson.[128]

There were so many elm avenues in Greater London[129] that they can only be considered selectively. The one at Harefield Place was the scene of Milton's *Arcades*, referred to in chapter 10. One of the Ham House avenues inspired Thomas Hood's lugubrious poem,[130] also mentioned in that chapter. The unusually orientated avenues at Canons, Little Stanmore, are described in chapter 9. The longest avenue in Greater London never existed. The central axis of Chelsea Hospital, when projected from the front elevation, passes through Kensington Palace. It is reputed that an avenue was planned to connect the two in the time of Queen Anne. The only trees, probably elm, that were planted were in Burton's Court, in front of the Hospital and in the road, still called Royal Avenue, immediately adjoining.[131] Two elm walks deserve a mention. Bishop's Walk ran along the Thames outside Lambeth Palace. Rupert's Walk, mentioned in chapter 9, was in Woolwich.

The region yet to be considered in MV7,[132] covering the eastern fringe of the county north of the

Fig. 146. Latimer's Elm, Monken Hadley (GL), by J. P. and W. H. Emalie after E. C. Wilde.[126]

Fig. 147. Chingford (GL) church, by J. C. Armytage after William Bartlett.[127]

Fig. 148. Home from sea, by Arthur Hughes.[127]

Thames. A small part of this extends into Essex. Both English Elm and Narrow-leaved Elm are frequent. The presence of elm at Havering atte Bower in the thirteenth century can be inferred from the surname de Ulmis, later atte Elmes, then current.[133] This is more likely to have been Narrow-leaved Elm, since English Elm is believed to be secondary in this area. Narrow-leaved Elm was certainly present at North Ockendon in the seventeenth century, being cited as *Ulmus folio glabro* in the second edition of Gerard's *Herball*. Elm is recorded from the fifteenth century onwards at Dagenham.[134] One of the notable elms in this region was the large specimen tree on Havering Green.[135]

There is very little Wych Elm in Greater London save in recent amenity plantings.

Settlement in prehistoric times was mainly along the Thames from London westward, in the Thames plain east of Dagenham, up the Lea and Roding valleys, and on the northern edge of the North Downs. The Roman period saw an intensification of settlement in the areas just mentioned, the foundation of London, and the occupation of the Cray valley, a southern tributary of the Thames.

The course of events after the Roman withdrawal is obscure. Except along the North Downs area, where there are a number of early Anglo-Saxon sites, the whole of Greater London is devoid of settlements going back to the initial stages of the Anglo-Saxon invasion. It is generally supposed that a Romano–British population remained during this period. This supposition is confirmed by the Celtic place name Penge [*pen coed*] south of the Thames. By the time that London is mentioned by Anglo-Saxon sources, it had become East Saxon territory. The Middle Saxons, who gave their name to Middlesex, had come and vanished as a separate group. It is often supposed that Surrey represented the southern half of an original Middle Saxon territory. Eventually, the land south of the Thames went to other Anglo-Saxon kingdoms, West Saxons in the west, Kent in the east. Possible traces of an extended Middlesex within modern Essex are considered in the Essex section.

Much of the eastern part of what is now Greater London was in Waltham Forest which came close to the city of London, rather as Galtres Forest came close to York. The citizens of London had certain hunting rights in Waltham Forest which have been held to indicate an ancient connection. Enfield Chase was in the north of Greater London and place names suggest that there was also wooded terrain in the west of the county. There was much woodland also to the immediate south of London.

The Narrow-leaved Elm of region MV7, like that in southern and eastern Essex, is related to elms growing today in northern France. It is suggested in chapter 3 that all these populations were introduced in the Iron Age. The distribution pattern is similar in all cases, relatively small populations centred on ancient settlement sites on the north side of the Thames estuary. MV7 corresponds with the Iron Age settlement area east of Dagenham.

English Elm was abundant around London by the twelfth century. The negative evidence of the Roman waterfront timbers suggests that it was not introduced before the Roman period. The failure, however, of Narrow-leaved Elm to spread either westward over the River Lea or southward over the Thames, suggests that another elm had meanwhile arrived. This would be English Elm from the west of England, which can hardly then have been introduced later than the Roman period. The linguistic evidence considered in chapter 4 certainly points to the conclusion that elm had already reached the eastern half of England when the Anglo-Saxon penetration began. It is highly probable that the relative disposition of English and Narrow-leaved Elm in the Thames estuary area had become established before the end of the Roman occupation.

Hertfordshire[136]

The recent changes in county boundaries have only slightly affected Hertfordshire. Barnet is now wholly in Greater London and a small strip of north Middlesex has been gained.

The principal upland of the county is the chalk outcrop of the Chilterns, and their extension beyond Hitchin, running from north-east to south-west. Part of the county projects west of this into the clay vale beneath the chalk scarp. To the south, the chalk dips under London Clay. There is much Boulder Clay masking older rocks and there are some sands and gravels in the south.

The east of the county is drained by tributaries of the Lea, itself an affluent of the Thames. These, from west to east, are the upper Lea, Maran, Beane, Rib, Ash and Stort. The main streams in the west of the county are the Gade and Ver, branches of the Colne, another

84

Fig. 149. Town Bottom Elm, Hatfield (He), by Alice Balfour.[139]

Thames tributary. In addition to the Lea valley, the Chilterns are breached by dry valleys at Hitchin and Tring.

English Elm is frequent in the west of the county, but is replaced by Narrow-leaved Elm in the east. The earlier designation for either kind was *wych*. This had been largely replaced by *elm* by the sixteenth century.

In much of the centre of the county elm is infrequent. This area is part of region Z3, which covers a considerable area in counties to the south-west of Hertfordshire. Much of the Chilterns lie within it and, in Hertfordshire, much terrain to the immediate south of them. English Elm is of scattered occurrence in the west of this area, and Narrow-leaved Elm in the east.

The presence of some elm in this area is suggested by the place name Wychdell in Knebworth[137] and Witch Hill in Northchurch.[138] The former is recorded in the fourteenth century, the latter in the seventeenth century. The elm on Preston village green reputedly commemorated the coronation of George III. The Town Bottom Elm in Hartfield Park was a notable specimen tree.[139] The elm avenue at Knebworth House was felled in 1957, and another existed in Berkhampstead Park. Ashridge House had its Elm Walk. The refectory table at Little Gaddesden Manor reputedly came from Ashridge Priory. The large elm that provided the timber is more likely to have grown in a nearby part of region V1 than in Z3.

A small region with little elm, Z21, runs southward from Great Mundon to Sacombe. It lies between the Beane valley, with English Elm, and the Rib valley, with mainly Narrow-leaved Elm.

V1,[140] the principal English Elm region, extends into Hertfordshire in four different places. It covers the parish of Puttenham, where the county projects westward into the clay vale beneath the chalk scarp face. It occupies the chalk hills north-east of Luton. A broad tongue enters the county from Greater London. St Albans lies in this section, which extends as far north as Flamstead in the heart of the Chilterns. The remaining part of V1 runs up the lower Lea valley, then northward along the Beane to join the second one at the southern end of the Hitchin dry valley. Occasional Wych Elm occurs in this area, mainly in the south-west Chilterns.

The earliest archeological evidence for elm in this area comes from an elm wattle hut of Roman age at Great Wymondley.[141] In the fifteenth century the manorial court rolls of Hoddesdon contain presentments for felling elm. The place name High Wyches, also in Hoddesdon, is first found at this time.[142] Elm floor boards in the tower of St Albans Abbey church, now the Cathedral, have been dated to the same century.[143] Elm is recorded at Hertingfordbury in the next century, and a Wytchfeild is mentioned in this parish in the seventeenth century.[144]

The Cavalier Elm stood at High Dawn, Pirton. A cavalier hiding in its hollow trunk was reputedly dragged therefrom and beheaded.[145] A prominent elm avenue ran across Redbourn Common. There were also elm avenues at Aldenham House.[146]

Eastern Hertfordshire, comprising the valleys of the Rib, Ash and Stort, is in region M1[147] with frequent Narrow-leaved Elm. This region covers extensive areas in Cambridgeshire and Essex to the north and east respectively. The Narrow-leaved Elms of the Hertfordshire part of region M1 show great variation. Three forms are readily distinguishable. The most frequent is in the west of the same area. Another has its centre of distribution at Eastwick, on the Stort. The third is centred on Braughing. Scattered English Elm occurs in the south of this area. The fact that information about elms may be gleaned in the most unlikely places is demonstrated by a marginal note in a copy of Bullinger's sermons to the effect that the elms in Anstey churchyard were lopped on 23 February 1581.[148]

There were elm avenues at Moor Park, Much Hadham and Thundridge Bury. The elmscape around Great Amwell features in the verse of John Scott of Amwell.[149]

A small region, M12,[150] with frequent Narrow-leaved Elm occurs around Digswell and Tewin. The presence of elm at Tewin in the seventeenth century is attested by the building accounts of Hatfield House.

Northern Hertfordshire, to the south of Royston, is in region MV11,[151] which also covers south-west Cambridgeshire. English Elm and Narrow-leaved Elm are both common.

The country between Bishops Stortford and Sawbridgeworth is in region HM1 with Narrow-leaved Elm and Hybrid Elms frequent. HM1 extends right across Essex to the Suffolk border. It is discussed in the Essex section. The place name High Wych may indicate the presence of elm in Sawbridgeworth in the sixteenth century. There was also an elm avenue at Hyde Hall in this parish.[152]

Wych Elm is not a common tree in Hertfordshire. It is probably native in the Chilterns south-west of Hitchin.

Prehistoric settlement was mainly along the Chiltern scarp face. In the Iron Age, settlements were also to be found along several of the river valleys, in particular the Ver, where St Albans is now, the upper Lea and the Maran, at Braughing, where the Quin runs into the Rib, and along the Stort. The settlement pattern in Roman times was very similar. There was an extension of the occupied area to the west of St Albans, and the Stort valley was more densely inhabited.

Anglo-Saxon entry was relatively late. One of the few assuredly identified Middle Anglian tribes was the Hicce who occupied the county around Hitchin. The west of the county, roughly as far as the Roman Ermine Street, was Middle Anglian terrain. East of Ermine Street was East Saxon. The division was perpetuated by the boundary between the medieval dioceses of Lincoln and London. It corresponds quite closely with the boundary between elm regions VI and MI.

Early place names indicative of wooded terrain and relatively late settlement are concentrated along the Cambridgeshire border and between Baldock and Luton.

The historical evidence suggests that the existing pattern of elm distribution is at least medieval in age. What is remarkable is the extent to which the pattern of Narrow-leaved Elm distribution reflects the Iron Age settlement pattern. There are relatively small Narrow-leaved populations centred on the Maran valley, at Braughing, and along the Stort. They are all different from one another and appear to mirror a settlement pattern of isolated riverain sites with limited communication between them. The elms appear to have been derived from northern France. As suggested in chapter 3, a date of introduction in the Iron Age is feasible. Whether this would have been direct or, in some cases, via other Iron Age sites in Essex or Kent, cannot be decided.

If the Romano–British hut of Great Wymondley were of English Elm, this would settle outright the Roman age of the Hertfordshire population of this elm. Though English Elm is far more frequent than any other kind at Great Wymondley today, one cannot dismiss the possibility that there was Wych Elm here earlier. It does not seem particularly likely, however, and it is argued in the Cambridgeshire section that

the closeness of the contact zone between English and Narrow-leaved Elm to the presumed centres of introduction of the latter in Cambridgeshire points to English Elm having arrived there in Roman times. If this were so, it is unlikely to be any later in Hertfordshire. The sparsity of English Elm in the Chilterns and the late settlement of the clay lowlands stretching away from the foot of the chalk scarp suggest that English Elm reached the Great Wymondley area via the lower Lea and Beane valleys. It is probably significant that a corridor of English Elm along the Beane should lie between Narrow-leaved Elm populations in the valleys of the Maran and Rib on either side of it. This would be explicable if there were Iron Age settlements with introduced elm in the last two rivers but no such settlement with elm along the Beane. This river valley was certainly entered during the Roman period and the Romano–British settlers might have obtained English Elm from the lower Lea.

The contact zone between English and Narrow-leaved Elm in eastern Hertfordshire also corresponds roughly to the later boundary between the territories of the Middle Angles to the west and the East Saxons to the east. The East Saxon sphere was in the enigmatic medieval Archdeaconry of Middlesex. If this did indeed represent an ancient arrangement, it suggests that it was Middle Saxons rather than East Saxons who advanced up the lower Lea and Stort. Perhaps they extended the area of the Narrow-leaved Elms they encountered. If so, it would not have been west of Ermine Street. This was Middle Anglian territory and the elm of the southern tribes of Middle Anglia was English Elm.

Kent[153]

Kent has retained its traditional extent except for the urbanized north-west which has gone to Greater London.

There are three upland areas: the chalk outcrop of the North Downs, the Greensand ridge to the south of and parallel to the chalk scarp, and the sandstone hills of the High Weald along the Sussex border. Tertiary beds lie north of the chalk, London Clay in the Hoo peninsula, the Isle of Sheppey and north of Canterbury, largely sandy rocks elsewhere. A narrow clay vale, the Holmesdale, separates the North Downs from the Greensand ridge. A wider clay plain, the Low Weald, separates the Greensand ridge from the High Weald. There is much coastal marshland along the Thames estuary and in Romney Marsh. Of the former offshore islands in the Thames estuary, the Isle of Grain and Thanet are now joined to the mainland. Elmley and Harty are now joined to Sheppey. The Thames estuary is the northern boundary. Two rivers cut through the North Downs, the Medway at Rochester and the Stour at Wye.

English Elm is common in the north-west of the county. Narrow-leaved Elm is frequent in the east. Both were called *elm* in earlier times, with *elvin* as an alternative. The two designations interchange with one another in two place names. Elmhurst in Brenchley was called Elvenhurst in the eighteenth century.[154] Elmton in Eythorne interchanges with Elventun.[155] An early form of this name suggests that it may have been derived from the personal name Aelfgyð. There are then two possibilities. One is that the place name was derived from the personal name but that later the derivation was forgotten and the name treated as an *elm* name with the variation to *elvin*. The other possibility is that it was originally an elm name but that the name of a later owner, Aelfgyð or some such, was used as an alternative designation.[156]

Roughly half of Kent lies in region Z4 where elm is infrequent. Z4 covers the High Weald, most of the Low Weald and the Greensand ridge, much of the North Downs, the lowlands north of the Downs between Chatham and Faversham, western Sheppey and most of Romney Marsh. English Elm occurs infrequently.[157] There is some Narrow-leaved Elm to the east and a little Wych Elm in the North Downs.

The presence of some elm in or near the region follows from the identification of the charcoal at Roman levels in Canterbury.[158] This would probably have been Narrow-leaved Elm. The parish of Lyminge lies within Z4. This place is first mentioned in the seventh century. Its name is thought to be derived from the River Limen, the earlier name for the Rother. Limen is usually said to be a derivative of Celtic *lem-*, meaning elm. Since the river is several kilometres distant from Lyminge, the name is not relevant to whether elm grew there. The river name is discussed below. Though elm is rare at Lyminge now, substantial quantities of it were obtained from the parish for building Sandgate Castle in the sixteenth century.[159]

The place name Elmhurst in Brenchley, mentioned above, is on record in the ninth century.[160] It may refer to the local English Elm, but it could also, as mentioned in chapter 4, be based on a personal name. The Elmton in Eythorne, also mentioned above, is in Domesday Book. The uncertainties surrounding this name have been discussed. If it does refer to elm, it would be Narrow-leaved Elm. Elmley Island, first mentioned in the thirteenth century and now joined to Sheppey, is almost devoid of elm today. Both English Elm and Narrow-leaved Elm grow nearby. Another Elmhurst, in Egerton, is on record in the same century.[161] Denge Marsh behind Dungeness has little elm today but in the sixteenth century watch was kept there at a place called Elmes.[162] Churchwardens' accounts at Canterbury show that elm board was being used in the sixteenth century.[163] There were small elms at Marden rectory in the seventeenth century.[164]

When the spa was opened at Tunbridge Wells, the walks were laid out under elm trees. The Upper Walk was originally a double row. One side was later felled and the surviving row long remained along the centre of what is now the Pantiles.[165] There was an elm avenue at St Clere, Kemsing.

Z16 is a small area of infrequent elm on the exposed chalk downland behind the South Foreland. Some Narrow-leaved Elm is present.

The largest area of frequent elm is the Kentish part of the English Elm region V1.[166] It covers the land immediately south of the Thames from the western border to the Medway. It extends into some of the less heavily wooded parts of the North Downs. It also goes up the valley of the Medway and some of its larger tributaries and so penetrates deeply into the Low Weald.

The principal early evidence for elm in this area comes from the pilework under the successive Rochester Bridges and from the bridgewardens' rolls.[167] The date of the first bridge is unknown but was probably late pre-Conquest. Its piles were mainly of oak but with some elm and beech.[168] The second bridge was built in the fourteenth century and all its piles were of elm. When this bridge was finally blown up to make way for the third and present bridge, it was estimated that some ten thousand elm piles had been driven in.

The bridgewardens' rolls contain a superb series of accounts, first in French, then in Latin, recording, year by year, the purchase of elms for piles and the sources of supply. From these, a virtually complete map can be drawn of the medieval distribution of English Elm in mid-Kent. It corresponds closely with the existing boundaries of this part of region V1. Elm for piles was obtained from Hoo and High Halstow in the Hoo peninsula, from Frindsbury, Strood, Cuxton and Snodland along the west bank of the lower Medway, and from Gillingham, Chatham, Rochester, Borstal, Wouldham and Burham along the opposite east bank. Further inland, elm was sent from Aylesford, Allington, Maidstone, Boxley, Ditton, from East Malling, Birling and Addington, and from Wateringbury, Yalding and Loose. One consignment came from Cobham. Each one of these places is in V1 except East Malling which is just outside it and has English Elm at lesser frequency. Elm piles were also used under Gravesend wharf in the fourteenth century.[169]

Elm water mains were laid in Rochester.[170] The 71 Court Tree on Boley Hill, Rochester, mentioned in chapter 9, was felled in 1831. It was under the elms called the Seven Sisters in the Vines at Rochester that Dickens strolled three days before his death, carrying with him to the grave the mystery of Edwin Drood for which the elms of Rochester Close provided the setting.[171] There was an elm avenue in Linton Park[172] and a Great Elm Avenue in Cobham Park.[173]

A small region, V9,[174] with frequent English Elm, covers the coastal lowland north-east of Faversham and the Isle of Harty, now joined to Sheppey. Elm was present at Graveney in the seventeenth century.[175]

In eastern Kent, there are two regions, M10 and M11, with frequent Narrow-leaved Elm. M10[176] covers a broad band of country behind the coast from just south of Deal to Hythe, and then north and behind the Royal Military Canal demarcating Romney Marsh to Warehorne. It also includes a narrow strip running down the east coast of Romney Marsh to New Romney.

The first place name of possible relevance is that of the River Limen, the old name of the branch of the Rother later made navigable as the Royal Military Canal. The river name is recorded in the seventh century, but the place name Lympne derived from it is cited in Roman times as Lemanae. If one could be confident of the usual derivation proposed for this name, from Celtic *lem-*, meaning elm, this would provide valuable evidence for the early presence here of, presumably, Narrow-leaved Elm.

The place name Elmstead permits no doubt. It is first recorded in the ninth century. The sixteenth century building accounts of Sandgate Castle enter purchases of elm timber from Dover.[159] The Rother was refurbished as the Royal Military Canal at the time of the Napoleonic Wars. It was planted with Wych Elm.[177]   150, 151

M11[178] lies behind the coast from Sandwich to Seasalter, and includes the Isle of Thanet. There is some English Elm towards the west. Elm piles were

Fig. 150. Royal Military Canal (Ke), by Paul Nash.[177]

used in the former port of Sandwich in the fifteenth century.[179] Surveys of ecclesiastical property in the seventeenth century record the existence of elm at Littlebourne, Reculver and Herne.[180] Elm was also present at this time at Whitstable and at a later date this port had elm water pipes.[181] There was elm too in the seventeenth century on the Archbishop of Canterbury's lands at Chislet and Nonington.[182]

Most of Sheppey Island proper is in region MV4[183] with English and Narrow-leaved Elm both frequent. A notable elm in this region was the Kingsborough Elm in Eastchurch. The functions performed beneath it are described in chapter 9.

Wych Elm is a rare tree in Kent. It is found occasionally in woodland on the chalk and in the High Weald. It may have survived in these areas since Boreal times.

Being the nearest part of England to the Continent, Kent has been settled since early times. Even so, prehistoric settlement was strongly localized. In the Iron Age the principal sites were the Isle of Thanet, the Stour valley up to Canterbury, and the coast around Deal and around Folkestone. The southern side of the Thames was settled, in particular about Swanscombe and near Sittingbourne. Inland, the major occupied areas were the Medway valley and the Greensand ridge south of Maidstone. The settlement pattern in Roman times was similar though more extensive. The east coast was continuously occupied between Reculver and Lympne. The Thamesside settlements expanded to give densely occupied terrain between Crayford and the western side of the Hoo peninsula, and between Rochester and Faversham. The Cray and Darent valleys, both running northward into the Thames, were entered. The inland settlement area on the Greenside ridge expanded westward as far as Ightham.

Kent was penetrated early by the Anglo-Saxon immigrants and the whole of the east of the county was settled then, except the heart of the chalk downs. They also took over the belt between Rochester and Faversham. Other sites occupied early were the Darent valley and the Holmdale between Otford and Lenham.

Though there was no royal forest in Kent, the Weald was recognized as a legal entity apart from the agricultural land surrounding it. Its agricultural development was medieval and later. Place names confirm its wooded state and late settlement. The Blean was an extensive wooded area north of Canterbury. Outside these two major wooded tracts, early place names indicate wooded terrain and late

70

Fig. 151. Royal Military Canal (Ke), by John Piper.[177]

settlement in the North Downs south of Gravesend, Gillingham and Faversham, and north of Folkestone.

The archeological and historical evidence indicates that elm was present in eastern Kent in Roman times. There is a high probability that this was Narrow-leaved Elm, now common in the area. It is argued in chapter 3 that this population was of Iron Age introduction, the source being northern France.

The English Elm of northern Kent and the Medway valley had arrived by the Norman Conquest. If it had arrived only shortly before, it would be difficult to explain why Narrow-leaved Elm is absent from the west of the county. It is more likely, therefore, that English Elm had been introduced into northern Kent from the London area in Roman times. Similar considerations led to the supposition that the English Elm of southern Essex was also introduced in the Roman period.

Leicestershire[184]

Modern Leicestershire amalgamates the former counties of Leicestershire and Rutland. Border rectifications only affected small areas.

The high ground in Leicester is mainly in the north of the county, namely, from west to east, the sandstone and Coal Measure upland centred on Ashby de la Zouch, the Precambrian hills of Charnwood Forest, the Boulder Clay Wolds north-east of Leicester, and the limestone hills in the east. The rest of the county is over clay. The part of the clay lowland at the foot of the limestone outcrop in the north-east of the county is the Vale of Belvoir which extends into Nottinghamshire. The principal river is the Soar, a tributary of the Trent. It has a major eastern tributary of its own, the Wreake. The Welland provides part of the southern county boundary.

Leicestershire is the centre of England for present purposes. The Wych Elm in the west of the county is at the southern limit of its main distribution. The English Elm in the south-west is at the northern limit of its main distribution. The Narrow-leaved Elm in the east is at its western limit. The name for each kind seems always to have been *elm*.

The main region of infrequent elm, Z1, covers three parts of Leicestershire. One part is contiguous with Z1 in northern Warwickshire. It comprises Charnwood Forest and the country between it and the Warwickshire border. The second part covered by Z1 is an extensive area in the eastern half of the county,

in particular the limestone hills. The third part is a small area around Castle Donington, on the north-west border of the county. Wych Elm, English Elm and Narrow-leaved Elm all occur occasionally.

There was some elm, probably Narrow-leaved Elm, at Stretton in the fourteenth century.[185] In the following century the building accounts of Kirby Muxloe Castle have an item on grubbing up elms.[186] The designation of St Mary in the Elms of Woodhouse church in Charnwood Forest appears to be recent. Belvoir Castle had an elm avenue.

Wych Elm is frequent in the hilly terrain in the west of the county. This is part of region G20,[187] which includes a small part of southern Derbyshire.

The main English Elm region V1[188] covers south-west Leicestershire and the environs of Leicester. It also goes up the Wreake valley as far as Rotherby. In addition to the prevalent English Elm, some Narrow-leaved Elm is present in the east.

It was presumably English Elm that was referred to in proceedings concerning ownership of an elm at Leicester in the fifteenth century.[189] The elm in the Saturday Market in this town was a well-known landmark.[190] Elm was amongst the trees growing at Rothley in the seventeenth century.[191] The qualifying phrase in the place name Sutton in the Elms is recent.

There are two areas where Narrow-leaved Elm is frequent, both in the north-east of the county. M17[192] covers the Wolds east and north of Melton Mowbray. Both English Elm and Lock's Elm are also present. M18, which is mainly in south-west Lincolnshire, covers the northern end of the Vale of Belvoir.

In several regions, two kinds of elms are frequent. In region GM2,[193] on the limestone hills north-east of Uppingham, Wych Elm and Narrow-leaved Elm are common. Felling of some kind of elm at Buddon Park, Drayton, is entered in a *coram rege* roll of the fourteenth century.[194] Region MV14,[195] where English Elm and Narrow-leaved Elm are frequent, lies at the head of the Wreake around Melton Mowbray. Elm of some sort was present at Melton Mowbray in the fourteenth century.[196] The same two elms coexist at high-frequency in region MV5 in the south of the county. This region is mainly in northern Northamptonshire. The earlier presence of elm in the Leicestershire part of it is attested by an Elm Field in Shawell, recorded in the seventeenth century.[197]

In one region, GMV1,[198] lying on both sides of

the border with Nottinghamshire, Wych Elm, English Elm and Narrow-leaved Elm are all frequent. There were three people called William Blount at Wymeswold in the seventeenth century. One was differentiated as William Blount at Elm, indicating the presence of one of these three kinds then.[199]

Wych Elm is common today in two upland regions: the western upland and the limestone hills north of Uppingham. It is probably native in these areas.

Prehistoric settlement in Leicestershire was sparse. There was one Iron Age site of consequence, Leicester itself. In Roman times settlement had extended along Fosse Way between Leicester and High Cross, where it intersects Watling Street. There was also extensive occupation of terrain along and to the immediate west of Ermine Street, which runs through the limestone upland in the east. A third area of occupation was the Leicestershire section of the Welland valley.

Anglo-Saxon penetration seems to have been up the Soar and thence up its major tributaries.

There was a royal forest, Rutland Forest, in the east of the county, mostly in region Z1. Charnwood Forest was not royal forest but lay outside the area of agricultural settlement. It is in another part of region Z1. There are early place names indicative of wooded terrain and late settlement in the Wolds region in the north of the county.

Both English Elm and Narrow-leaved Elm had arrived before the end of the Middle Ages. The dearth of prehistoric settlement makes it unlikely that either was around before the Roman period. The Leicestershire part of region V1 is a wedge of country lying over the Fosse as far north as Leicester. North of Leicester it goes up the Wreake valley. This distribution is explained most easily on the assumption that English Elm had been taken along the Fosse in Roman times from Warwickshire as far as Leicester. Then when the Anglo-Saxon invaders had reached Leicester via the Trent and Soar, they took the English Elm they encountered at Leicester with them as they advanced up the Wreake.

Narrow-leaved Elm seems to have come to Leicestershire from two sources: from south-west Lincolnshire, and along the Ise valley, a tributary of the Nene, from northern Northamptonshire. If the conclusion be correct that English Elm was taken up the Wreake in Anglo-Saxon period, the fact that it got

no further than Melton Mowbray suggests that Narrow-leaved Elm had advanced concurrently into the north-east of the county. In chapter 12, it was argued that Narrow-leaved Elm had reached the south-west upland of Lincolnshire in Roman times. Its expansion westward into the previously unoccupied country of north-east Leicestershire would seem to be an incidental of Anglo-Saxon movements. The date of entry of Narrow-leaved Elm in the south of the county is presumably somewhat after it had reached northern Northamptonshire, that is, no earlier than the Anglo-Saxon period.

The introduction of the scattered tree of Lock's Elm is not likely to go back more than a few centuries. South-west Lincolnshire is the likely source.

Norfolk[200]

The only change that this large county has recently undergone is the absorption of the country immediately south of Great Yarmouth, formerly in Suffolk.

There are no dramatic topographical features in Norfolk and no really high ground. Gently undulating country over clay is the rule. To the north-west there is a Greensand outcrop and south of this are the sandy soils of the Norfolk part of the Breckland. There is a slight eminence, the Cromer Ridge, to the north-east. There is much marshland. In the west, a large part of the Fenland is in Norfolk. In the east are the Norfolk Broads, which surround a small area of slightly higher ground, the district of Flegg.

One set of rivers radiates from the Breydon Water behind Great Yarmouth. They are the Bure, Yare, with its tributary the Wensum, and the Waveney. The last provides the boundary with Suffolk for most of its length. Other rivers, the Nar, Wissey and Little Ouse, are tributaries of the Great Ouse. The Little Ouse, which rises a few metres from the Waveney, continues the Suffolk boundary in the west. The Stiffkey is one of the few rivers running northward.

Narrow-leaved Elm is widely but irregularly distributed throughout the county. It appears always to have been known as *elm*.

More than half the county lies in region Z1 where elm is infrequent. This covers most of the Fenland and Breckland and most of north-east Norfolk, including the Broads. Narrow-leaved Elm occurs sporadically throughout this area.[201] Wych Elm *152* is found rarely in the centre of the county. There is a little English Elm in the east.

Very near the centre of Norfolk and in the centre of the Norfolk part of region Z1, is the village of North Elmham. In the Suffolk section it is argued that this is a transferred name, like Melbourne, Australia, named after the South Elmham district of Suffolk. There is little elm at North Elmham today. There was little in the Tudor period when all the trees mentioned in the North Elmham churchwardens' accounts were *ooks*.[202]

Neither is elm frequent at Norwich except in relatively recent amenity plantings. The underwater piles on the Norwich waterfront, whatever their date, are mainly of oak.[203] The elm of the medieval coffin excavated beneath Norwich Cathedral might have been obtained from region M2, not far to the south.[204] The elm tree after which Elm Hill was named, the survivor of a pair in Tudor times, is notable as an unusual rather than a typical tree.[205] There was, however, sufficient elm in the neighbourhood of Norwich in the sixteenth century to make gunstocks.[206]

Elm plank was used for constructional work on Wroxham Bridge in the sixteenth century, but the piles beneath it were of alder and aspen.[207] References to elm are to be found in the sixteenth century accounts of the Guild of Our Lady of Wymondham.[208]

There are nine regions with frequent Narrow-leaved Elm. The largest of these, M2,[209] is in south-east Norfolk. It also occupies an extensive area in eastern Suffolk and in north-east Essex. The place name Wicklewood, mentioned in Domesday Book, has been derived from *wice*.[210] However, *wych* appears not to have been used for elm in Norfolk so this etymology is dubious. Deopham High Tree was an elm visible for some 10 km around Deopham.[211] M22[212] is a small region centred on Kings Lynn. The extensive medieval underwater piling of the waterfront of this port is of alder. Some of its houses were on oak piles. The only elm artefact that excavators have recovered is a turned bowl of thirteenth–fourteenth century date.[213] The

Fig. 152. Elm at Haveringland Park (Nf), by Henry Ninham.[201]

presumption is that there was little elm here in the Middle Ages. The churchwardens of Tilney All Saints, however, were able to purchase *ellmyng* timber in the sixteenth century.[214]

Three of the regions with frequent Narrow-leaved Elm are coastal. M23[215] lies behind Hunstanton. M24[216] is in the Stiffkey valley. M16,[217] which is mainly in north-east Suffolk, extends into Norfolk between Beccles and Lowestoft. The four inland regions where Narrow-leaved Elm is common are: M21,[218] on either side of Peddar's Way, the principal Roman road of north-west Norfolk; M15 in the Norfolk Breckland; M26[219] in the valley of the Tud, a tributary of the Wensum; M15[220] in the Norfolk Breckland; and M27[221] in the heart of the Broads.

Around Lowestoft, Narrow-leaved Elm and English Elm are both frequent. This is part of region MV13[222] which runs down a short distance along the Suffolk coast.

Wych Elm is a rare tree in Norfolk. Nor are there extensive populations of Hybrid Elm as in Suffolk and Essex. The presumption is that Wych Elm disappeared over most of the county at the Elm Decline.

Prehistoric settlement was mainly peripheral. There were Iron Age sites along the lower reaches of the Great Ouse, on the eastern shores of the Wash and up the Stiffkey valley. The principal inland settlements were along the Little Ouse around Thetford and around East Harling, and along the Yare and Wensum just above their confluence. The Roman pattern was an extension of the earlier one. Most of the land west of Peddar's Way in north-west Norfolk appears to have been occupied. New coastal sites were settled near Sheringham and between Caister on Sea and Burgh Castle. More land was taken in along the Little Ouse. Other inland sites were developed, notably at North Elmham on the upper Wensum, and the upper Bure valley to the south-east of Aylsham.

Early Anglo-Saxon sites were more sparse. In west Norfolk they have been found in the Hunstanton area and along the valleys of the Nar and Wissey. The Stiffkey valley continued in occupation but not the Sheringham site. The coastal strip between Caister on Sea and Burgh Castle was taken over. Of the inland sites, the centre of occupation of the Little Ouse site shifted south into Suffolk. Settlement continued around Norwich, North Elmham and Aylsham.

There was no royal forest in Norfolk. In several areas early place names suggest woodland cover and presumably late settlement, in particular to the north-east of Kings Lynn, to the north of North Walsham, to the west of North Elmham, along the Yare valley west of Reedham, to the west of Wymondham, and along the border between Fenland and Breckland around Methwold. Most of these areas are wholly or partly in region Z1.

The form of Narrow-leaved Elm occurring in Norfolk is similar to that occurring further south and much more frequently in Suffolk. The overall pattern of distribution in the two counties suggests that Norfolk derived its elm from Suffolk. Some at least of the nine regions of Narrow-leaved Elm must have been established by the end of the Middle Ages. Several, namely M22, M26 and M27, do not coincide with early settlement sites and for these a medieval origin is likely. M2 and M16 are mainly in Suffolk and are discussed further in the Suffolk section.

The four remaining regions, M15, M21, M23 and M24, all in the western half of Norfolk, mirror the Romano–British settlement pattern, which suggests that the introduction of elm in these regions was during the Roman period. The source would have been Suffolk.

The introduction of English Elm into region MV13, so far from the nearest part of V1, the main English Elm region, suggests a relatively recent date of introduction, probably post-Medieval. It is also likely that there was little Narrow-leaved Elm here earlier, otherwise there would have been little motive for introducing another elm.

Northamptonshire[223]

Except for very minor rectifications, Northamptonshire has retained its old boundaries.

Northamptonshire lies along the outcrop of Jurassic rocks and is consequently largely an undulating limestone area of moderate elevation. Lower terrain, mostly clay, is found along both its western and eastern borders.

The major river is the Nene which runs down the centre of the county from south-west to north-east. The western lowland is drained in the north by the Welland, for some distance the county boundary, and in the south by the Avon and its tributary the Leam. The Cherwell drains the south-west of the county and the Great Ouse and its tributary the Tove the south-east.

English Elm is frequent in the southern half of the county. Narrow-leaved Elm and some Hybrid Elms occur in the northern half. The older designation of these was *wych*. This was replaced by *elm* from the seventeenth century onward.

There is little elm in the north-east of the county which is covered by region Z1. What elm there is, is mainly English Elm in the west and Narrow-leaved Elm in the east. Wych Elm is also present. Lock's Elm is frequent at Laxton. The place names possibly indicating the earlier presence of elm are irregular and the evidence they provide is inconclusive. Weekly, recorded first in the tenth century, is supposedly based on *wice*.[224] Elmington, known from the eleventh century, has been thought to be derived from *elm*. Some elm, most likely Narrow-leaved Elm, was present at Rockingham in the seventeenth century.[225] A vast system of avenues, elm and lime, was planted for John, second Duke of Montagu, in Boughton Park. Elms for pipes were also purchased at Boughton Park[226] in the early eighteenth century.[227] There was also an avenue of English Elm at Biggin Hall, Benefield.[228]

Region Z20, another area with infrequent elm, lies over both sides of the border with Buckinghamshire.

The southern half of Northamptonshire is in region V1,[229] the main English Elm region. Some Narrow-leaved Elm is present, especially along the Warwickshire border. The river name Leam, supposedly based on Celtic *lem-, meaning elm, is discussed in the preceding chapter. It would probably be English Elm that was present at Northampton and at Puxley, in Potterspury,[230] in the sixteenth century. A seventeenth century manorial court roll records unauthorized felling of elm at Badby.[231] Elm was also present at this time at Croughton.[232]

The avenues at Castle Ashby Park were originally laid out in elm and lime. The lime was selectively felled in the early nineteenth century. Capability Brown reputedly reduced the number of rows in the avenues.[233] He left unscathed the elm avenues in Althorp Park. John Wesley, as mentioned in chapter 9, remarked on an elm on the road between Northampton and Towcester.

Narrow-leaved Elm is frequent along the Cambridgeshire border to the south-east of Oundle. This area is in region M1[234] which is widespread in southern Cambridgeshire.

Between regions V1 and Z1 is a band of country with frequent English Elm and Narrow-leaved Elm. This is region MV5.[235] An early seventeenth century

lease stipulated that a lessee in Lowick should plant oak, elm or ash,[236] but this might be no more than a legal formula. Around Wellingborough is region HV1 with frequent English Elm and Hybrid Elms. Wych Elm and Narrow-leaved Elm are also present. This region also extends into northern Bedfordshire. The Tinker's Tree of Mears Ashby was a well-known specimen elm.[237] Finedon Hall had an elm avenue.

Wych Elm occurs sporadically on the limestone hills in the north of the county. Its indigenous status here is feasible. It was presumably wild Wych Elm that was parent to the Hybrid Elms in region HV1.

Prehistoric settlement in Northamptonshire was sparse. Iron Age sites are largely confined to the Nene valley, near Oundle and near Northampton. A great influx took place in Roman times. The whole of the Nene valley was occupied as far west as its crossing by Watling Street. Settlement was also dense along its northern tributary the Ise, as far upstream as Kettering. Romano–British sites also occur along the road linking Leicester with Godmanchester.

Early Anglo-Saxon occupation was lighter. It was principally in the Nene valley, but in two sections, one downstream from Wellingborough, the other along the upper Nene and its tributaries. Settlement along the Ise shifted upstream.

Northamptonshire had much royal forest. Rockingham Forest lay mostly in region Z1 north-east of Kettering. Salcey Forest, largely coincident with Z20, was situated on the Buckinghamshire border south-east of Northampton. Whittlewood Forest lay further south along the same border. Place names indicative of woodland and late settlement are found, in addition to those in the forest areas just listed, in the angle between the Nene and Ise, and south of Daventry.

The light Iron Age settlement makes it improbable that either English Elm or Narrow-leaved Elm entered the county before Roman times. English Elm is not likely to have come from Buckinghamshire since the boundary with this county was heavily forested. It is unlikely to have moved up the Cherwell valley from Oxfordshire, the northern part of which was also heavily wooded in early times. The most probable source is Warwickshire via the Leam valley. This was an important Anglo-Saxon route. It seems most likely then that English Elm entered Northamptonshire some time in the Anglo-Saxon period.

The Narrow-leaved Elm in the Northamptonshire part of region M1 is on the edge of its distribution. It was postulated in the Cambridgeshire section that this elm was introduced into the valley of the Great Ouse in the Later Bronze Age, but was not taken into the wooded country on either side till the Anglo-Saxon period. If this were so, this would be the earliest date at which Narrow-leaved Elm could have entered Northamptonshire. It would also be the earliest date for hybridization with native Wych Elm in region HV1. This region lies between Rockingham and Salcey Forests and is likely to have been heavily wooded too in earlier times. Hybridization between Narrow-leaved Elm and Wych Elm in region HM1 in Essex is also believed to have been a consequence of Anglo-Saxon penetrations into previously unoccupied woodland.

Suffolk[238]

The only recent changes in the boundaries of Suffolk have been the transfer of the north-east corner of the former county to Norfolk.

Suffolk is devoid of conspicuous variation in relief. The centre is mainly gently undulating lowland over clay. The Fenland enters Suffolk at its north-west corner. To the east of the Fenland are the light sandy soils of the Breckland. Sandy soils occur also behind the North Sea coastline. There are marshes along the lower reaches of the river Waveney.

Three major rivers radiate from the neighbourhood of Harwich, the Stour, for much of its course the border with Essex, the Gipping and the Deben. The Waveney running into the North Sea, and the Little Ouse running westward, provide the boundary with Norfolk. The Lark is another westward running river which, like the Little Ouse, feeds the Great Ouse.

The characteristic elm of Suffolk is Narrow-leaved Elm. Hybrid elms occur in south-central Suffolk. Both are usually known as *elm*.

In two areas of the county, both in region Z1, elm is infrequent. One is in the north-west, covering the Suffolk Fenland and a large part of the Breckland. The other is a band of country crossing the Norfolk border between Bungay and Beccles and running first southward and then south-east to meet the sea between Aldeburgh and Orford. The latter area comprises the seven South Elmham parishes: South Elmham All Saints, South Elmham St Cross, and so on. Elmham in Suffolk is first mentioned as such in Domesday Book.[239] One of the two bishops of East Anglia was however Bishop of Elmham, a title first

encountered in the ninth century. The existence of two bishops in East Anglia goes back to the seventh century and there is no reason against supposing that one of the seventh century sees was at Elmham. The difficulty is that both North Elmham in Norfolk and South Elmham in Suffolk have been claimed as the site of the see. With local patriotic zeal, Norfolk antiquaries have argued in favour of Norfolk[240] and Suffolk antiquaries for Suffolk.[241]

In favour of North Elmham is the existence of a late pre-Conquest church and an episcopal manor in the parish, and the fact that, in the fourteenth century, North Elmham was believed to be the site of the see. In favour of South Elmham is the existence of another pre-Conquest church, the so-called Old Minster in a Roman earthwork, the location of this within a ring of seven churches, the present South Elmham parish churches, an arrangement reminiscent of Coptic and Irish custom, the grouping of the South Elmham parishes into a separate deanery with a privileged ecclesiastical jurisdiction of its own, and the possession of extensive property in the South Elmhams by the Bishop of Norwich.

This extraordinary coincidence of similar circumstances in the two Elmhams suggests that their history is intertwined. Latterly, it has been urged as likely, or admitted as a possibility, that the see was founded at South Elmham and moved to North Elmham later, probably in response to the Viking raids.[242]

What has been left completely out of account is the significance of elm distribution in Suffolk and Norfolk. Elm is not frequent in the South Elmhams today though regions M2 and M16 where Narrow-leaved Elm is frequent, are nearby. Elm grew on copyholds in South Elmham St Peter in the sixteenth century.[243] In North Elmham, however, there is much less elm and, as noted in the Norfolk section, negative evidence against its earlier occurrence. The conclusion that follows most naturally is that the original Elmham was at South Elmham, named after locally growing elm. Here, one presumes, the second East Anglian see was established. At some later date, perhaps in the troublous times of the Viking descents, as already suggested, the see would have been removed to the more secluded situation of North Elmham to which the name of the original see was transferred.

Two Lakenheath documents of the sixteenth and seventeenth centuries respectively, refer to the local occurrence of *wych*, the latter to *elm* and *wych*.[244] This is the only place, on present evidence, where the term *wych* appears to have been current in East Anglia. Lakenheath is on the western border of the county, however, and it may also be significant that the lord of the manor was the Prior, later the Dean and chapter of Ely, in Cambridgeshire, where *wych* was long the local word for elm.

Elsewhere in Z1 there is a seventeenth century reference to several Elm Fields in Ickworth parish.[245] Concurrent with this is a mention of elm for constructing a curble for the bottom of a well at Bury St Edmunds.[246]

The south-west of Suffolk is in region M1[247] which extends through Cambridgeshire to Northamptonshire. Narrow-leaved Elm with large-toothed leaves as in eastern Cambridgeshire is frequent. Elm is known to have been present at Wickhambrook in the seventeenth century.[248] Local legend held that ten knights were buried beneath the elm avenue leading to Kedington church.[249] St Mildred's Well, Exning, was in a hollow surrounded by elms.

Most of eastern Suffolk is in region M2.[250] Narrow-leaved Elm of similar form to that in the Suffolk part of M1 is frequent. The presence of elm in earlier times is attested by the place names Elmsett, recorded in the tenth century, and Elmswell, known from the fourteenth century. The church of St Mary at Elms in Ipswich has been so designated since the early thirteenth century.[251] Elm was present at Walsham le Willows[252] and Witnesham[253] in the sixteenth century. The White Elm Inn at Copdock, mentioned in chapter 10, can be traced back to the early seventeenth century.[254] There was an elm avenue at Campsea Ash.

Immediately behind the North Sea coast and separated from region M2 by region Z1, is region M16[255] which extends a short distance into Norfolk. The presence of elm at Yoxford in the fifteenth century is indicated by the mention of an Elmwege.[256] The court rolls of Hinton manor in Blythburgh[257] and a manorial survey in Sternfield,[258] attest the local occurrence of elm in the sixteenth century.

There are two areas where Narrow-leaved Elm and English Elm are both frequent. One, MV13,[259] which lies partly in Norfolk, extends down the coast from the county boundary to Kessingland. An elm at Lowestoft associated with Oliver Cromwell is mentioned

in chapter 9. The other, MV12,[260] is centred on Kersey in southern Suffolk.

Regions M1 and M2 are separated by a tract of country in which Narrow-leaved Elm and Hybrid Elms are frequent. This is part of region HM1 which runs across Essex from eastern Hertfordshire. Buck's Elm, a landmark on Bergholt Heath, is mentioned in the letters of John Constable.[261] HM1 includes the lower Stour valley which provided the Constables with the 96, 97 sites for a number of major elm paintings and 98, 99 drawings.

Wych Elm is seldom encountered in Suffolk. It seems that there may have been some once in the centre of the county but that, as in Essex, it has been hybridized out of existence.

Suffolk has a long history of human settlement. In the Iron Age there were coastal settlements near Orford, around Ipswich and, as noted in the Essex section, along the middle section of the Stour valley. Of inland sites, the most intensive occupation was the country around the headwaters of the Ouse, and along the River Lark.

In Roman times, all these sites, except that at Orford, remained important. The Ipswich area was further developed. In the Lark valley there was perhaps a shift downstream. The lower Waveney was also occupied.

Early Anglo-Saxon occupation was largely in or near Roman sites. The Ipswich settlement expanded to include the Deben estuary. Settlement in the Lark moved upstream again. The middle Stour was rather neglected.

As elsewhere in East Anglia, there were no royal forests in Suffolk. Place names suggestive of wooded terrain and late settlement occur south of Bury St Edmunds, to the east of Sudbury, and in east Suffolk between Halesworth and Eye. Much of the last area is in Z1.

The historical evidence provides satisfactory evidence for the presence of elm in the seventh century. This can hardly be later than Roman in origin. The overall distribution pattern, a western, M1, and an eastern, M2, population of Narrow-leaved Elm, separated by a zone with little elm in the north, Z1, and a hybridization zone in the south, MH1, reflects the bipolar settlement pattern that persisted from Bronze Age to late Anglo-Saxon times. The Narrow-leaved Elm of Suffolk shows only limited variation, which suggests that it may all have arisen from a single introduced stock.

The western population M1 is continuous with that in the Cam valley of Cambridgeshire which, it was suggested in chapter 3, was introduced from central Europe in the Later Bronze Age. It may be supposed that elm from this centre of dispersal would have reached Suffolk, perhaps in the Iron Age, certainly by Roman times.

The eastern population M2 is more of a problem. It seems to have spread out from the Ipswich area. It might have been introduced direct from central Europe concurrently with the introduction of similar elm into the Cam valley. Alternatively, it may have been imported at a later date from Cambridgeshire. The second possibility seems the more likely since, if the Ipswich area had imported its elms directly from central Europe, some significant differences might have been expected between the elms in the Suffolk parts of M1 and M2. No significant differences have been detected. It is certainly necessary to suppose that the Narrow-leaved Elm in M2 had arrived before the end of the Roman period. The fact that South Elmham appears to have received its *elm* name no later than the seventh century, suggests that M2 had reached the Waveney in Roman times. The elms in M16 had arrived by the end of the Middle Ages. The sparsity of elm to the west of M16 and the fact that English Elm was introduced into region MV13, which it nearly surrounds, tell against there being much elm here before the Norman Conquest.

Hybridization between presumed native Wych Elm in central Suffolk and Narrow-leaved Elm brought into this area both from the west and the east would account for the Suffolk part of region HM1. It covers the first two of the areas mentioned above whose place names indicate wooded terrain. Penetration into this area would have largely taken place in the later stages of Anglo-Saxon expansion and this seems likely to be the date when Narrow-leaved Elm was introduced and when hybridization with Wych Elm occurred.

The two regions with frequent English Elm are many kilometres from the main English Elm region V1. Its introduction is likely to be late, perhaps post-medieval.

15
Conservation

The severity of the second Dutch elm disease epidemic has been such that mature elms have been eliminated from a large part of the English landscape. Clearly, a conservation policy is urgently required.[1]

The first point to be taken into consideration is the area of devastation. It does not cover the whole of the British Isles since the disease vectors, the bark-boring beetles, do not occur, or only sporadically, at the higher altitudes and greater elevations. As a rough generalization, the Humber can be taken as marking the critical latitude. South of the Humber, the epidemic has been near universal. North of the Humber, outbreaks have been more local and much elm has remained free of the disease.

Conservation measures based on disease control have been largely abandoned in the southern area as useless, save for injection of a few specimen trees. Injection costs prohibit its use on a wide scale. In the northern area, felling of isolated diseased trees may restrain the spread of the disease under conditions suboptimum for it.

The first item then in a conservation programme is to identify the area where the disease is present or likely to enter. The next consideration is whether, in the area of devastation, elms are worth conserving. Hopefully, this book provides material for an affirmative answer. Reasons in favour of conservation can be assembled under a number of heads. Landscape aesthetics can be taken first. The English elmscapes were appreciated and enjoyed by the majority of those in contact with them. Their disappearance has been felt as a loss. They were, in various respects, unique.

In no other country was exactly the same experience possible. It has naturally followed that elmscapes have come to be regarded as one of the essential elements of the English landscape. When patriotism is analysed it usually has a landscape component. It is natural to identify with landscape and to identify with the English landscape has often meant to identify with an English elmscape. Much of what is said in chapter 10 demonstrates that this has been so. If, then, there is a sort of patriotism that is balanced, reasonable and a right outlet of deep feeling, it involves, by implication, the conservation of elmscape.

English elmscapes have also a special place in English culture consequent on their role in literature and the visual arts. Chapters 10 and 11 respectively illustrate how extensive this role has been. A thorough grasp of many facets of English literature and fine art will elude those who have never experienced an elmscape at first hand. Any conservation programme should identify the most significant associations between major writers and artists and the elmscapes that have most influenced them. These can then receive such conservation treatment as best illuminates these associations.

The temper of scientific conservation is rather different. The assumption now is that anything that has evolved should be kept in being. Or, at least, man should not be responsible for extinction of an organism, either by directly exterminating it, or by failing to take measures that would counter whatever puts the organism at risk. In the case of elm, not only have the two species to be considered, but also the various kinds of each which can be differentiated.

Further, not only elm is at risk. It supports, as shown in chapter 6, a large number of associated organisms whose existence depends obligatorily upon it. It also supports many other organisms, owls, for instance, which can go elsewhere. The disappearance of elm has, however, interfered drastically with their overall environment.[2]

It is true that man is not responsible for the evolution of the fungus causing Dutch elm disease. He cannot be absolved from responsibility for introducing it into most of the regions where it has caused most devastation. Certainly, if the conservation of as much of the world flora and fauna as has survived to the present day is a good thing, then the conservation of the English elms and their associated organisms is a good thing.

In chapter 8, the uses to which elm has been put are surveyed. Many of these are now of historical significance only. Elm timber, however, retains its value and now that its seasoning is well understood is likely to increase in esteem, especially as it is likely to become in short supply. No timber, either home grown or foreign, has a grain with all the attractions of that of elm. A sound elm grown to maturity is most unlikely to prove a bad investment.

Chapters 3 and 5 investigate the significance of elm for elucidating prehistoric and historical problems. If the arguments advanced there have any validity, the significance of elm in these areas is very great, and with improved investigative techniques, could be considerably greater. Elm is the only plant introduced by man that has weathered two millenia of history without, in some stocks, suffering genetic change. It may have much more to reveal about man in the first millenium BC, in the Dark Ages, and in later periods right up to the present day.

Once it is accepted that elm conservation in a devastated area or an area at risk is worthwhile, it is prudent to determine how intensively conservation measures should be applied. The maps accompanying chapters 12–14 are of assistance here. Elm does not contribute much to the landscape of the Z regions, and conservation in these is likely to be justifiable mainly when some particular association, historical, literary or artistic is involved. In all the other regions, elm has been an important landscape component and conservation measures merit consideration.

Different approaches are appropriate for Wych Elm, much of which is native, and for the various Field Elms, all of which are believed to be introduced. The two species have different reproductive systems, the former reproducing by seed, and the latter mainly by suckering from the roots.

Wych Elm flourishes best along streams, in shady ravines, and to a lesser degree in some categories of woodland, principally on calcareous soils. By establishing such sites as nature reserves, or at least managing them to some extent as nature reserves, it is likely that the Wych Elm in them, even if attacked by disease, will produce sufficient seed to survive.

The Field Elms owe their present sites to man and are of longest standing in the hedges and closes around inhabited places, whether nucleate settlements or, where the settlement pattern is dispersed, single houses. Conservation, in these cases, depends on conserving the usually very mixed hedges in which elms principally occur. Since reproduction is largely by suckering, all that needs to be done is to allow such suckers as arise spontaneously to continue growth. Recent observations indicate that there is much elm regeneration by suckering in devastated areas.[3]

The assumption is that conservation should be done on the spot. A quite different approach would be to collect a representative range of elms and grow them under quarantine conditions in some distant and isolated site where Dutch elm disease could, hopefully, be kept out. This second approach has two fatal shortcomings. First, it would be prohibitively expensive. Secondly, a great deal of the rationale of elm conservation rests on the significance of particular kinds of elm in particular places. When an elm is removed from where it has long grown, much of its overall significance goes.

The intensity of conservation appropriate in any region varies with the kind of elm. As regards Wych Elm, there is considerable variation, both between stands and, in the case of isoperoxidase content, within stands. The number of its conservation areas should be in proportion to the variability between stands, and the stands should contain sufficient elm trees to conserve variation within stands.

With Field Elm, there are wide differences in variation both between and within stands. In English Elm, there is little of either category of variation and it would be reasonable to conserve relatively lightly, say two or three hedges per settlement unit. In other Field Elms and the Hybrid Elms, variation can be very great, both between and within stands. When this is so, it would be appropriate to conserve as much hedgerow containing elms as possible.

The sort of conservation envisaged is essentially directed at preservation of local kind of elms rather than of individual trees or particular plantings such as avenues. Both these last are too vulnerable to be conserved as a rule. However, elms in a clump, say in a Capability Brown landscape, are far less vulnerable, considered in aggregate. Measures to conserve elms in such a planting pattern, where the whole landscape is a conservation objective, are not unreasonable.

In any conserved stand, elm can be largely left to look after itself. It may be useful, however, in places where Dutch elm disease has infected the roots, to cut root connections so that infection along a hedgerow is impeded.

Replacement of elms by other species is clearly not a conservation measure and is not considered here. Suggestions for partial replacement have also been made, as for example replacing elm by the nearest related genus *Zelkova* or replacing local kinds by new hopefully resistant varieties. In either case, the effect on the landscape is likely to be very different from what it was, and the historical significance of the local elms goes by the board. Partial replacement is not recommended. The general pattern of severe disease epidemics is one in which a combination of natural selective mechanisms normally results in eventual amelioration of virulence, as indeed happened in the earlier outbreak of Dutch elm disease. The conservation programme outlined, taken with the high variability of the English elm population as a whole, which may extend also to disease resistance, and the prospect that the disease will eventually attenuate, offers a reasonable hope that the elms will come through.

Key to species and varieties of elm

Note: all references to leaf apply to the second leaf from the apex of a mature dwarf shoot.

1   Leaves large (7–16 cm), usually rough above; leaf stalk short, less than 0.3 cm
                    Ulmus glabra: Wych Elm
    Not as above   2
2   Leaves over 6.5 cm, product of length × breadth over 28 cm   *U. × hollandica*: Hybrid Elm 3
    Leaves less than 7.0 cm, product of length × breadth less than 28 cm   *U. minor*: Field Elm 4
3   Habit spreading, branching crooked and irregular, long shoots corky
            U. × hollandica nm. *hollandica*: Dutch Elm
    Habit fan-shaped, branches mostly straight
        U. × hollandica nm. *vegeta*: Huntingdon Elm
    Not as above
                U. × hollandica: Other notomorphs
4   Leaves relatively broad (length/breadth ratio over 0.75), usually rough above, eventually dark green
            U. minor var. *vulgaris*: English Elm
    Leaves relatively narrow (length/breadth ratio under 0.75) usually smooth above, usually light green   5
5   Trunk very straight, silhouette narrow, crown flat, total tooth number of leaf margin below 70
        U. minor var. *cornubiensis*: Cornish Elm
    Trunk very straight, silhouette pyramidal, crown pointed   *U. minor* var. *sarniensis*: Guernsey Elm
    Not as above   6
6   Leading shoot drooping, long shoots conspicuous in the mature canopy, leaves less than 5 cm
                U. minor var. *lockii*: Lock's Elm
    Habit not as above, mature canopy composed of short shoots, leaves up to 7 cm
            U. minor var. *minor*: Narrow-leaved Elm

APPENDIX 2

Key to mines in elm leaves

1   Leaf surface not pleated above mine   2
    Leaf surface pleated above mine   8
2   Mine long and narrow (length/breadth ratio above 7)   3
    Mine extending in all directions (length/breadth ratio below 3)   5
3   Mine unbranched   4
    Mine with several blind ends
                    Bucculatrix albedinella
4   Mine pursuing an irregular course in the leaf blade or following a vein, not closely convoluted
                    Stigmella ulmivora
    Mine closely convoluted in a small area of the leaf, no leaf discoloration   *Stigmella viscerella*
    Mine closely following the leaf margin, or, at first closely convoluted in a fleck of yellow discoloration, then following an irregular course
                Stigmella marginicolella
5   Mine small (less than 1 cm), larva leaving it to live in a cigar-shaped case of a piece of folded leaf   6
    Not as above   7
6   Case over 1 cm   *Coleophora limosipennella*
    Case less than 1 cm   *Coleophora badiipennella*
7   Mine narrow at first, from a vein near the leaf apex, then expanding to form a large blotch
                    Rhynchaenus alni
    Mine with no initial narrow section   *Fenusa ulmi*
8   Mine elongate (length/breadth ratio over 4), lying between two leaf veins   *Phyllonorycter tristrigella*
    Mine oval (length/breadth ratio less than 2)
              Phyllonorycter schreberella

APPENDIX 3

Key to elm galls

1   Gall on overground shoots   2
     Gall on young roots   *Mimeura ulmiphila*
2   Gall a large bladder (3–8 cm) involving distortion
     of a whole shoot   *Eriosoma lanuginosum*
     Not as above   3
3   Leaf strongly puckered   4
     Leaf not strongly puckered   5
4   Single leaves on mature shoots puckered and
     rolled   *Eriosoma ulmi*
     Successive leaves of juvenile shoots puckered into
     rosettes   *Eriosoma patchiae*
5   Gall projecting conspicuously from the leaf
     surface   7
     Gall not projecting conspicuously from the leaf
     surface   6
6   Gall a circular spot (2–4 mm), first yellowish
     green, then brown, finally breached by a slit
     Physemocecis ulmi
     Gall brown and scab-like on the under surface of
     the leaf. Not apparent at first on the upper surface but
     ultimately evident above as a black 'tar spot'
     Eriophyes filiformis
7   Gall large (more than 1 cm), 1–4 present   8
     Gall small (less than 1 cm), usually many present   9
8   Gall a stalkless egg-like structure over the central
     vein   *Kaltenbachiella pallida*
     Gall a stalked body arising from the upper leaf
     surface away from the central vein
     Tetraneura ulmi
9   Gall a small flask-like structure on a vein on the
     under surface of the leaf   *Janetiella lemeei*
     Gall a small pimple-like body projecting from the
     upper surface of the leaf   *Eriophyes campestricola*

Notes

Preliminaries

1 Frontispiece is Michael Young. Relaxing in Doolittle country, ph. repr. *Times* (1901), 18 April.
2 J. N. L. Myres. *Anglo-Saxon pottery and the settlement of England* (1969), p. 119.

1: Introduction

1 Earlier general articles on elm have been written by Baker (1934), Edlin (1965), Evelyn (1964), Hadfield (1960), Lees (1874*a*, *b*), Richens (1956, 1972), Webster (1926).
2 Authorities for Latin names are cited in chapter 7.
3 Fig. 1 is Worthington Smith. *The great wych elm at Moor Court* (HW), woodcut. In *Gardeners' Chron.* (1876), **5**, 435.
4 *Ulmus minor* sensu latissimo Richens (1968).
5 Fig. 2 is A. Sinclair. *Swaffham Prior* (Ca), ph. repr. *E. Anglian Mag.* (1965), **24**, 383.
6 Fig. 3 is Samuel Read. *Sunday morning*, woodcut by Edmund Evans, 34 × 23 cm. In *Illustrated London News* (1856), **28**, 388.
7 Fig. 4 is *Cornish elms at Coldrenick* (Co), one of the classic elm photographs, repr. Elwes and Henry (1913), plate 397.
8 Fig. 5 is Mary Neal. *Lock's elm at Laxton* (Np), ph. repr.
9 Fig. 6 is Nigel Luckhurst. *Guernsey Elm*, ph. repr.
10 Fig. 7 is Nigel Luckhurst. *Dutch Elm*, ph. repr.
11 Fig. 8 is Nigel Luckhurst, *Huntingdon Elm, along Gonville Place, Cambridge:* ph. repr.

2: Botany

1 Though there is general agreement that the Ulmaceae divide into two tribes, there is disagreement among taxonomists as to which tribe some genera pertain. The treatment adopted here is that of Grudzinskaya (1968) who, however, treats the two tribes as separate families.
2 *Ulmus* Linnaeus (1753). No advantage attaches to segregating some elms in other genera such as *Chaetoptelea* Liebmann (1850) and *Microptelea* Spach (1844).
3 D. F. Cutler. Survey and identification of tree roots. *Arboricultural J.* (1948), **3**, 243–6.
4 J. Cullen. A preliminary survey of ptyxis (vernation) in the angiosperms. *Not. Roy. Bot. Gard. Edinburgh* (1938), **37**, 161–214.
5 P. J. A. Howard and D. M. Howard. Microbial decomposition of tree and shrub leaf litter. I. Weight loss and chemical composition of decomposing litter. *Oikos* (1974), **25**, 341–52.
6 The suggestion of Tanai and Wolfe (1977) that the fine venation of *Ulmus* leaves can be distinguished from that of *Zelkova* by a high frequency of percurrent tertiary veins could not be substantiated.
7 Popular description by Keate (1912).
8 Vinje and Vinje (1955).
9 Mature elm pollen is trinucleate according to H. D. Behnke and R. Dahlgren. The distribution of characters within an angiosperm system. 2. Sieve-element plastids. *Bot. Not.* (1976), **129**, 287–95. Artificial germination of elm pollen has been studied by Greguss (1966).
10 The criteria proposed by Stockmarr (1970, 1974) for discriminating between the pollen of different elm species are not reliable.
11 N. T. Moar. Two pollen diagrams from the Mainland, Orkney Islands. *New Phytologist* (1969), **68**, 201–8.
12 B. Fredskild. Postglacial plant succession and climatic changes in a west Greenland bog. *Rev. Palaeobot. Palynol.* (1967), **4**, 113–27; J. C. Ritchie and S. Lichti-Federovich. Pollen dispersal phenomena in arctic–subarctic Canada. *Rev. Palaeobot. Palynol.* (1967), **3**, 255–66.
13 H.-D. Behnke. Sieve-tube plastids in relation to angiosperm systematics – an attempt towards a classification by ultrastructural analysis. *Bot. Rev.* (1972), **38**, 155–97; H.-D. Behnke. P-type sieve-element plastids: a correlative ultrastructural and ultrahistochemical study on the diversity and uniformity of a new reliable character in seed plant systematics. *Protoplasma* (1975), **83**, 91–101; H.-D. Benhke and R. Dahlgren, see note 9.
14 The embryo sac in *Ulmus* is tetrasporic. It may develop according to the *Adoxa*, *Drusa* or *Chrysanthemum cinerariaefolium* types. Descriptions of embryological development are given by Crété *et al.* (1966), D'Amato (1940), Ekdahl (1941), Guignard and Mestre (1966), Hjelmqvist and Grazi (1965).
15 The production of dihaploid calli in *Ulmus americana* which, as so often in trees, developed no further, by the technique of anther culture, is reported by Redenbaugh *et al.* (1981).
16 Chromosomes of *Ulmus* spp. are described by Ehrenberg (1949), Grudzinskaya and Zahar'eva (1967), Leliveld (1933), Sax (1933).
17 Anderson (1933), Gill *et al.* (1939, 1946), Hirst *et al.* (1951), Hough *et al.* (1950).
18 Sørensen and Søltoft (1958).
19 The following interspecific hybrids have been produced involving the English elm species: *Ulmus glabra × U. minor, U. glabra × U. pumila. U. glabra × U. rubra, U. glabra × U. wallichiana,*

U. glabra × *U. parvifolia*, *U. minor* × *U. pumila*, *U. minor* × *U. rubra* and *U. minor* × *U. wallichiana*. The principal papers on these hybrids are by Hans (1981), Went (1954, 1955), Santamour (1972*b*).

20 Santamour (1972*b*).

21 Endtmann (1967), Townsend (1975).

22 Some cytological observations have been reported in putative hybrids between *Ulmus minor* and *U. pumila* (Chernyavskaya, 1978, 1979).

23 Germinability of elm seed under natural or artificial conditions has been investigated by Buszewicz and Holmes (1961–2), Enescu and Enescu (1957), Krstić (1950), Lupe (1956), Manaresi (1935), Popovski (1972), Vincent (1958), Went (1954).

24 The principal kinds of elm are described briefly in the previous chapter and in detail in chapter 7.

25 The incidence of cork in British elm is shortly discussed by Rivers (1855).

26 The foliar characters just described are those used in the biosystematic description of elm populations by Richens (1955 onwards) and in the computer-aided multivariate analysis of these and further populations by Jeffers and Richens (1970) and Richens and Jeffers (1975, 1978). One additional character, tooth length, was also used but this is so strongly correlated with tooth breadth that it can be ignored for general descriptive purposes. Almost all other morphological characters, whether of the leaf or other organs, are correlated to varying degrees with the foliar characters mentioned or some function of them. The principal characters not so correlated are the chemical ones.

A concise leaf formula is useful for representing the biometrical leaf characters of any elm or of any population of elms. The range of each character is divided into three segments, a central mid-range segment, roughly one-fifth of the range, and the segments on either side of this. If a measurement falls within the central segment it is ignored in the formula. If it is within the segment with higher numerical values, it receives a capital letter representing the character. If it is within the segment with lower numerical values, it receives an appropriate lower-case letter. The range of the segments for each character and the relevant letters are as follows:

Character	Inner segment designation	Midrange segment (not designated)	Upper segment designation
leaf length	l	50–59 mm	L
tooth breadth	b	5.0–5.5 mm	B
tooth depth	d	2 mm	D
relative breadth	w	0.60–0.69	W
relative petiole length	p	0.08–0.11	P
relative asymmetry	a	0.08–0.11	A
tooth number	n	90–109	N

To take an example, an English Elm collected at Lewknor (Ox) had leaves whose measurements for the characters listed above were respectively 56 mm, 5.0 mm, 2.5 mm, 0.82, 0.09, 0.10, 89. The leaf is designated DWn. Leaf breadth, tooth breadth, relative

petiole length and relative asymmetry fall within the midrange segment and are therefore omitted in the formula.

When a leaf formula is quoted for a population it is that which occurs most frequently.

27 The morphology and evolution of the elm inflorescence has been investigated by Grudzinskaya (1966), Grudzinskaya and Chernik (1976).

28 C. Chalk, C. and B. J. Rendle. British hardwoods: their structure and identification. *Bull. For. Products Res.* (1929), no. 3.

29 Le Sueur (1941) has suggested that branch dropping may be a consequence of cross tensions resulting from the cross grain.

30 Santamour (1972*a*).

31 Bate-Smith and Richens (1973).

32 The preliminary results of a mainly unpublished series of observations on elm isoperoxidases are reported by Pearce and Richens (1977).

33 Rowe *et al.* (1972).

34 R. Kanerva. Pollenanalytische Studien über die spätquartäre Wald- und Klimageschichte von Hyrynsalmi in NO-Finnland. *Ann. Acad. Sci. Fenn., A* (1956), no. 46.

35 Brett (1978*a, b*).

36 J. R. Lowell. *Letters* (ed. C. E. Norton), 1966, AMS Press, vol. 3, p. 257.

37 Bancroft (1936).

38 Data on flowering dates of various elm species over periods of years are presented by Armitage (1917, 1922), Christy (1909–10, 1922), T. Gray. Correspondence (ed. P. Toynbee and L. Whibley, 1971), Oxford University Press, 681–809.

39 According to Daumann (1975), the initial stage of female receptivity lasts 1–4 days, the ensuing bisexual stage 2–10 days, and the final female phase after the stamens have shrivelled 1–4 days.

40 Anon. (1908), Henry (1910), Ley (1909), Melville (1909), Went (1954).

41 Greguss (1972), Heybroek (1965), Winieski (1959).

42 Inconclusive results on segregation of leaf insertion in elm seedlings were published by Henry (1910), Melville (1937). Some further information on Henry's seedlings was given by Backhouse (1934). Cytoplasmic inheritance of variegation in the cultivated variety Argenteo-variegata of English Elm was inferred by Carter (1946).

3: Prehistory

1 *Ulmipollenites undulosus* Wolff (1934). Synonym: *Polyporopollenites undulosus* (Wolff, 1934) Pflug (1953).

2 Distinctions between the leaves of *Ulmus* and *Zelkova* are discussed in chapter 2.

3 The sections of *Ulmus* are described in chapter 6.

4 *Ulmus carpinoides* Goeppert (1855).

5 *Ulmus braunii* Heer (1856).

6 The fossil leaves contained quercitin-3-glycosides as in contemporary *Ulmus* species. Their steroid composition seemed also to correspond with that of *Ulmus* today. This work is reported by Giannasi and Niklas (1977); also K. J. Niklas and D. E. Giannasi. Angiosperm paleobiochemistry of the Succor Creek

Flora (Miocene) Oregon, USA. *Amer. J. Bot.* (1978), **65**, 943–52.

7 *Ulmus longifolia* Unger (1847).

8 Suggested most nearly related modern species are *U. castaneifolia*, *U. lanceifolia* or *U. parvifolia* in the Far East, or *U. alata* in North America. Possible affinities are discussed by Berger (1957), Mädler (1939), Szafer (1961).

9 The principal Spanish pollen profile is that published by F. Florschütz, J. Menéndez and T. A. Wijmstra. Palynology of a thick Quaternary succession in southern Spain. *Palaeogeogr. Palaeoclimat. Palaeoecol.* (1972), **10**, 233–64.

10 The evidence summarized in this paragraph was assembled by Grosset (1967). The hairy-fruited elm on Gotland was described by Johansson (1921) as *U. glabra* var. *trautvetteri*. The similar form on the Swedish mainland was found by Lindquist (1930, 1932), who segregated it as subvar. *dasycarpa*. The hairy-fruited Hybrid Elm of Öland was described by Johansson (1922), who called it *U. foliacea* var. *pilosula*.

11 The significance of the distribution of these two insects was first pointed out by Richens (1963a).

12 Iversen (1941), Morrison (1959), Pilcher (1969), Smith and Willis (1961–2), ten Hove (1968) Troels-Smith (1960).

13 W. Stokes (ed.) *The martyrology of Oengus the Culdee* (1905), pp. 204–5; W. Stokes (ed.) *The tripartite life of Patrick* (1887), pp. 84–5; *Ancient Laws of Ireland* (1879), vol. 4, pp. 146–7. See also chapter 8 under Elm in medicine.

14 *Calendar of the State Papers relating to Ireland, 1615–1625* (1880), p. 572.

15 There is no evidence in support of the theory (Moss, 1914) that Field Elm is a survivor from alluvial forest.

16 G. O. Clark. Neolithic bows from Somerset, England, and the prehistory of the archery in north-western Europe. *Proc. Prehist. Soc.* (1963), **29**, 50–98; G. Rausing. *The bow* (1967); Troels-Smith (1959).

17 V. G. Childe. The first waggons and carts – from the Tigris to the Severn. *Proc. Prehist. Soc.* (1951), **17**, 177–94; V. G. Childe. The diffusion of wheeled vehicles. *Ethnograph.-Arch. Forsch.* (1954), **2**, 117.

18 M. Ventris and J. Chadwick. *Documents in Mycenaean Greek*, 2nd edn (1973), Cambridge University Press, pp. 369–72.

19 J. Curle. *A Roman frontier post and its people* (1909), p. 292.

20 G. Macdonald. The Roman forts on the Bar Hill, Dumbartonshire, excavated by Mr Alexander Whitelaw of Gartshore. *Proc. Soc. Antiq. Scot.* (1906), **40**, 403–546.

21 P. Vouga. La Tène (1923), p. 56.

22 Troels-Smith (1960).

23 A. Steensberg. *Ancient harvesting implements* (1943), pp. 179–209.

24 B. Pettersson. Gotland and Öland. Two limestone islands compared. *Acta Phytogeograph. Suecia*, no. 50, pp. 131–40; E. Sjögren. Epiphytische Moosvegetation in Laubwäldern der Insel Öland (Schweden). *Acta Phytogeograph. Suecia*, no. 44.

25 See chapter 13 under Wiltshire.

26 Pliny (1938–62) vol. 4, pp. 474–5.

27 The lines of argument that follow were first developed by Richens and Jeffers (1978).

28 A. Dauzat and C. Rostaing. *Dictionnaire étymologique des noms de lieux en France*, (1963) Librairie Larousse.

29 Richens and Jeffers (1975).

30 Dodoens (1554, 1557), de Lobel (1576).

31 Gysseling (1960).

32 The Germanic linguistic roots for elm are considered in the next chapter.

33 See chapter 13 under Avon.

34 See chapter 14 under Greater London.

35 V. G. Childe. Double-looped palstaves in Britain. *Antiq. J.* (1939), **19**, 320–3; H. St G. Gray. Double-looped palstaves found at Curland near Taunton. *Antiq. J.* (1937), **17**, 63–9; C. F. C. Hawkes. Las relaciones en el bronce final, entre la Península Ibérica y las Islas Británicas con respecto a Francia y la Europa Central y Mediterránea. *Ampurias* (1951), **13**, 81–119; H. N. Savory. The Atlantic Bronze Age in south-west Europe. *Proc. Prehist. Soc.* (1949), **15**, 128–55; H. N. Savory. *Spain and Portugal* (1968), pp. 221–7. The argument advanced here is not affected by the possibility that these palstaves may have originated in Britain, as suggested by H. N. Savory. A double-looped bronze palstave possibly from Central Wales. *Bull. Board Celt. Stud.* (1967), **22**, 407–10.

36 The late celticization of Brittany is asserted by P. R. Giot. *Brittany* (1960), pp. 173–4. The existence of people speaking a non-Indoeuropean language in north-west Germany and north-east France is argued by R. Hackmann, G. Kossack and H. Kuhn. *Völker zwischen Germanen und Kelten* (1962); H. Kuhn. *Namenforschung* (1968), vol. 3, pp. 311–34.

37 J. M. Coles, S. V. E. Heal and B. J. Orme. The use and character of wood in prehistoric Britain and Ireland. *Proc. Prehist. Soc.* (1978), **44**, 1–45.

38 See chapter 13 under Somerset.

39 See chapter 14 under Greater London.

40 G. Clark. *World prehistory in new perspective.* 3rd edn, (1977), Cambridge University Press, p. 182; J. Maluquer de Motes. Late Bronze and Early Iron in the valley of the Ebro. In J. Boardman, M. A. Brown and T. G. E. Powell (eds.). *The European community in later prehistory* (1971), Rowman, pp. 105–20; H. N. Savory. *Spain and Portugal* (1968), pp. 227–32.

41 P. Bosch-Gimpera. Two Celtic waves in Spain. *Proc. Brit. Acad.* (1942), **26**, 3–126; P. Bosch-Gimpera. Les mouvements celtiques. *Ét. Celt.* (1950–1), **5**, 352–400; **6**, 71–126, 328–55; (1955–56), **7**, 147–69; P. Bosch-Gimpera. Íberes, basques, celtes. *Orbis* (1956), **5**, 329–38; (1957), **6**, 126–34.

42 J. J. Butler. Bronze Age connections across the North Sea. *Palaeohistoria* (1963), vol. 9.

43 C. F. C. Hawkes. North Germany, Britain and the Fengate pin. *Antiq. J.* (1976), **56**, 234–5.

44 Richens and Jeffers (1978).

45 D. Allen. The origin of coinage in Britain: a reappraisal. In S. S. Frere (ed.). *Problems of the Iron Age in southern Britain* (1958), pp. 97–308.

46 B. Edeine. La technique de fabrication du sel marin

dans les sauneries protohistoriques. *Ann. Bretagne* (1970), **77**, 95–133; P.-L. Gouletquer and J.-P. Pinot. Les briquetages du Trégor. *Ann. Bretagne* (1968), **75**, 142–8; P.-L. Gouletquer. Études sur les briquetages. IV. *Ann. Bretagne* (1969), **76**, 119–47; P.-L. Gouletquer. Briquetages et sauneries. *Ann. Bretagne* (1970), **77**, 135–53; S. Piggott. Ancient Europe (1965), Edinburgh University Press, pp. 170–1, 248–9.

47 Gerard (1633).

48 A. T. Lloyd. The salterns of the Lymington area. *Proc. Hants. Field Club Arch. Soc.* (1967), **24**, 86–102.

49 S. Piggott. A glance at Cornish tin. In. V. Markotic (ed.) *Ancient Europe and the Mediterranean* (1977), Aris and Phillips, pp. 141–5.

4: Vernacular names

1 C. de Smedt and J. de Backer. *Acta sanctorum Hiberniae ex codice Salmanticensi* (1888), p. 203; W. Stokes. *Lives of the saints from the Book of Lismore* (1890), p. 80.

2 The form *tilh* is cited in an Old Breton medical treatise published by W. Stokes. A Celtic leechbook. *Z. celt. Phil.* (1897), I, 17–25; the dictionary was published by E. Ernault. *Dictionnaire étymologique du Breton moyen* (1888).

3 Bartholomeus Anglicus. *Liber de proprietatibus* (1505) lib. xvii, cap. cxcii.

4 Serra (1951).

5 Duhamel du Monceau (1755); du Mont de Coursey (1811).

6 Theophrastus (1916).

7 de Lobel (1576).

8 F. La Flesche. A dictionary of the Osage language. *Bull. Smithsonian Inst. Bur. Amer. Ethn.* (1932), no. 109.

9 Dodoens (1557).

10 Lyte (1578).

11 Gerard (1597).

12 Gerard (1633).

13 J. Bodaeus à Stapel. In Theophrastus (1644), p. 318.

14 Poederlé (1792).

15 As stated in chapter 1, linguistic roots are treated here merely as codes for a presumed ancestral word. There is no intention of reproducing any conjecture as to the exact form this may have taken.

16 G. Cardona, H. M. Hoenigswald and A. Senn. *Indo-European and the Indo-Europeans* (1970), Pennsylvania University Press, pp. 89–111.

17 Richens (1958, 1961b).

18 This type of inference is discussed by W. F. Wyatt, Jr. The Indo-Europeanization of Greece. See Cardona *et al.*, note 16, pp. 89–111.

19 P. Trépos. Le saints bretons dans la toponymie. *Ann. Bretagne* (1954), **61**, 372–406.

20 P. Friedrich (1970a, b).

21 T. Wright and R. P. Wülcker. *Anglo-Saxon and Old English vocabularies*, 2nd edn (1884).

22 C. Lindberg (ed.). MS. Bodley 959. *Stockholm Stud. English* (1969), vol. 20.

23 See chapter 3, note 36, last two references.

24 *Geiriadur prifysgol Cymru* (1950–67), Gwasg Prifysgol Cymru, vol. I, p. 1207.

25 W. de G. Birch. *Cartularium saxonicum* (1897), vol. 2, p. 386.

26 The first and traditional view is presented by R. Haberl. Zur Kenntnis des Gallischen. *Z. celt. Phil.* (1910–12), **8**, 82–101. Ribezzo (1950) developed the fifth view.

27 J. Corominas. *Diccionario crítico etimológico de la lengua castellana* (1954) Editorial Gredos, vol. I, p. 78.

28 A. Dauzat. *La toponymie française* (1946), pp. 192–3; T. F. O'Rahilly. *Early Irish history and mythology* (1946), pp. 453–4; J. Pokorny. Zur Urgeschichte der Kelten und Illyrier. *Z. celt. Phil.* (1940), **21**, 83, 119–21.

29 M. Vasmer. *Z. slav. Phil.* (1938), **15**, 448–55.

30 M. Gimbutas. *The Slavs* (1971).

31 The Lincolnshire occurrence is given in The witch of Grasby. *Folk-lore* (1934), **45**, p. 255. The *wicken* of John Clare's poetry is generally interpreted as rowan (*Sorbus aucuparia*). This is unlikely since rooks do not nest in this tree. The place name High Wicks (Bd) is probably an analogue of the not uncommon High Wyches and High Elms.

32 J. A. Raflis and M. P. Hogan. *Early Huntingdonshire lay subsidy rolls* (1976), pp. 187, 247.

33 M. Gelling. English place names derived from the compound wīchām. *Med. Arch.* (1967), **11**, 87–104.

34 E. Moór. Die slawischen Ortsnamen der Theissebene. *Z. Ortsnamenforsch.* (1930), **6**, 1–37, 105–40.

35 Eilers and Mayrhofer (1961).

36 N. Jokl. Zur Vorgeschichte des Albanischen und der Albaner. *Wörter und Sachen* (1929), **12**, 63–91.

37 Considered improbable by G. Alessio. La base preindoeuropea *kar(r)a 'pietra' II. *Stud. Etruschi* (1936), **10**, 165–89.

38 R. Trautmann. Die slavischen Orstnamen Mecklenburgs und Holsteins. *Abhand. sächsischen Akad. Wissenschaften Leipzig, Phil.-hist. Klasse* (1950), **45**(3).

39 I. Iordan. *Rūmanische Toponomastik* (1924), vol. I, p. 33.

40 S. Bugge. Beiträge zur etymologischen Erläuterung der armenischen Sprache. *Z. vergleich. Sprachforsch. Indogerman. Sprachen* (1893), **32**, 1–87; E. Boisacq. *Dictionnaire étymologique de la langue grecque*, 4th edn (1950) Carl Winter; P. Chantraine. *Dictionnaire étymologique de la langue greque* (1974) Éditions Klincksieck, vol. 3; P. Friedrich (1970a, b); J. B. Hofmann. *Etymologisches Wörterbuch des Griechischen* (1971), R. Oldenbourg; H. Hübschmann. *Armenische Grammatik* (1897), vol. I, pp. 374–5, 449.

41 The etymology of Caucasian elm words is discussed by Eilers and Mayrhofer (1961).

42 E. Lhuyd. *Archaeologia britannica* (1707), pp. 175, 275; W. Pryce. *Archaeologia cornu-britannica* (1790).

43 T. Vuorela. The Finno-Ugric peoples. *Indiana Univ. Pub., Uralic Altaic Ser.* (1964), **39**.

44 C. Battiste. *Sostrate e parastrati nell' Italia preistorica* (1959), p. 261.

45 V. Bertoldi. Relitti etrusco-campani. *Stud. estruschi* (1933), **7**, 279–93.

46 F. Le Quesne. Private communication.
47 Richens (1958, 1961*b*).
48 F. Le Maistre. *Dictionnaire Jersiasis-Français* (1966), Don Balleine Trust.

5: History

1 J. D. Pheifer. *Old English glosses in the Épinal-Erfurt glossary* (1974), Oxford University Press; H. Sweet. *The oldest English texts* (1885); T. Wright and R. P. Wülcker. *Anglo-Saxon and Old English vocabularies*, 2nd edn (1884).
2 Particular use has been made of P. H. Blair. *An introduction to Anglo-Saxon England*, 2nd edn (1977), Cambridge University Press; J. N. L. Myres. Romano-Saxon pottery. In D. B. Harden (ed.), *Dark-Age Britain* (1956), pp. 16–39; J. N. L. Myres. *Anglo-Saxon pottery and the settlement of England* (1969); W. Piroth. Studies on place-names and Anglo-Saxon migration: a comparison of *-ingas, inga-* names in England with their parallels on the European Mainland. *Nomina* (1977), 1, 27–31.
3 F. M. Stenton. *Anglo-Saxon England*, 3rd edn (1971), Oxford University Press, p. 6.
4 Bede. *Historia ecclesiastica* (ed. C. Plummer, 1896) p. 30–3.
5 Richens and Jeffers (1978).
6 They are noted in the early seventeenth century by A. Standish. *New directions of experience authorized by the Kings most excellent Majesty, as may appeare, for the increasing of timber and fire-wood* (1616).
7 C. Vancouver. *General view of the agriculture in the county of Cambridge* (1794), pp. 76–7.
8 Whittlesford (Ca) Manorial Court Roll (1391). Cambridge, County Record Office. Huddleston Coll.
9 See chapter 12 under Staffordshire, chapter 13 under Oxfordshire and chapter 14 under Bedfordshire.
10 See chapter 12 under Staffordshire.
11 See chapter 14 under Bedfordshire and Cambridgeshire.
12 See chapter 14 under Cambridgeshire and Hertfordshire.
13 J. Aubrey. In *The natural history of Wiltshire* (ed. J. Button, 1847), p. 54.
14 R. F. Kilvert. *Diary* (ed. W. Plomer, 1938–40).
15 T. F. Kirby. *Annals of Winchester College* (1892), p. 371.
16 Plukenet (1696).
17 D. B. Green. *Gardener to Queen Anne* (1956), p. 43.
18 A. Pope. *Correspondence* (ed. S. Sherburn, 1956), vol. 2, pp. 13–15.
19 Miller (1748).
20 H., H. Selections from the Monymusk papers. *Publ. Scot. Hist. Soc.* (1945), 39, 97, 113.
21 Richens and Jeffers (1975, 1978).
22 R. A. Cotgrave. *A dictionarie of the French and English tongues* (1611).
23 W. V. Wartburg. *Französisches etymologisches Wörterbuch* (1961), Helbing & Lichtenhahn, vol. 14, pp. 655.
24 H. I. Triggs. *Garden craft in Europe* (1913), p. 191.
25 H. R. Rankin. Cattle droving from Wales to England. *Agriculture* (1955), 62, 218–21.
26 J. Ingram. (1847).
27 T. Quayle. *General view of the agriculture and present state of the islands on the coast of Normandy* (1815), p. 271.
28 Loddiges (1836).
29 Loudon (1838).
30 Rogers (1869).
31 L. Wilson. *Portrait of the Isle of Wight* (1965), p. 160.
32 J. T. Dillon. *Travels through Spain* (1780), p. 183; Evelyn (1664); A. Fanshawe. *Memoirs* (1905), p. 178; H. Swinburne. *Travels through Spain* (1779), pp. 327–32. The Aranjuez avenues were indifferently drawn and engraved by Henry Swinburne, *ibid.*, p. 327.
33 English Elm was being propagated in an American nursery near Boston in the eighteenth century (Anon., 1911).
34 Armfield (1939).

6: Associated organisms

1 The general theory of evolutionary parallelism between elms and their associates has been elaborated by Richens (1963*a*). The particular relations between Himalayan and European elms and between their respective associates have also been investigated by Richens (1962).
2 Wetwood is described further by Peace (1962).
3 Westerdijk and Buisman (1929).
4 P. Pesson, C. Toumanoff and C. Hararas. Étude des épizooties bactériennes observées dans les élevages d'insectes xylophages. *Ann. Épiphyties* (1955), 6, 315–28.
5 J. M. Trappe. Fungus associates of ectotropic mycorrhizae. *Bot. Rev.* (1962), 28, 538–606.
6 Imperfect state: *Graphium ulmi*.
7 As stated in the preface, it is not feasible in a general account such as this to list all papers on Dutch elm disease. Some of the principal are: Brasier (1979); Brasier and Gibbs (1973, 1975); Clinton and McCormick (1936); Fransen (1939); Gibbs (1978*a*, *b*); Gibbs and Brasier (1973); Gibbs and Howell (1972); Jeffers (1972); Laut *et al.* (1979); Peace (1960); Takai (1974); Went (1954); Westerdijk and Buisman (1929).
8 Imperfect state: *Cytospora carbonacea*.
9 Imperfect state: *Libertella dissepta*.
10 Imperfect state: *Macrodiplodia ulmi*.
11 Imperfect state: *Piggotia astroidea*.
12 The following *Phyllosticta* species have been described from elm: *Ph. bellunensis, Ph. confertissima, Ph. lacerans, Ph. melaleuca, Ph. ulmaria, Ph. ulmi* and *Ph. ulmicola*. The relations of the European elm *Phyllosticta* species have been considered by Săvulescu and Eliade (1962) and Cejp (1967, 1969).
13 Troels-Smith (1960); B. Walldén. Misteln vid dess nordgräns. *Svensk Bot. Tid.* (1961), 55, 427–549. Far Eastern varieties of *Viscum album*, in contrast to those in Europe, have elm as a preferred host according to A. Hosie. *Manchuria* (1901), pp. 14, 149, and A. de C. Sowerby. *Sport and science on the Sino–Mongolian frontier* (1918), pp. 233, 243.
14 H. E. Durham. On mistletoe. *Trans. Woolhope Nat. Field Club* (1933–5), 140–53; P. Sims. *A history of Saltford village* (1976), p. 99; Wrest Park. II. *Country Life* (1904), 16, 90–8.
15 J. J. Lipa. *Stempellia scolyti* (Weiser) comb. nov. and

Nosema scolyti sp. n., microsporidian parasites of four species of *Scolytus* (Coleoptera). *Acta Protozool.* (1968), **6**, 69–77.

16 K. Purrini. Zur Kenntnis der Krankheiten des grossen Ulmenssplintkäfers, *Scolytus scolytus* F. im Gebiet Kosova, Jugoslavien. *Anzeiger Schädlingskunde Pflanzenschutz Umweltschutz* (1975), **48**, 154–6; J. Weiser. *Plistophora scolyti* sp. n. (Prot. Microsporidia), a new parasite of *Scolytus scolytus* F. (Col. Scolytidae). *Folia Parasitol.* (1968), **15**, 11–14.

17 D. J. Hunt and N. G. M. Hague. A redescription of *Parasitaphelenchus oldhami* Rühm. *Nematologica* (1974), **20**, 174–80; J. N. Oldham. On the infestation of elm bark-beetles (Scolytidae) by a nematode, *Parasitylenchus scolyti* n. sp. *J. Helminth.* (1930), **8**, 239–48.

18 D. J. Hunt and N. G. M. Hague. The bionomics of *Crystophelenchoides scolyti* n. comb., syn. *Ektaphelenchus scolyti* Rühm., 1956 (Nematoda: Aphelenchoididae) a nematode associate of *Scolytus scolytus* (Coleoptera: Scolytidae). *Nematologica* (1976), **22**, 212–16.

19 *T. cruciata*, *T. hippocastani* and *T. plebeja*. The distinguishing characters are described by W. E. China. New and little-known species of British Typhlocybidae (Homoptera). *Trans. Soc. Brit. Ent.* (1943), **8**, 111–53.

20 Heslop Harrison (1920).

21 Richards (1967). The elm associates are *T. platani* in Europe, *T. saltans* in Central Asia, *T. ulmi-parvifoliae* in the Far East and *T. ulmifolii* in North America.

22 *E. ulmi* is not treated here as specifically distinct from *E. anncharlotteae* (Danielsson, 1979).

23 *E. pyricola* is not treated here as specifically distinct from *E. lanuginosum*.

24 Hille Ris Lambers (1968–9).

25 The wide geographical range of *Tetraneura* species on grass roots does not establish the hypothesis of Mordvilko (1929) that elms once had a correspondingly wide range.

26 General accounts: Beaver (1966a), Fisher (1928), Laidlaw (1932), Munro (1926). Taxonomic revision of *Scolytus*: Michalski (1973).

27 Fisher (1931, 1937).

28 MacDougall (1900).

29 Meyer and Norris (1967).

30 Doskotch et al. (1970, 1973).

31 Pearce et al. (1975), Peacock et al. (1971).

32 Blight et al. (1977, 1978, 1979).

33 Beaver (1966b).

34 Allen (1975), Donisthorpe (1941), Elton (1970), Marshall (1978), Pool (1904), Verdcourt (1947).

35 Gordon (1943).

36 Carson (1958).

37 Miles (1978).

38 Illustrated but not identified by Horne (1938).

39 Hering (1931).

40 Kuznetsov (1962).

41 Richens (1962).

42 Braun (1963).

43 Taxonomic revision: Emmett (1973–4); Klimesch (1975). English stigmellid mines were described by Richens (1963b).

44 *Stigmella fulvomacula*. Described by Skala (1936, 1941).

45 *Stigmella ulmicola* and *S. ulmifoliae* respectively.

Described by Hering (1931, 1932) and Jacobs (1962).

46 Klimesch (1941).

47 *Stigmella ulmiphaga* (Preissecker, 1942) is only known with certainty from the Danube. Its presence in England inferred by E. M. Hering from a mine in an Essex elm leaf described by Richens (1963b) seems unlikely. *S. suberosella* (Toll, 1934) is doubtfully distinct from *S. marginicolella*. *S. tauromeniella* (Groschke, 1944), from southern Italy, is closely related to *S. viscerella*. *S. gracilivora* (Skala, 1942) is a completely uncertain entity.

48 Skala (1933, 1934, 1939, 1941).

49 Identified in the author's mine herbarium by A. G. Carolsfeld-Krausé.

50 T. Kumata. Private communication.

51 *Stigmella apicialbella* and *Stigmella ulmella*.

52 Kumata (1959, 1963, 1967).

53 Benson (1936).

54 Verzhutskiĭ et al. (1976).

55 P. Starý. Biology of *Aeropraon lepelleyi* (Wat.), a parasite of some eriosomatid aphids (Hymenoptera, Aphidiidae). *Acta Ent. Bohemoslavaca* (1976) **73**, 312–17.

56 First recorded in England by Richens (1961c).

57 Keifer (1939).

58 Keifer (1959).

59 Farkas (1960).

60 Keifer (1962).

61 Keifer (1940).

62 *Aculops neokonoella* Keifer (1975).

63 *Aculops verapasi* Keifer (1971).

64 There is some uncertainty as to the mite species responsible for the galls on *U. laevis*. Almost always *E. brevipunctatus* is present together with another species, *E. multistriatus*, which is related to *E. filiformis*, discussed below.

65 Keifer (1965).

66 See note 62.

67 Keifer (1975).

68 H. H. Keifer. Private communication.

69 They have five-rayed feather claws. *E. ulmi* has a three-rayed feather claw and *E. campestricola* a two-rayed equivalent.

70 I. Newton. Bud-eating by bullfinches in relation to the natural food supply. *J. Appl. Ecol.* (1964), **1**, 265–79.

71 M. Christy. *The birds of Essex.* (1890), p. 136.

7: Botanical classification

1 *International Code of Botanical Nomenclature* (1978), Bohn, Scheltema & Holkema.

2 Theophrastus (1916).

3 Throughout this chapter, 'form' is used in a general sense as a subordinate category to kind. It is not used for the botanical rank designated forma.

4 Vernacular designations with initial capitals are used with the same significance in this chapter as in earlier ones. They are defined precisely in succeeding sections.

5 Columella (1968).

6 Pliny (1938–62).

7 Theophrastus (1644), p. 214.

8 Ruellius (1536).

9 Dodoens (1554).

10 Dodoens (1557).
11 Lyte (1578).
12 Dodoens (1583).
13 Gerard (1597).
14 Dalechamps (1586).
15 Bauhin (1623).
16 Bock (1539).
17 There are no grounds for equating Bock's *Ulmus lata* with *U. montana* of Bauhin (1623) as the latter supposed. Nor is the view (Hoppe, 1969) that Bock's *Ulmus lata* is the modern *U. laevis* and his *U. procera* is Wych Elm (*U. glabra*) any better supported.
18 de Lobel (1581).
19 Gerard (1633).
20 Parkinson (1640).
21 Plot (1677). Fig. 42 is Michaell Burghers. *Ulmus folio angusto glabro*, copper engraving, *loc. cit.*
22 Plukenet (1696).
23 Tournefort (1700).
24 Miller (1731). Fig. 43 is Georg Ehret. *Ulmus folio glabro*, water-colour, 53 × 35 cm. London, British Museum (National History).
25 Miller (1739).
26 Miller (1759).
27 Little (1923).
28 There are a large number of general accounts of British elm taxonomy. Many present little new beyond nomenclatural revision. The principal papers not considered further elsewhere include: Bancroft (1934, 1935, 1936, 1937); Boulger (1879), Clapham *et al.* (1952, 1962), Ley (1910b), Richens (1981), Schneider (1916b), Stearn and Gilmour (1932).
29 Linnaeus (1737).
30 Linnaeus (1753).
31 Few botanists have followed Linnaeus in combining Wych Elm (*Ulmus glabra*) and Field Elm (*U. minor*) into a single species, *U. campestris*. One who did so was Hudson (1778).
32 Hudson (1762).
33 Miller (1768).
34 Miller's confusion over *Ulmus sativa* led to its being diversely interpreted by later botanists. Melville (1939b), for this reason, proposed that it be discarded as a *nomen ambiguum*.
35 Weston (1770). Five years later, (Weston, 1775), he published a second treatment of the British elms, with much difference in detail.
36 Moench (1785).
37 There has been much confusion about corky elms in the past, especially with corky Field Elms and Dutch Elm. The supposition (Burbidge, 1896) that some British corky elms pertained to *Ulmus alata* was dispelled by Elwes (1907).
38 Afanasijev (1957) and Andronov (1969) supported specific status for *U. suberosa*. Grudzinskaya (1956) presented a strong case for rejecting it.
39 Sowerby and Smith (1790–1814), vol. 31, p. 2161. Elucidated further by Garry (1904).
40 Syme (1868).
41 Lindley (1829).
42 Loudon (1838).
43 Loddiges (1836).
44 Druce (1907).
45 Druce (1911a).
46 Henry (1913).
47 Moss (1914).
48 Horwood (1933).
49 Bancroft (1934 onwards).
50 Melville (1938a–75).
51 Synonymous with the type var. *coritana*.
52 Melville (1937d, 1939d) used a biometrical method to define elm leaf shape. This was used for improving the precision of description of type or selected specimens. The application of numerical taxonomic methods to elm classification is found in papers by Richens (1955–67). Results of computer-aided analysis of biometrical data on elm leaf variation have been published by Jeffers and Richens (1970) and Richens and Jeffers (1975, 1978).
53 The pros and cons of the two approaches are argued by Melville (1978) in favour of the traditional approach, and by Richens (1980) in favour of the biometrical.
54 Surveys of the world array: Planchon (1848, 1873); Rehder (1940); Schneider (1916a).
55 *Microptelea* (Spach, 1841a) Schneider 1916a). *Ulmus crassifolia* Nuttall (1837), put by Schneider into this section, is not closely related to *U. parvifolia*. It is best transferred to the next section.
56 *Ulmus parvifolia* Jacquin (1798).
57 *Trichoptelea* Schneider (1916a).
58 *Chaetoptelea* (Liebmann, 1850) Schneider (1916a).
59 *Ulmus villosa* Gamble (1902).
60 *Blepharocarpus* Dumortier (1827) has priority over section *Oreoptelea* of Spach (1841b). Subgenus *Oreoptelea* of Planchon (1848) comprises the same species as section *Blepharocarpus*. Since no name has priority outside its own rank, should anyone wish to subdivide *Ulmus* into subgenera rather than sections, then *Oreoptelea* would have to be used.
61 *Ulmus americana* Linnaeus (1753).
62 *Ulmus divaricata* Mueller (1936).
63 *Ulmus laevis* Pallas (1784).
64 Synonyms of section *Ulmus* are section *Madocarpus* of Dumortier (1827) and section *Dryoptelea* of Spach (1841b). If subgenera are used, the correct designation is subgenus *Ulmus*, of which *Dryoptelea* of Planchon (1848) is a synonym. Cheng (1963) erected a section *Trichocarpus* for the Chinese species *U. glaucescens* Franchet (1884) and *U. kunmingensis* Cheng (1963). These are better retained in section *Ulmus*.
65 Moss (1914) erected three series. Series *Nitentes* was roughly equivalent to Field Elm (*Ulmus minor* var. *minor*). Series *Campestres* comprised English Elm (*U. minor* var. *vulgaris*). Series *Glabrae* corresponded to Wych Elm (*U. glabra*). Schneider (1916a) took over series *Nitentes* and added the various Far Eastern elms with the seed displaced to the apex of the fruit. Series *Campestris* he ignored. He erected two new series: series *Pumilae* for *U. pumila* and *U. glaucescens* and series *Lanceaefoliae* for *U. lanceifolia* Wallich (1831). The three series so constituted he placed under subsection *Foliaceae*. Moss's series *Glabrae* was converted into series *Euglabrae* and the Chinese species with glabrous fruits were added. A series *Wallichianae* was erected for oriental elms with hairs over the entire fruit surface. Elms with

hairs only over the seeds, considered in the next section, so-called *U. elliptica* from the USSR, and *U. rubra* from North America, were put into a series *Fulvae*. These last three sections were placed under subsection *Glabrae*. Under the current rules, series *Euglabrae*, if retained, would have to be changed back to series *Glabrae*.

66 *Ulmus rubra* Muhlenberg (1793).

67 *Ulmus wallichiana* Planchon (1848).

68 *Ulmus chumlia* Melville and Heybroek (1972). Synonym: *U. androsowii* Litvinov (1922) var. *virgata* (Planchon, 1873) Grudzinskaya (1971*a*).

69 *Ulmus davidiana* Planchon (1873); *Ulmus wilsoniana* Schneider (1912).

70 *Ulmus pumila* Linnaeus (1753); *Ulmus macrocarpa* Hance (1868).

71 The two papers referred to in note 52 also present the cases, respectively, for recognizing many or few species of European elms.

72 The correct name of Wych Elm is *Ulmus glabra* Hudson (1762) sensu Moss (1912). Some, chiefly Continental, botanists have used the name *U. campestris* Linnaeus (1753) sensu Pallas (1784) for this elm, since the elm labelled *U. campestris* in Linnaeus' herbarium is this species. Linnaeus, however, did not designate type specimens and his writings make it clear that he regarded the Field Elm (*U. minor*) as the typical representative of what he called *U. campestris*. Rejected synonyms at specific rank are *U. latifolia* (Weston, 1770) Cullum (1774), *U. montana* Stokes (1787) and *U. scabra* Miller (1768). Rejected synonyms at varietal rank are *U. campestris* Linnaeus (1753) var. *campestris* sensu Druce (1908), *U. campestris* Linnaeus (1753) var. *latifolia* (Weston, 1770) Aiton (1789), *U. glabra* Hudson (1762) var. *glabra* sensu Hudson (1762), *U. glabra* Hudson (1762) var. *latifolia* Weston (1770), *U. montana* Withering (1787) var. *genuina* Syme (1787), *U. montana* Stokes (1787) var. *montana* sensu Smith (1800) and *U. scabra* Miller (1768) var. *scabra* sensu Gürke (1897). Earlier description of the Wych Elm (*U. glabra*) are given by Moss (1912, 1914) as *U. glabra*, and by Bancroft (1934), Elwes and Henry (1913) and Sowerby and Smith (1790–1814), **27**, p. 1887 as *U. montana*.

Fig. 44 is Edward Hunnybun. *U. glabra*, drawing, repr. Moss (1914). Other botanical illus.: Mary Ferguson. *U. glabra*, Butcher (1961), Stella Ross-Craig. *U. glabra*, Ross-Craig (1970), both drawing repr.; James Sowerby. *U. montana*, coloured copper engraving. In Sowerby and Smith, *loc. cit.*

73 *Ulmus bergmanniana* Schneider (1912).

74 *Ulmus uyematsui* Hayata (1913).

75 *Ulmus changii* Cheng (1936).

76 *Ulmus gaussenii* Cheng (1939).

77 The elms described under the names *U. elliptica* Koch (1849), *U. excelsa* Borkhausen (1800), *U. laciniata* (Trautvetter, 1859) Mayr (1906), *U. podolica* (Wilczinski, 1921) Klokov (1963) and *U. sukaczewii* Andronov (1955) are not regarded as distinct from Wych Elm (*U. glabra*) at specific level. Foliar biometrical data do not support a distinction in England between *U. glabra* Hudson (1762) var. *glabra* (= subsp. *glabra*, · = var. *scabra* Lindquist

(1929–31)) and *U. glabra* Hudson (1762) var. *montana* Lindquist (1929–31) (= subsp. *montana* (Lindquist (1929–31) Tutin)).

78 *U. glabra* Hudson (1762) var. *trautvetteri* Johansson (1921). This variety has often been treated as a species under the name *U. elliptica*.

79 With the taxonomic treatment favoured in this book, the correct name for the Field Elm is *Ulmus minor* Miller (1768) sensu latissimo Richens (1968). *U. campestris* Linnaeus (1753) sensu Stokes (1787) would have been the obvious name. However, its rejection as a nomen ambiguum has been argued by Stearn and Gilmour (1933) and Melville (1938*b*), largely on the grounds that it has been applied to Wych Elm (*U. glabra*) as well as to the Field Elm (*U. minor*). Their arguments have been generally accepted though Grudzinskaya (1957, 1971*b*) found them unsatisfactory. The earliest substitute name available is *U. minor* Miller (1768) with an enlarged circumscription. Melville (1939*c*) has shown that, with a narrow circumscription, *U. minor* (1768) is also a nomen ambiguum. With a broad circumscription it is not ambiguous. Another name with equal priority to *U. minor* Miller (1768) is *U. sativa* Miller (1768). This is also a nomen ambiguum and reasons against taking it up have been sufficiently argued by Melville (1939*b*).

Rejected synonyms at specific rank, all later than *U. minor* Miller (1768), are *U. carpinifolia* Gleditsch (1773), *U. carpinifolia* (Gleditsch, 1773) Suckow (1777), *U. foliacea* Gilibert (1792), *U. glabra* Miller (1768), *U. nitens* Moench (1794), *U. suberosa* Moench (1785) sensu Hooker and Arnott (1855) and *U. surculosa* Stokes (1812). *U. carpinifolia* Gleditsch (1773) was based on the pre-Linnaean *U. carpini folio seu cortice arboris albido* of Ruppius (1745), a German Field Elm. Gleditsch's species is discussed by Melville (1946, 1956) and Rehder (1938). An earlier description of the Field Elm is given by Richens (1976).

80 Isoperoxidases with Rf values of 28 and 48 respectively according to Pearce and Richens (1977).

81 *Ulmus castaneifolia* Hemsley (1889–1902).

82 The following elms, described at specific rank, are all regarded as coming within the circumscription of the Field Elm (*Ulmus minor*) as understood here: *U. angustifolia* (Weston, 1770) Weston (1775), *U. araxina* Takhtadzhyan (1945), *U. asperrima* Simonkai (1890), *U. boissieri* Grudzinskaya (1977), *U. canescens* Melville (1957), *U. carpinifolia* Gleditsch (1773), *U. carpinifolia* (Gleditsch, 1773) Suckow (1777), *U. georgica* Shkhiyan (1953), *U. germanica* Hartig (1851), *U. globifera* Hartig (1851), *U. micranthera* Kittel (1844), *U. modiolina* du Mont de Coursey (1811), *U. nemorosa* Borkhausen (1800), *U. nitens* Moench (1794), *U. pilifera* (Borbás, 1881) Borbás (1891), *U. sativa* Miller (1768) sensu Moss (1912), *U. suberosa* Moench (1785), *U. tetrandra* Schkuhr (1791), *U. tortuosa* Host (1827), *U. uzbekistanica* Drobov (1953) and *U. wyssotzky* Kotov (1940).

Rejected names at infraspecific ranks are: *U. campestris* Linnaeus (1753) subsp. *campestris* sensu Arcangeli (1882), *U. campestris* Linnaeus (1753) subsp. *suberosa* (Moench, 1785) Arcangeli (1882),

86 J. R. Walbran. Memorials of the abbey of St Mary of Fountains. *Publ. Surtees Soc.* (1862), **42**, 34.

87 Spofforth manorial court rolls transcript. Leeds, Yorkshire Archaeological Society. MS880, p. 656.

88 Specimen preserved at Skipton, Craven Museum.

89 T. Horsfall. *Notes on the manor of Welland Snape* (1912), p. 239.

90 LF: LDwpaN.

91 LF: LbDpaN.

92 LF: LDwpaN.

93 LF: DWn.

94 LF: DWAn.

95 The elms of Nottinghamshire have been previously described by Howitt and Howitt (1963).

96 LF: LDwpaN.

97 LF: LDpaN.

98 Accounts of the Nottingham elm have been published by A. B. (1855); and J. Holland Walker. An itinerary of Nottingham. *Trans. Thoroton Soc. Notts.* (1930), **34**, 1–52.

99 LF: DWAn.

100 LF: LDwPAn.

101 LF: Wych Elm LDpaN; English Elm DWn. Illus.: Samuel Oscroft. *Early morning – the Trent from the Willow Holt, Wilford*; *Evening, Wilford Church*; *Wilford Church*; *Evening, Wilford Green*; *Wilford Church*; *Old cottages, Wilford*; all water colours. Nottingham, Castle Museum.

102 D. G. Wilson. Plant remains from a Roman well at Bunny, Nottinghamshire. *Trans. Thoroton Soc. Notts.* (1968), **72**, 42–9.

103 LF: Wych Elm LDwpN; English Elm DWn.

104 LF: Wych Elm LDwpN; Narrow-leaved Elm LBPAn; English Elm DWn.

105 J. Ogilby. *Road maps of England and Wales* (1971), p. 50.

106 Ph. repr., *Country Life* (1947), **102**, 1310–13.

107 LF: LDpN.

108 C. H. Drinkwater. Seven Shrewsbury gild merchant rolls of the 14th century. *Trans. Shropshire Arch. Soc. Nat. Hist. Soc.* (1903), **3**, 47–98.

109 LF: LDpaN.

110 LF: LDpaN.

111 Lease (1656). Shrewsbury, Shropshire Record Office, 38/231.

112 LF: BDWPA.

113 See note 19.

114 Place names and surnames not otherwise attributed are from Smith, see note 14.

115 LF: LDpaN.

116 Articles of agreement (1657). Leeds, Yorkshire Archaeological Society. DD/70/54. Articles of agreement (1675). Leeds, Yorkshire Archaeological Society. DD/70/127.

117 LF: Wych Elm LDpN; English Elm DWn.

118 The elms of Staffordshire have been previously described by Edees (1972).

119 Ekwall (1960).

120 Illus.: Thomas Peploe Wood. *Norbury elm*, ink and wash. Stafford, William Salt Library.

121 G. Wrottesley. Extracts from the plea rolls of the reign of Edward II. *Coll. Hist. Staffs.* (1889), **10**, 17–18.

122 G. Wrottesley. The subsidy roll of A.D. 1327. *Coll. Hist. Staffs.* (1886), **7**, 193–255.

123 R. Plot. *The natural history of Stafford-shire* (1686), 210–11.

124 LF: LDpaN.

125 G. Wrottesley. Extracts from the plea rolls of Edward III and Richard II. *Coll. Hist. Staffs.* (1892), **13**, 164.

126 LF: LDpaN.

127 G. Wrottesley. The subsidy roll of Edward III, A.D. 1332–3. (1889), vol. 10, pp. 77–132.

128 See note 121, p. 12.

129 G. Wrottesley. Extracts from the *coram rege* rolls of Edward III and Richard II. *Coll. Hist. Staffs.* (1893), **14**, 75.

130 LF: LDpaN.

131 Fig. 120 is Jacob Strutt. *The Tutbury wych elm*, etching. In Strutt (1822), plate 6; Fig. 121 is Thomas Peploe Wood. *Tutbury elm*, ink. In Stafford, Museum and Art Gallery. Another illus.: Thomas Peploe Wood. *Tutbury elm*, ink, Stafford, William Salt Library.

132 G. Wrottesley. Extracts from the plea rolls of the reigns of Richard II and Henry IV. *Coll. Hist. Staffs.* (1894), **15**, 64.

133 LF: DWn.

134 LF: LDpaN.

135 The elms of West Midlands have been previously described by Cadbury *et al.* (1971).

136 LF: LDpaN.

137 J. E. B. Gover, A. Mawer and F. M. Stenton. *The place-names of Warwickshire* (1936), Cambridge University Press.

138 G. Wrottesley. Extracts from the plea rolls. Temp. Edward IV, Edward V and Richard III. *Coll. Hist. Staffs.* (1903), **6**, 161.

139 LF: DWn.

140 LF: LDpaN. Fig. 122 is *Old elm, church, lych-gate, and village cross, High Ackworth*, woodcut. In W. Smith. *Old Yorkshire* (1884), vol. 5, p. 171.

141 Place names and surnames not otherwise attributed are from A. H. Smith, see note 14.

142 W. P. Baildon. Court rolls of the manor of Wakefield. II. 1297–1309. *Rec. Ser. Yorks. Arch. Soc.* (1906), **36**; J. W. Walker. Court rolls of the manor of Wakefield. V. 1322–1331. *Rec. Ser. Yorks. Arch. Soc.* (1945), **109**.

143 Abstract of the earliest remaining court roll of the rectory manor of Dewsbury. *Yorks. Arch. Topograph. J.* (1911), **21**, 393–478.

144 S. J. Chadwick. Kirklees Priory. *Yorks. Arch. Topograph. J.* **16**, 347.

145 Specimen pipes preserved at Halifax, Bankfield Museum.

146 T. W. Woodhead. *History of Huddersfield water supplies* (1939). Specimen pipes preserved at Huddersfield, Tolson Memorial Museum.

13: Elm in the southern counties

1 The elms of Avon have been previously described by White (1912).

2 LF: DW. The large roadside elm in Abbots Leigh is the subject of a sonnet by Hardwicke Rawnsley (1877). Illus.: Samuel Jackson. *Brislington church and village*, water colour. Bristol, City Art Gallery, repr. S. Stoddard. *Mr Braikenridge's Brislington* (1981),

Bristol Museum & Art Gallery, cover; Lonsdale Ragg. *The great elm in the Botanic Garden, Bath. Tree Lover* (1939–42), vol. 3, p. 11, and *Big elm in Victoria Park, Bath. Tree Lover* (1941–4), **4**, 81, both pencil repr.

3 W. de G. Birch. *Cartularium saxonicum* (1883–93), vol. 2, p. 575. G. B. Grundy. The Saxon charters of Somerset. VII. *Proc. Somerset Arch. Nat. Hist. Soc.* (1933), 79, Appendix I, 193–224.

4 Place names and surnames not otherwise attributed are from A. H. Smith. *The place-names of Gloucestershire* (1964–5), Cambridge University Press.

5 F. H. Dickinson. Exchequer lay subsidies 169/5. [*Publ.*] *Somerset Rec. Soc.* (1889), **3**, 79–281.

6 E. G. C. F. Atchley. On the parish records of the church of All Saints, Bristol. *Trans. Bristol Gloucester. Arch. Soc.* (1904), **27**, 267.

7 Survey of timber sold (1566). Taunton, Somerset Record Office. DD/SE.

8 R. C. S. Walters. *The ancient well, springs and holy wells of Gloucestershire* (1928), pp. 143–52.

9 Ph. repr. *Country Life* (1900), **7**, 401.

10 LF: Wych Elm LDpaN; English Elm DWn.

11 Admeasurement of timber (1775). Taunton, Somerset Record Office. DD/PO.L.1.

12 A. Braine. *A history of Kingswood Forest* (1891), p. 1.

13 The elms of Berkshire have been previously described by Druce (1897, 1927).

14 Place names and surnames not otherwise attributed are from M. Gelling. *The place-names of Berkshire* (1973–6), Cambridge University Press. There is some doubt about the etymology of Wickenholt, but a similar name, Wikeholt, now Wiggonholt, occurs in W. Sussex.

15 A. L. Humphreys. *Bucklebury* (1932), p. 125.

16 LF: DW. Illus.: Leonard Richmond. *Elm trees at Windsor*, water colour, repr. L. Richmond. *The art of landscape painting* (1965), p. 53; Paul or Thomas Sandby. *Windsor Castle from Datchet Lane*, water colour and pencil. Windsor Castle, repr. A. P. Oppé. *The drawings of Paul and Thomas Sandby at Windsor Castle* (1947), plate 33.

17 L. F. Salzman. *Building in England down to 1540*, (1967), p. 250.

18 G. Wrottesley. Extracts from the plea rolls of the reigns of Henry V and Henry VI. *Coll. Hist. Staffs.* (1896), **17**, 125.

19 S. Barfield. *Thatcham, Berks, and its manors* (1901), vol. 2, pp. 96, 99, 100, 102.

20 W. M. Childs. *The town of Reading during the early part of the nineteenth century* (1910), p. 33.

21 See note 19, p. 290.

22 J. Shepherd. *Old days of Eton parish* (1908), p. 32.

23 O. Hedley. *Windsor Castle* (1967), pp. 116, 180. Fig. 124 is William Heath. *A short ride in the Long Walk*, engraving (1824), 23 × 35 cm. The driver is George IV, his companion Lady Conyngham; Fig. 125 is Sir Alfred Munnings. *The return from Ascot*, oil. Lode, Anglesey Abbey. Other illus.: James Harding. *Long Walk*, steel engraving by J. C. Varrall. In L. Ritchie. *Windsor Castle and its environs*, 2nd edn (1848), p. 206; Paul or Thomas Sandby. *The Castle from the Long Walk*, water colour and pencil. Windsor Castle, repr. A. P. Oppé, see note 16, plate 43; William

Westell. *King George IV Gate*, lithograph, repr. C. Hibbert. *The court of Windsor*, 1st edn (1964), p. 193.

24 The elms of Buckinghamshire have been previously described by Druce (1926).

25 J. G. Jenkins. The cartulary of Missenden Abbey. [*Publ.*] *Bucks. Arch. Soc. Rec. Branch* (1938), **2**, 197.

26 J. W. Garrett-Pegge. Richard Bowle's book. *Rec. Bucks.* (1916), **10**, 6.

27 For a woodcut of the churchyard elm of Olney see T. Wright. *The town of Cowper*, 2nd edn (1893), p. 22. This elm was far from the present church. It reputedly marked the site of an earlier one. There is a woodcut of one of the former three market place elms of Olney in the same work (p. 13). For an earlier illustration of them, see J. and H. Storer. *Cowper's house, Olney*, copper engraving. In J. Storer. *The rural walks of Cowper* (n.d.), p. 64.

28 LF: DWA. Fig. 126 is Frederick Griggs. *Boveney church*, woodcut. In C. Shorter. *Highways and byways in Buckinghamshire* (1910), p. 216. Other illus.: Birket Foster. *Milton's cottage at Chalfont St Giles*, water colour. Newcastle, Laing Art Gallery; John Nash. *The Aylesbury plain*, oil, repr. John Nash (1925); Ian Strang. *Walnut tree farm, Walton*, etching, repr. F. Markham. *A history of Milton Keynes* (1973), White Crescent, vol. 1, p. 148.

29 W. H. Ward and K. S. Block. *A history of the manor and parish of Iver* (1933), p. 87.

30 See chapter 10 under Symbol.

31 W. Bradbrook. Aston Abbotts: parish account book. *Rec. Bucks.* (1916), **10**, 34–50.

32 Account of timber sales (1634). Aylesbury, Buckinghamshire Record Office, D13/12.

33 F. E. Hyde and S. F. Markham. *A history of Stony Stratford* (1948), p. 80.

34 J. Yoarks. Wood accounts (1698–1705). Aylesbury, Buckinghamshire Record Office. D/C/4/12.

35 Hampden Estate wood book (1696–1706). Aylesbury, Buckinghamshire Record Office. D/MH/30/5.

36 G. Bland. The annual progress of New College. *Rec. Bucks.* (1935), **13**, 115.

37 Note book (1786–92). Aylesbury, Buckinghamshire Record Office. D/C/4/65/1; Note book (1776–9) *ibid.* D/C/4/65/2.

38 Ph. repr. *Country Life* (1949), **106**, 110–13.

39 R. Gibbs. *A history of Aylesbury* (1885), p. 606.

40 The elms of Cornwall have been previously described by Davey (1909), Thurston (1930), M. Turk (1981).

41 J. Leland. *Itinerary* (ed. L. T. Smith, 1907), vol. 1, p. 316.

42 Quoted without date by G. Grigson. *Freedom of the parish* (1954), p. 17. The form *witchhals* suggests the seventeenth century.

43 A. K. Hamilton Jenkin. Lluyd manuscripts in the Bodleian Library. I. The Duchess of Cornwall's progress. *J. Roy. Inst. Cornwall* (1925), **21**, 401–13.

44 Place names are from J. E. B. Gover. *The place names of Cornwall* (1948). Truro, Royal Institute of Cornwall; P. A. S. Pool. *The place-names of West Penwith* (1973); O. Padel, personal communication.

45 J. H. Matthews. *A history of the parishes of St Ives, Lelant, Towednach and Zennor* (1892), p. 144.

46 This mixture of trees was also recommended for the

Scilly Isles. See W. Borlase. *Observations on the ancient and present state of the islands of Scilly* (1966), p. 28.

47 Fig. 127 is Holman Hunt. *Helston, Cornwall*, water colour, 19 × 26 cm. Manchester, Whitworth Art Gallery.

48 LF: bwan.

49 St Breock churchwardens' accounts. Truro, County Record Office. DDP19/5/1.

50 C. H. Henderson. *Essays in Cornish history* (1935), p. 144.

51 W. Beckford. *Travel-diaries* (ed. G. Chapman, 1928), vol. 2, p. 5; S. E. Gay, *Old Falmouth* (1903), p. 190.

52 St Stephen by Saltash churchwardens' accounts. Truro, County Record Office. DDP214/5/2.

53 LF: bwan.

54 R. Potts. A calendar of Cornish glebe terriers, 1673–1735. [Publ.] *Dev. Cornwall Rec. Soc.* (1974), 19.

55 E. Peacock. On the churchwardens' accounts of the parish of Stratton, in the county of Cornwall. *Archaeologia* (1880), 46, 235.

56 The elms of Devon have been previously described by Martin and Frazer (1939).

57 Illus.: John Tucker. *Near Chumleigh*, oil. Exeter, Royal Albert Memorial Museum.

58 Place names and surnames not otherwise attributed are from J. E. B. Gover, A. Mawer and F. M. Stenton. *The place-names of Devon* (1931–2), Cambridge University Press.

59 LF: DWA. Fig. 128 is Robert Bevan. *The smithy, Luppitt*, lithograph (1920), 24 × 31 cm. Other illus.: John Gendall. *Exeter*, water colour. Exeter, Royal Albert Memorial Museum; Joseph Pennell. *A Devonshire lane*, repr. H. James. *English hours* (1905), p. 87.

60 E. S. Chalk. *A history of the church of St Peter Tiverton* (1903), p. 136.

61 E. Lega-Weekes. *Some studies in the topography of the cathedral close, Exeter* (1915).

62 D. St Leger-Gordon. *Portrait of Devon* (1963), p. 80.

63 See chapter 9 under Cult object.

64 LF: DWA.

65 LF: DWn.

66 LF: bwan.

67 I. L. Gregory. *Hartland church accounts, 1597–1706* (1950), p. 237.

68 Glebe terriers. Exeter, Devon Record Office. Exeter diocesan records.

69 Survey (1574). Exeter, Devon Record Office 123M/E72.

70 Survey of trees (1732). Taunton, Somerset Record Office. DD/WHb3108.

71 Place names and surnames not otherwise attributed are from A. D. Mills. *The place-names of Dorset* (1977–80), Cambridge University Press.

72 The history of the planting of the elms is related by A. Pope. The walks and avenues of Dorchester. *Proc. Dorset Nat. Hist. Antiq. Field Club* (1918), 38, 23–33. The principal allusions by Hardy are: T. Hardy. *The life and death of the mayor of Casterbridge* (1975), Macmillan; *idem*. The third kissing-gate. In *Complete poems* (1976), Macmillan, pp. 904–5.

73 A. D. Mills. Dorset lay subsidy roll, 1332. *Publ. Dorset Rec. Soc.* (1971), no. 4.

74 Illus.: Lonsdale Ragg. *Tall elm at Shaftesbury*, pencil, repr. E. Hughes-Gibb. *Trees and men* (1938), p. 76.

75 LF: DWA.

76 View of Gillingham Manor. *NQ Somerset Dorset* (1923), 17, 93–7.

77 An early 17th century building account. *NQ Somerset Dorset* (1942), 23, 121–6.

78 The church bells of Dorset. *Proc. Dorset Nat. Hist. Antiq. Field Club* (1906), 27, 118.

79 F. J. Furnivall. *The fifty earliest English wills* (1882), p. 27.

80 LF: DWA.

81 See note 41, p. 252.

82 LF: DWn. Fig. 129 is Charles Cheston. *Norman church, Studland*, etching (1909), 12 × 18 cm.

83 C. Reid. The origin of the British flora (1899), p. 59.

84 The elms of Gloucestershire have been previously described by Riddelsdell *et al.* (1948).

85 The accounts of St Katharine's Hospital, Ledbury, 1584–95. *Trans. Woolhope Nat. Field Club* (1952–4), 34, 88–132.

86 Place names and surnames not otherwise attributed are from Smith, see note 4.

87 Calendar of patent rolls. Edward VI (1925), vol. 3, p. 119.

88 Folklore Survey. London, University College.

89 LF: DWn. Illus.: Lonsdale Ragg. *The great elm of West End House, Fairford*, pencil, repr. *Tree Lover* (1941–4), 4, 182; *Wych elm, Oakley Park* [*Cirencester*], woodcut. In *Trans. Worc. Nat.* (1897–9), suppl. 14.

90 The pile is described by W. H. Knowles and L. E. W. O. Fullbrook-Leggatt. Report on the excavations during 1934 on the Barbican and Bon Marché sites, Gloucester. *Trans. Bristol Gloucester. Arch. Soc.* (1934), 56, 65–81. The suggestion by Richens and Jeffers (1978) that the pile may have been English Elm now seems unlikely.

91 S. E. Bartlett. The manor and borough of Chipping Camden. *Trans. Bristol Gloucester. Arch. Soc.* (1884–5), 9, 134–95.

92 W. H. Stevenson. *Calendar of the records of the corporation of Gloucester* (1893), pp. 123, 337.

93 H. Barkly. The Berkeleys of Dursley during the 13th and 14th centuries. *Trans. Bristol Gloucester. Arch. Soc.* (1884–5), 9, 227–76.

94 G. Cross. Select cases from the coroners' rolls, A.D. 1265–1413. [*Publ.*] *Seldon Soc.* (1896), 9, 50.

95 H. T. Lilley. *A history of Standish* (1932), p. 93.

96 J. Ogilby. *Road maps of England and Wales* (1971), p. 59.

97 E. Conder. Pauntley manor and the Pauntley custom. *Trans. Bristol Gloucester. Arch. Soc.* (1917), 40, 115–31.

98 Defoe (1704), pp. 95, 129.

99 E. Humphries and E. C. Willoughby. *At Cheltenham Spa* (1928), pp. 51–4.

100 W. Marshall. *On planting and rural ornament*, 3rd edn (1803), vol. 1, pp. 423–33.

101 A. C. Painter. Notes on some old Gloucestershire maps. *Trans. Bristol Gloucester. Arch. Soc.* (1929), 51, 79–93.

102 Ph. repr. *Country Life* (1950), 107, 1880–4.

103 Specimen pipes preserved at Gloucester, Folk Museum.

104 G. Hart. *A history of Cheltenham* (1965), p. 125.

105 C. Hussey. Sezincote, Gloucester. I. *Country Life* (1939), **85**, 502–6.

106 LF: LDwpaN.

107 The elms of Hampshire have been previously described by Townsend (1904).

108 Place names and surnames not otherwise attributed are from J. E. B. Gover. *Hampshire place names*. Winchester, Hampshire Record Office.

109 J. F. Williams. *The early churchwardens' accounts of Hampshire* (1913), p. 100.

110 C. D. Stooks. *A history of Crondall and Yateley* (1905), p. 28.

111 Ph. repr. *Country Life* (1899), **5**, 464–7.

112 Mrs Suckling. Some notes on the manor of Stanbridge Earls in the parish of Romsey Extra. *Pap. Proc. Hants. Field Club Arch. Soc.* (1907–10), pp. 6, 41–64.

113 LF: DWn.

114 A. A. Ruddock. The method of handling the cargoes of medieval merchant galleys. *Bull. Inst. Hist. Res.* (1942–3), **19**, 140–8.

115 J. Vaughan. *Winchester Cathedral close* (1914), p. 140.

116 F. J. Baigent and J. E. Millard. *A history of the ancient town and manor of Basingstoke* (1889), p. 279.

117 Reports from local secretaries. *Pap. Proc. Hants. Field Club Arch. Soc.* (1914), **7**(1), 113–6.

118 Specimen pipes preserved at Portsmouth, Eastney Pumping Station.

119 J. P. M. Pannell. *Old Southampton shores* (1967), p. 68.

120 The history of this tree is recounted by A. F. Leach. *A history of Winchester College* (1899), p. 452, and by I. T. 'ad penates'. *Wykehamist* (1910), no. 476, 247. It is celebrated in verse by G. Huddesford (1804).

121 R. Mudie. *Hampshire* (1840), vol. 2, p. 299; R. Pococke. *Travels through England* (ed. J. J. Cartwright, 1888), vol. 2, p. 129.

122 LF: Narrow-leaved Elm bdwan; English Elm DW.

123 Gerard (1633). The detailed distribution of this population of Narrow-leaved Elm was established by Melville (1938a).

124 The elms of Malvern Chace have been described previously by Lees (1853–70).

125 Illus.: *Distorted pollard wych elm near Cradley*, woodcut. In *Trans. Worc. Nat. Club* (1897–9), suppl. 15.

126 See note 85.

127 See note 96, p. 44.

128 LF: DWn. Elm in Worcester is the subject of a general paper by Barker (1916). Illus.: *Hollow wych elm near Knightsford Bridge*, woodcut. In Lees (1953–70), p. 89; *Elm with swollen base, Stanbrook Park, Powick*; *Old elm near Upton*; *Tall hollow elm, Holt Castle*; and *Stump of pollard elm near Shrawley Wood*, woodcuts, *ibid.*, note 125, pp. 10, 12, 13, 17.

129 Place names and surnames not otherwise attributed for the former county of Worcester are from A. Mawer and F. M. Stenton. *The place-names of Worcestershire* (1927), Cambridge University Press: those for the former county of Hertfordshire are from Ekwall (1960).

130 Ekwall (1960); Mawer and Stenton, see note 129,

offer a tentative etymology based on the tribal name Hwicce.

131 F. J. Eld. Lay subsidy roll for the county of Worcester, I Edward III. [*Publ.*] *Worc. Hist. Soc.* (1895).

132 F. B. Andrews. *Memorials of old Worcestershire* (1911), p. 176.

133 See note 96, p. 72.

134 Folklore Survey. London, University College. Fig. 130 is *The Friar's Elm*, woodcut by Thomas Armstrong. In Lees (1853–70), p. 101.

135 Ph. repr. *Country Life* (1960), **143**, 18.

136 LF: DWn. Illus.: Ladmore and Son. *Rectory elm, Stretton Sugwas*, ph. In *Trans. Woolhope Nat. Field Club* (1870), viii; *Hagley Park elm*, ph. In *ibid.*, p. 160; T. H. Winterbourn. *Elm at King's Acre, Hereford*, ph. repr., *ibid.* (1898–9), 112.

137 F. C. Morgan. The steward's accounts of John, First Viscount Scudamore of Sligo (1601–1671) for the year 1632. *Trans. Woolhope Nat. Field Club* (1949–51), **33**, 155–84.

138 LF: LDpaN.

139 LF: LDpN.

140 LF: Wych Elm LDwpN; English Elm DWn.

141 LF: Wych Elm LDpaN; English Elm DWn.

142 Ph. repr. *Country Life* (1900), **8**, p. 814.

143 LF: DWn.

144 Fig. 131 is W. I. Walton. *The old church at Bonchurch*, engraving by A. Willmore. In W. H. D. Adams. *The history, topography and antiquities of the Isle of Wight* (1856), p. 76. Other illus.: G. Brannon. *The parish church of Bonchurch*, steel engraving. In G. Brannon. *Picture of the Isle of Wight* (1846), p. 68; William Buck. *Bonchurch*, water colour. Oxford, Ashmolean Museum; G. S. Shepherd. *Bonchurch*, steel engraving by J. Shury and Son. In Mudie, *Hampshire*, **3**, 215.

145 The elms of Oxfordshire have been previously described by Druce (1927).

146 Place names and surnames not otherwise attributed are from M. Gelling. *The place-names of Oxfordshire* (1953–4), Cambridge University Press and Smith, see note 14.

147 LF: DWA. Illus.: Ralph Chubb. *Berkshire farm*, oil. Manchester, City Art Gallery; William Delamotte. *Water Perry*, oil. London, Tate Gallery; Frederick Griggs. *East Hendred* and *Godstow Bridge*, woodcuts. In J. E. Vincent. *Highways and byways in Berkshire* (1906), pp. 25, 56; Lonsdale Ragg. *Churchyard elms, Sutton Courtenay*, pencil, repr. *Tree Lover* (1941–4), **4**, 185; Jacob Strutt. *The elms at Mongewell*, etching. In Strutt (1822), plate 16.

148 A. Wood. Survey of the antiquities of the city of Oxford, (ed. A. Clark). [*Publ.*] *Oxf. Hist. Soc.* (1889), **15**.

149 H. E. Salter. Survey of Oxford. [*Publ.*] *Oxf. Hist. Soc.* (1960), **14**, 198.

150 E. Stone. Oxfordshire hundred rolls of 1279. The Hundred of Brampton. [*Publ.*] *Oxf. Hist. Soc.* (1968), **46**, 19, 42, 45, 48.

151 P. D. A. Harvey. *A medieval Oxfordshire village* (1965), p. 25.

152 M. Dickens. A history of Hook Norton (1928), p. 54.

153 T. F. Hobson. Adderbury 'rectoria'. [*Publ.*] *Oxf. Rec. Soc.* (1926), **8**, 118.

154 G. Wrottesley. Extracts from the plea rolls of the reigns of Edward III and Richard II. *Coll. Hist. Staffs.* (1892), **13**, 121.

155 C. R. J. Currie and J. M. Fletcher. Two early cruck houses in north Berkshire identified by radiocarbon. *Med. Arch.* (1972), **16**, 136–42.

156 R. E. G. Kirk. Accounts of the obedientiars of Abingdon Abbey (1892), 75; B. Challenor. *Selections from the municipal chronicles of the borough of Abingdon* (1898), p. xxxi.

157 J. R. H. Weaver and A. Beardwood. Some Oxfordshire wills. [*Publ.*] *Oxf. Rec. Soc.* (1958), **39**, 64–5.

158 J. E. Field. Notes of the topography of the parish of North Moreton, Berks. *Berks. Bucks. Oxon. Arch. J.* (1913), **19**, 15–19.

159 E. Corbett. *A history of Spelsbury* (1962), p. 78.

160 See note 41, p. 112.

161 Plot (1677).

162 R. L. Rickard. The progress notes of Warden Woodward round the Oxfordshire estates of New College, Oxford. [*Publ.*] *Oxf. Rec. Soc.* (1949), **27**.

163 J. E. Field. Brightwell, Berks. *Berks. Bucks. Oxon. Arch. J.* (1911), **17**, 88–9.

164 J. Hunt. Valuation of the woods at West Wickham etc. (1696). Aylesbury, Buckinghamshire Record Office. D/D/14/55.

165 W. C. Costin. The history of St John's College, Oxford, 1598–1860. [*Publ.*] *Oxf. Rec. Soc.* (1958), **12**, 93.

166 A. Wood. *Diaries* (ed. A. Clark). [*Publ.*] *Oxf. Hist. Soc.* (1894), **26**, 485.

167 Illus.: W. Delamotte. *Joe Pullen's tree*, woodcut by Orlando Jewitt. In J. Ingram. *Memorials of Oxford* (1837), vol. 3; also an etching by Delamotte (1834).

168 See note 96, pp. 85, 114.

169 J. S. Burn. *A history of Henley-on-Thames* (1861), pp. 310–2.

170 Ph. repr. *Country Life* (1954), **115**, 216–9.

171 Ph. repr. *Country Life* (1900), **7**, 83.

172 Ph. repr. Elwes and Henry (1913).

173 Matthew Arnold. The scholar gypsy. In *Poems* (1913), pp. 230–7. Illus.: ph. repr. W. (1948) and Woodman (1957).

174 Murray's Handbook for Berkshire (1902), p. 125.

175 Scott-Snell (1940).

176 A. Williams. *Villages of the White Horse* (1913).

177 LF: DWn.

178 Place names and surnames not otherwise attributed are from Ekwall (1960).

179 W. de G. Birch, see note 3, vol. 2, p. 74; G. B. Grundy. The Saxon charters of Somerset. I. *Proc. Somerset Arch. Soc.* (1927), **73**, appendix I, 25.

180 A. Kirke. North Curry chest. *NQ Somerset Dorset* (1930–2), **20**, 121.

181 D. D. Shilton and R. Holworthy. Wells city charters. [*Publ.*] *Somerset Rec. Soc.* (1932), **46**, 20.

182 A. Watkin. The great chartulary of Glastonbury. [*Publ.*] *Somerset Rec. Soc.* (1952), **63**, 424.

183 F. H. Dickinson. Exchequer lay subsidies 169/5. [*Publ.*] *Somerset Rec. Soc.* (1889), **3**, 79–281.

184 F. Hancock. *Wifela's Combe* (1911), p. 267.

185 T. B. Dilks. Bridgwater borough archives, 1377–99. [*Publ.*] *Somerset Rec. Soc.* (1938), **53**, 242.

186 G. Bradford. Somerset Star Chamber cases, 1485–1547. [*Publ.*] *Somerset Rec. Soc.* (1911), **27**, 85.

187 Bishop Hobhouse. Church-wardens' accounts of Tintinhull. [*Publ.*] *Somerset Rec. Soc.* (1890), **4**, 179–207.

188 A. L. Humphreys. *Materials for the history of the town and parish of Wellington* (1908), p. 180.

189 Field name problem. *NQ Somerset Dorset* (1955), **26**, 98, 119, 136.

190 *Loc. cit.*, note 41, pp. 156, 160.

191 Lord Hylton. *Notes on the history of the parish of Kilmersdon* (1910), p. 183.

192 Perambulation and customs of the manor of Muddesley, Somerset. *NQ Somerset Dorset* (1920), 16, 163–8, 206–8.

193 Survey of timber sold (1566). Taunton, Somerset Record Office. DD/SE.

194 F. W. W. Manor of Castle Cary. *NQ Somerset Dorset* (1917), **15**, 88–91.

195 H. C. Maxwell Lyte. The Lytes of Lytescary. *Proc. Somerset Arch. Nat. Hist. Soc.* (1892), **38**(2), 1–110.

196 Anon. (1926).

197 C. A. Johns. *The forest trees of Britain* (1849), vol. 2, pp. 118–9.

198 Ph. repr. *Country Life* (1898), **4**, 528; and *ibid.* (1911), **29**, 924–34.

199 LF: Wych Elm LDpaN; English Elm LBDW.

200 E. Dwelly. *Directory of Somerset* (1929), p. 65.

201 A. Bulleid and H. St G. Gray. *The Glastonbury lake village* (1911–7).

202 A. Bulleid and H. St G. Gray. *The Meare lake village* (1948–66).

203 The elms of Surrey have been previously described by Lousley (1976) and Salmon (1931).

204 Fig. 132 is Henry Dawson. *Large elms, St Annes Hill*, water colour, 46 × 33 cm. Nottingham, Art Gallery. Also at Nottingham by the same artist: *Brook and trees, Thorpe*, pencil.

205 E. Toms. Chertsey Abbey court rolls abstract. [*Publ.*] *Surrey Rec. Soc.* (1954), **21**, 25, 77.

206 Chertsey cartularies. [*Publ.*] *Surrey Rec. Soc.* (1933), **12**, 101.

207 LF: DWn. Fig. 133 is William Henry Hunt. *Pollard elm and village pound, Hambledon*, water colour, 76 × 53 cm. Preston, Harris Museum and Art Gallery. Fig. 134 is Birket Foster, *Lane near Dorking* or *The green lane*, water colour, woodcut by Dalziel Bros. In M. B. Foster. *Pictures of rustic landscape* (1896), p. 91. The two background elms are an improvement over the original, repr. in F. Lewis. *Myles Burket Foster* (1973). Other illus.: George Barnard, *ibid.*, chapter 11, note 66; P. Sandby. *Ember Court* [Thames Ditton], copper engraving by M. A. Rooker. In P. Sandby. *The virutosi's museum* (1778), plate 63.

208 Place names and surnames not otherwise attributed are from J. E. B. Gover, A. Mawer and F. M. Stenton. *The place-names of Surrey* (1934), Cambridge Univesity Press.

209 W. de G. Birch, see note 3, vol. 1, p. 58.

210 Surrey taxation returns. [*Publ.*] *Surrey Rec. Soc.* (1932), **10**.

211 S. J. Madge. The economy of the rural estates in

Middlesex. *Trans. London Middlesex Arch. Soc.* (1918–22), **4**, 404.

212 Timber inventory. Guildford, Surrey Record Office. LM 1947; Timber accounts. *Idem.* 85/16/17.

213 Evelyn (1959), p. 484.

214 N. C. C. Barrell. *The history of the Guildford waterworks*, 2nd edn (1952). Guildford, Surrey Record Office. There are some specimen elm pipes in Guildford Museum.

215 G. Home. *Epsom, its history and surroundings* (1901), p. 45.

216 The elms of Warwickshire have been previously described by Cadbury *et al.* (1971).

217 Survey of the manor of Griss and Coton (1684). Warwick, County Record Office. CR/136/V12.

218 LF: DW. Illus.: J. E. Duggins. *A bye-lane: Ettington* and *Warwickshire Elms*, repr. M. D. Harris. *Unknown Warwickshire* (1924), pp. 66, 222.

219 Place names and surnames not otherwise attributed are from J. E. B. Gover, A. Mawer and F. M. Stenton. *The place-names of Warwickshire* (1936), Cambridge University Press.

220 W. F. Carter. The lay subsidy roll for Warwickshire of Edward III (1332). [*Publ.*] *Dugdale Soc.* (1926), **6**.

221 J. H. Bloom. *The register of the gild of the Holy Cross, the Blessed Mary and St John the Baptist, of Stratford-upon-Avon* (1907), p. 36.

222 Abstract of the bailiff's accounts of monastic and other estates in the county of Warwick, 1547. [*Publ.*] *Dugdale Soc.* (1923), **2**, 19, 48.

223 W. Cooper. *Wootton Wawen: its history and records* (1936), p. 13.

224 S.H.A.S. *Ladbroke and its owners* (1914), p. 279.

225 Minutes and accounts of the corporation of Stratford-upon-Avon. [*Publ.*] *Dugdale Soc.* (1926), **5**, 105–9; (1929), **10**, 129.

226 T. Waryng. Survey of timber on the Archer estates. Stratford-upon-Avon, Shakespeare Birthplace Library.

227 D. M. Barratt. Ecclesiastical terriers of Warwickshire parishes. [*Publ.*] *Dugdale Soc.* (1955), **22**, 101.

228 *Victoria history of the county of Warwick* (1969), vol. 8, p. 93.

229 Specimen pipes preserved at Warwick, St John's House.

230 Ph. repr. *Country Life* (1952), **111**, 1080–3, 1164–7, 1328–31. There is also a repr. of an early view of the avenue, possibly by John Stevens.

231 Ph. repr. *Country Life* (1906), **19**, 54–65.

232 The elms of Sussex have been previously described by Wooley-Dod (1937).

233 W. Hudson. The three earliest subsidies for the county of Sussex. [*Publ.*] *Sussex Rec. Soc.* (1909), **10**.

234 Lord Leconfield. *Sutton and Duncton manors* (1956), pp. 8, 78.

235 Illus.: Jacob Strutt. *The Crawley elm*, etching. In Strutt (1822), plate 32.

236 J. Dale. Extracts from the churchwardens' accounts and other matters belonging to the parish of Bolney. *Sussex Arch. Coll.* (1853), **6**, 244–52.

237 Ph. repr. *Country Life* (1942), **91**, 1134–7.

238 LF: DW. Illus.: Frederick Griggs. *Boxgrove from the south* and *Steyning church*, woodcuts. In E. V. Lucas. *Highways and byways in Sussex*, 2nd edn (1935), pp. 35, 154.

239 Place names and surnames not otherwise attributed are from A. Mawer, F. M. Stenton and J. E. B. Gover. *The place-names of Sussex* (1929–30), Cambridge University Press. There is some doubt about Wiggonholt. The occurrence of Wikenholt in Lambourn (Br) in the twelfth century suggests that both are based on *wice* and not on some other word.

240 W. D. Peckham. The acts of the Dean and chapter of the cathedral church of Chichester, 1472–1544. [*Publ.*] *Sussex Rec. Soc.* (1951–2), **52**, 79.

241 J. R. Daniel-Tyssen. The parliamentary surveys of the county of Sussex. *Sussex Arch. Coll.* (1872), **24**, 246.

242 B. C. Redwood. Quarter sessions order book, 1642–9. [*Publ.*] *Sussex Rec. Soc.* (1954), **54**, 199.

243 T. G. Willis. *Records of Chichester* (1928), pp. 33, 73, 88. Illus.: J. Broughton. *North Walls*, ink, repr. *Chichester Pap.* (1963), no. 37.

244 LF: DW.

245 The elms of Wiltshire have been previously described by Grose (1957) and Preston (1888).

246 Place names and surnames not otherwise attributed are from J. E. B. Gover, A. Mawer and F. M. Stenton. *The place-names of Wiltshire* (1939), Cambridge University Press.

247 B. H. Cunnington. *Records of the county of Wilts* (1932), p. 58.

248 Earl of Cardigan. The wardens of Savernake Forest. IV. The Brudenell wardens. *Wilts. Arch. Nat. Hist. Mag.* (1950), **53**, 1–62.

249 LF: DWAn.

250 E. M. Forster. *The longest journey* (1947), Arnold, p. 129.

251 R. C. C. Clay. Pre-Roman coffin burials with particular reference to one from a barrow at Fovant. *Wilts. Arch. Nat. Hist. Mag.* (1927–9), **44**, 101–5.

252 C. Wordsworth. *The fifteenth century cartulary of St Nicholas' hospital, Salisbury* (1902), p. 63.

253 See chapter 8 under Timber. I.

254 See note 17.

255 F. Goddard. Notes on Clyffe Pypard and Broad Town. *Wilts. Arch. Nat. Hist. Mag.* (1927–9), **44**, 143–70.

256 Calendar of the patent rolls. Edward VI (1926), vol. 4, p. 160.

257 D. H. Robertson. Notes on some buildings in the city and close of Salisbury connected with the education and maintenance of the cathedral choristers. *Wilts. Arch. Nat. Hist. Mag.* (1937–9), **48**, pp. 1–30.

258 R. L. Rickard. Progress notes of Warden Woodward for the Wiltshire estates of New College, Oxford. [*Publ.*] *Wilts. Arch. Nat. Hist. Soc. Rec. Branch* (1957), **13**, pp. 4, 26–37.

259 J. Aubrey. *The natural history of Wiltshire* (ed. J. Britton, 1847), pp. 54–5.

260 E. Coward. William Gaby, his book. 1656. II. *Wilts. Arch. Nat. Hist. Mag.* (1932–4), **46**, 336–49.

261 Ph. repr. *Country Life* (1900), **8**, 277; (1937), **82**, 516–21.

262 R. F. Kilvert. *Diary* (ed. W. Plomer, 1938–40), Cape.

263 H. Tanner. *Wiltshire village* (1939), R. Garton.

14: Elm in the eastern counties

1 The elms of Bedfordshire have been described previously by Richens. (1955, 1961b). Sources for information not otherwise attributed are given in these papers. Another earlier description is by Dony (1953).

2 LF: DWn.

3 *Victoria history of the county of Bedford* (1904–14), vol. 2, p. 348.

4 J. E. Farmiloe and R. Nixseaman. Elizabethan churchwardens' accounts. *Publ. Beds. Hist. Rec. Soc.* (1953), **33**, pp. 60, 93.

5 See note 3, vol. 3, p. 98.

6 J. Aubrey. *The natural history of Wiltshire* (ed. J. Britton, 1847), pp. 54–5.

7 A true report of certaine wonderful overflowing of waters, now lately in Summersetshire, Norfolk, and other places of England (ed. E. E. Baker, 1884), p. 33.

8 Ph. repr. *Country Life* (1904), **16**, pp. 54–64, 90–8.

9 LF: bwPa.

10 LF: Wych Elm LbDwpN; English Elm DWn.

11 A particular of the rents of the mannor of Husborne Crawley. Bedford, Bedfordshire Record Office. AD1033.

12 LF: Narrow-leaved Elm LbwPA; English Elm DWn.

13 The elms of Cambridgeshire have been described previously by Richens (1955, 1958, 1961a, 1965). Sources for information not otherwise attributed are given in these papers.

14 Illus.: Hallam Murray. *Elm, Elton Hall*, water colour, repr. *Tree Lover* (1932–5), **1**, 217.

15 Place names and surnames not otherwise attributed are from A. Mawer and F. M. Stenton. *The place-names of Bedfordshire and Huntingdonshire* (1926), Cambridge University Press; P. H. Reaney. *The place-names of Cambridgeshire and the Isle of Ely* (1943), Cambridge University Press; and Ekwall (1960).

16 W. T. Mellows. The book of William Morton. *Publ. Northants. Rec. Soc.* (1952–3), **16**, pp. 2, 4.

17 J. Ravensdale. *Liable to floods* (1974), Cambridge University Press, p. 39.

18 J. Bodger. *A chart of the beautiful fishery of Whittlesea Mere* (1786).

19 LF: bwPa and LBDn. Illus.: Lonsdale Ragg. *Hinton elms*, pencil, repr. M. Sackville. *The lyrical woodland* (1945), p. 29.

20 E. K. Purnell. *Magdalene College* (1904), p. 14.

21 W. Jones. Extracts from the parish registers, Wendy, Cambs. *E. Anglian* (1903), **10**, 134–8.

22 *Cambridge Evening News* (1979), 12 February.

23 Fig. 136 is *Wimpole Hall and Park*, ph. repr. Cambridge University Aerial Photography Collection. Other illus.: Leonard Knyff. *Wimpole Hall*, engraving by Johannes Kip. In *Britannia Illustrata* (1707), plate 32.

24 LF: bwPan and LBDPA.

25 LF: LBDn.

26 LF: Narrow-leaved Elm bwPn and LDwa; English Elm DWn.

27 W. M. Palmer. *The neighbourhood of Melbourn and Meldreth* (1923), p. 3. Fig. 137 is Richard Bankes Harraden. *Melbourn elm*, water colour, 30 × 40 cm.

An ink drawing of this tree is in Cambridge, Folk Museum, repr. Richens (1972).

28 W. M. Palmer, see note 27, p. 31.

29 *Victoria history of the county of Huntingdon* (1926–38), vol. 3, p. 69.

30 The elms of Sussex have been previously described by Wooley-Dod (1937).

31 Place names and surnames not otherwise attributed are from A. Mawer, F. M. Stenton and J. E. B. Gover. *The place-names of Sussex* (1929–30), Cambridge University Press.

32 J. M. Baines. *Historic Hastings*, 2nd edn (1963), p. 166.

33 Ledger. Lewes, East Sussex Record Office. ASH1178, p. 4.

34 LF: DWpn.

35 C. Thomas-Stanford. An abstract of the court rolls of the manor of Preston. [Publ.] *Sussex Rec. Soc.* (1921), **27**, p. 24.

36 Jevington manorial court rolls transcript. Lewes, Sussex Archaeological Society. Budgen Collection vol. 115, p. 24.

37 Eastbourne Burton manorial court rolls transcript. Lewes, Sussex Archaeological Society. Budgen Collection vol. 106, p. 5; Folkington manorial court rolls transcript. *Ibid.*, Budgen Collection vol. 123.

38 Some seventeenth and eighteenth century Sussex tradesmen's accounts. *Sussex Arch. Coll.* (1896), **40**, 274–7.

39 C. Musgrave. *Royal pavilion*, 2nd edn (1959), p. 43.

40 LF: lbWan.

41 J. R. Daniel-Tysson. The parliamentary surveys of the county of Sussex. *Sussex Arch. Coll.* (1872), **24**, 268.

42 L. A. Vidler. *A new history of Rye* (1934), p. 95.

43 The elms of Essex have been previously described by Richens (1955, 1967). Sources for information not otherwise attributed are given in these papers.

44 W. C. Waller. Essex field-names. VI. *Trans. Essex Arch. Soc.* (1903), **8**, 199–222.

45 F. G. Emmison. The 'very naughty ways' of Elizabethan Essex. *Essex Rev.* (1955), **64**, 85–91.

46 W. R. Fisher. *The Forest of Essex* (1887), p. 400–3.

47 Place names and surnames not otherwise attributed are from P. H. Reaney. *The place-names of Essex* (1935), Cambridge University Press.

48 G. M. Benton. Wivenhoe records. *Essex Rev.* (1928), **37**, 156–69.

49 LF: DW, DWAn, DWAN. Illus.: Charles Taylor. *To Childerditch*, woodcut (1929), repr. *Print. Coll. Quart.* (1936), **23**, 131.

50 P. J. Huggins. Excavation of a medieval bridge at Waltham Abbey, Essex, in 1968. *Med. Arch.* (1970), **40**, 126–47.

51 L. F. Salzman. *Building in England down to 1540* (1967), p. 84.

52 J. C. Shenstone. The woodlands of Essex. *Essex. Nat.* (1907–8), **15**, 105–15.

53 P. Benton. *History of Rochford hundred* (1867–88), p. 323.

54 Fig. 138 is John Varley. *Waltham Abbey*, water colour, 30 × 49 cm. London, Victoria and Albert Museum. Fig. 139 is George Campion. *Waltham Abbey*, steel engraving by J. Rogers. In T. Wright. *The history and topography of the county of Essex* (1831–5), vol. 2, p. 450.

55 LF: lbPan, bdwPn, bwPan, bP, LbwPaN. Illus.: Frederick Griggs. *Grace's Walk* [Little Baddow] and *Tolleshunt Major*, woodcuts. In C. Bax. *Highways and byways in Essex* (1939), pp. 50, 183.

56 A. Clark. *Great Waltham five centuries ago* (1904), p. 65.

57 L. C. Sier. Manor of East Hall, East Mersea. *Essex Rev.* (1945), **54**, 7–11.

58 G. Eland. *At the courts of Great Canfield, Essex* (1949), p. 62.

59 R. Holinshed. *Chronicles* (1807–8), vol. 1, pp. 357–8.

60 S. Taylor. *History and antiquities of Harwich and Dovercourt* (ed. S. Dale, 1730), pp. 87–8.

61 G. F. Beaumont. *A history of Coggeshall* (1890), p. 113.

62 H. Smith. Two perambulations of White Roding – 1615 and 1810. *Essex Rev.* (1935), **44**, 222–7.

63 K. Walker. Jay Wick – an ancient Essex homestead. *Essex Rev.* (1950), **59**, 15–24.

64 Illus.: William Bartlett. *Boreham House*, steel engraving by J. Rogers. In T. Wright, see note 54, vol. 1, p. 108; photograph repr. *Country Life* (1914), **36**, p. 54.

65 Ph. repr. *Country Life* (1942), **91**, 68–71.

66 E. P. Dickin. A note on the county records at Chelmsford. *Trans. Essex Arch. Soc.* (1930), **19**, 289–94.

67 LF: lban, LbdwPa, LwaN, LBDwa.

68 T. Wright, see note 54, vol. 2, p. 82.

69 LF: Wych Elm LDwpaN; Narrow-leaved Elm bWn.

70 LF: Narrow-leaved Elm lbPan; English Elm DW.

71 LF: Narrow-leaved Elm bDPn; English Elm DWN.

72 LF: Narrow-leaved Elm bdP; English Elm DWA. Illus.: Frederick Griggs. *Stow Maries, near Purleigh*, woodcut. See note 55, p. 302.

73 Survey of manors of Graces, St Cleers and Herons. Chelmsford, Essex Record Office. D/DGe M135.

74 LF: Narrow-leaved Elm lbPan; English Elm DWAn.

75 Rivers (1861).

76 The frequency of Hybrid Elms in central Essex had been noted by Greville (1949).

77 J. W. Keaworthy. A supposed Neolithic settlement at Skitt's Hill, Braintree, Essex. *Essex Nat.* (1899–1900), **11**, 94–121; F. W. Reader. Further notes on the pile-dwelling site at Skitt's Hill, Braintree, Essex. *Essex Nat.* (1905–6), **14**, pp. 137–47.

78 G. Rickword. Colchester in the twelfth and thirteenth centuries. *Trans. Essex Arch. Soc.* (1900), **7**, pp. 115–35; W. G. Benham. *The red paper book of Colchester* (1902), p. 162.

79 Articles of agreement between Sir Thomas Davies, and Jonas Shish, Henry Wooden and James Goldsmith (1676). Chelmsford, Essex Record Office. D/DU191/31.

80 Specimen pipes preserved at Colchester Castle.

81 Ph. repr. Elwes and Henry (1913).

82 *Victoria history of the county of Essex* (1903–78), vol. 2, p. 81.

83 The elms of the part of Greater London formerly in Essex and Hertfordshire have been previously described by Richens (1959, 1967). Sources for information not otherwise attributed are given in these papers. Other earlier descriptions are by Kent (1975) and Trimen and Thistleton Dyer (1869).

84 Gerard (1633).

85 A survey of the manor of Westham (1649). London, Public Record Office. E317 Essex 17.

86 LF: DWA. Fig. 141 is *Paddington church*, etching. In D. Lysons. *The environs of London* (1795), vol. 3, p. 332. Other illus.: Joseph Farington. *Barne elms* [Barnes], ink, Oxford, Ashmolean Museum; P. Sandby. *A view of Charlton*, copper engraving by M. A. Rooker. In P. Sandby. *The virtuosi's museum* (1778), plate 26; J. Spencer. *Ham Common*, oil, repr. M. H. Grant. *A chronological history of the old English landscape painters* (1959), vol. 5, plate 188. See also chapter 11, note 39.

87 C. Welch. *History of the Tower Bridge* (1894), p. 28.

88 A. Marks. *Tyburn tree* (1908).

89 C. J. Feret. *Fulham old and new* (1900), vol. 3, p. 264.

90 *Liber custumarum* (ed. H. T. Riley, 1860), vol. 1, p. 150.

91 B. Hobley and J. Schofield. Excavations in the city of London. First interim report, 1974–5. *Antiq. J.* (1977), **57**, 31–66; G. Milne and C. Milne. Excavations on the Thames waterfront at Trig Lane, London, 1974–76. *Med. Arch.* (1978), **22**, 84–104; D. M. Wilson and D. G. Hurst. Medieval Britain in 1964. *Med. Arch.* (1965), **9**, 200.

92 See note 88, vol. 2, p. 230; *Court rolls of the manors of Bruces, Dawbeneys, Pembrokes (Tottenham) 1 Richard II to 1 Henry IV* (1961), p. 240.

93 Ekwall (1960). Place names and surnames not otherwise attributed are from J. E. B. Gover, A. Mawer and F. M. Stenton. *The place-names of Surrey* (1934), Cambridge University Press; J. E. B. Gover, A. Mawer and F. M. Stenton. *The place-names of Middlesex* (1942), Cambridge University Press; J. K. Wallenberg. *The place-names of Kent* (1934); see note 47.

94 See note 51, p. 250.

95 W. St John Hope. *The history of the London Charterhouse* (1955), p. 127; A. P. Stanley. *Historical memorials of Westminster Abbey* 5th edn (1882), p. 354.

96 T. Faulkner. *The history and antiquities of Brentford, Ealing and Chiswick* (1845), p. 300.

97 C. Welch. *History of the Tower Bridge* (1894), p. 55.

98 Court rolls of Tooting Bec manor (1909), vol. 1, p. 132.

99 J. Stow. *A survey of London* (ed. C. L. Kingsford, 1971), Oxford University Press, vol. 1, pp. 127, 144; vol. 2, p. 71.

100 G. J. Turner. Chancery Lane: Sir Nicholas Bacon's estate. *Trans. London Middlesex Arch. Soc.* (1923–8), **5**, 486–502.

101 W. R. Douthwaite. *Gray's Inn* (1886), pp. 183, 185.

102 T. Faulkner. *An historical and topographical description of Chelsea* (1829), vol. 1, p. 151.

103 E. Law. *The history of Hampton Court Palace* (1885), vol. 1, pp. 5, 372. King Charles' swing is engraved as *Great elm tree in the home park*, woodcut, p. 5.

104 W. E. Morden. *The history of Tooting-Graveney* (1897), p. 10; M. Sharpe. *Some account of bygone Hanwell* (1924), p. 27.

105 R. R. C. Gregory. *The story of royal Eltham* (1909), p. 186.

106 E. Walford. *Greater London* (n.d.), vol. 1, p. 534.

107 P. Cunningham. *Hand-book of London*, 2nd edn (1850), p. 372. There is a woodcut of the elm on the site of St Paul's cross in B. Holmes. *The London burial grounds* (1896), p. 60.

108 B. Jonson. *The divell is an asse* (1967), p. 20.

109 H. L. L. Bellot. *The Inner and Middle Temple* (1902), pp. 248–9.

110 Evelyn (1664), introduction.

111 Fig. 142 is *Rosamond's pond*, oil. Its attribution is discussed by M. H. Grant, see note 85, vol. 2, pp. 85, 115. It is not likely that William Hogarth was the artist.

112 E. Beresford Chancellor. *The history and antiquities of Richmond* (1894), p. 61.

113 Fig. 143 is Wenceslaus Hollar. *The great hollow tree*, etching, 18 × 19 cm.

114 W. J. Roe. *Ancient Tottenham* (n.d.), p. 115. Fig. 144 is *Seven sisters*, copper engraving. In W. Robinson. *History and antiquities of the parish of Tottenham High Cross* (1818). Fig. 145 is Hans Unger. *Seven sisters*, ceramic tiles. London, Seven Sisters Underground station. Another illus.: *Seven sisters*, woodcut. In E. Walford. *Old and new London* (n.d.), vol. 5, p. 373.

115 S. J. Madge. Rural Middlesex under the Commonwealth. *Trans. London Middlesex Arch. Soc.* (1918–22), **4**, pp. 308, 453; W. McB. Marcham and F. Marcham. Court rolls of the Bishop of London's manor of Hornsey, 1603–1701 (1929), p. xx; R. de Salis. *Hillingdon through eleven centuries* (1926), pp. 63, 66; A. Saunders. *Regent's Park* (1969), p. 223.

116 C. R. L. Fletcher. All Souls College v. Lady Jane Stafford. [*Publ.*] *Oxford Hist. Soc.* (1885), **5**, pp. 179–247.

117 W. C. Waller. *Loughton in Essex* (1889–1900), p. 59.

118 Evelyn (1959), p. 459.

119 W. T. Vincent. *The records of the Woolwich district* (1889), p. 498.

120 B. Platts. *A history of Greenwich* (1973); A. D. Webster. *Greenwich Park* (1902).

121 A. Saunders, see note 115.

122 Kalm (1966), p. 338.

123 Kalm (1966), pp. 334–5; W. Wroth. *Cremorne and the later London gardens* (1907), p. 23.

124 C. A. White. *Sweet Hampstead and its associations* (1900), pp. 137, 172, 187, 188.

125 J. C. Wall. *Traditionary site of Warwick's death*, woodcut. In J. Tavenor-Perry. *Memorials of old Middlesex* (1909), p. 112.

126 Fig. 146 is E. C. Wilde. *Old elm tree on Hadley Common*, lithograph by J. P. and W. H. Emalie. In F. C. Cass. *Monken Hadley* (1880), p. 8.

127 Fig. 147 is William Bartlett. *Chingford church*, steel engraving by J. C. Armytage. In T. Wright, see note 54, vol. 2, p. 472; Fig. 148 is Arthur Hughes. *Home from sea*, oil, 51 × 65 cm. Oxford, Ashmolean Museum.

128 Illus.: Wingfield. *Nine elms*, water colour. London, Greater London County Hall, Print Collection.

129 There were elm avenues at: Addington Palace, Battersea Park, Claysmore in Enfield, Bishop's Avenue at Fulham, Belsize House and Vane House in Hampstead, Hanwell Park, Swakeleys in Ickenham, Campden House, Holland House and Earls Court in Kensington, Peckham Manor House, Pinner Grove House, Broomfield Park in Southgate, Queen Elizabeth's Walk at Stoke Newington, and Tooting Bec Common.

130 Other illus.: see chapter 10, note 179.

131 J. Tavenor-Perry, see note 125, p. 229.

132 LF: Narrow-leaved Elm bP; English Elm DWA.

133 H. F. Westlake. *Hornchurch Priory* (1923).

134 J. P. Shawcross. *A history of Dagenham* (1904), p. 6.

135 Illus.: S. R. Badmin. *Havering-atte-Bower: village stocks*, woodcut, see note 55, p. 128.

136 The elms of Hertfordshire have been described previously by Richens (1955, 1959). Sources for information not otherwise attributed are given in these papers.

137 Place names and surnames not otherwise attributed are from J. E. B. Gover, A. Mawer and F. M. Stenton. *The place-names of Hertfordshire* (1938), Cambridge University Press.

138 Church terriers. *Herts. Genealogist Antiq.* (1899), **3**, p. 299.

139 Fig. 149 is Alice Balfour. *Town Bottom Elm*, water colour, 35 × 24 cm. Hatfield House.

140 LF: DWn. Illus.: Frederick Griggs. *St Ippolyts church* (Ippolits), *Wymondley Magna* (Great Wymondley) – *morning*, *Wymondley Paroa* (Little Wymondley) and *The Icknield Way at Cadwell*, woodcuts. In H. W. Tompkins. *Highways and byways in Hertfordshire* (1902), pp. 187, 206, 208, 242.

141 W. P. Westell. Excavation of an uncharted Romano–British occupation site at Great Wymondley, Herts. *Trans. E. Herts. Arch. Soc.* (1937), **10**, pp. 11–5.

142 A. Tregelles. *History of Hoddesdon* (1908), pp. 127, 309.

143 J. C. Rogers. The great tower of St Albans Abbey church. *Trans. St Albans Herts. Architect. Arch. Soc.* (1930), pp. 56–64.

144 See note 138, p. 178.

145 Nicoll (1944).

146 E. Beckett. The avenues at Aldenham. *Country Life* (1920), **47**, pp. 103–5. Ph. repr. *Country Life* (1924), **55**, 282–90.

147 LF: lbDPan, bdwan. Illus.: Frederick Griggs. *Gilston elms*, in Tompkins, see note 140, p. 24; John Linnell. *The valley of the River Lea*, water colour, London, British Museum, repr. K. Crouan. *John Linnell* (1982), Cambridge University Press, p. 63.

148 F. R. Williams. *Anstey* (1929), pp. 12–15.

149 J. Scott. Amwell. In *Poetical works* (1782), pp. 57–89. This edition has illustrations of the Amwell elms engraved after drawings by J. Feary.

150 LF: wan.

151 LF: Narrow-leaved Elm bwPan and PAn; English Elm DWn.

152 See note 135, where an etymology based on *wic* is proposed.

153 The elms of Kent have been described previously by Hanbury and Marshall (1899).

154 E. Hasted. *The history and topographical survey of the county of Kent* (1797–1801), vol. 5, p. 281.

155 Place names and surnames not otherwise attributed

are from K. Wallenberg. *The place-names of Kent* (1934).

156 The suggestion of Wallenberg, see note 155, that there were two different but adjacent places with very similar names is unlikely.

157 Illus.: Jacob Strutt. *Chipstead elm*, etching. In Strutt (1822), plate 5.

158 F. Jenkins. Canterbury. Excavations in Burgate Street, 1946–8. *Arch. Cantiana* (1950), **63**, pp. 82–118.

159 W. L. Rutton. Sandgate castle, A.D. 1539–40. *Arch. Cantiana* (1893), **20**, pp. 228–57.

160 K. P. Witney. *The Jutish forest* (1976), Athlone, p. 236.

161 See note 160, p. 259.

162 W. J. Lightfoot. Documents relating to a dispute between the Seven Hundreds and Lydd concerning the watch at Denge Marsh. *Arch. Cantiana* (1872), **8**, pp. 299–310.

163 Accounts of St Dunstan's, Canterbury, from A.D. 1485 to A.D. 1625. *Arch. Cantiana* (1923), **36**, p. 85.

164 Parliamentary surveys. London, Lambeth Palace. XIIa/22, p. 400.

165 M. Barton. *Tunbridge Wells* (1937), p. 117; A. Savidge. *Royal Tunbridge Wells* (1975).

166 LF: DW. Illus.: Charles Hullmandel. *Rochester castle from the Medway*, lithograph. In W. Westall, C. Hullmandel and J. D. Harding. *Britannia delineata. Kent* (1922); Samuel Williams. *Beckenham church*, woodcut. In T. Miller. *Pictures of country life* (1847), p. 65.

167 Bridge Wardens' accounts, transcribed by M. J. Becker. Rochester, Bridge Chamber. Some excerpts from these have been published by M. J. Becker. *Rochester Bridge: 1387–1856* (1930).

168 A. A. Arnold. The earliest Rochester bridge. Was it built by the Romans? *Arch. Cantiana* (1921), **35**, pp. 127–44.

169 See note 51.

170 F. F. Smith. *A history of Rochester* (1928), p. 432.

171 C. Dickens. *The mystery of Edwin Drood* (1974), Penguin.

172 Ph. repr. *Country Life* (1945), **99**, 578–81.

173 Ph. repr. *Country Life* (1943), **94**, 1124–7.

174 LF: DWn.

175 Surveys of timber and underwood on the estates of the Archbishop of Canterbury. London, Lambeth Palace. T52, p. 104a.

176 LF: lbda.

177 Fig. 150 is Paul Nash. *Canal under Lympne*, water colour; Fig. 151 is John Piper. *The Royal Military Canal at Ruckinge*, repr. J. Piper. *Romney Marsh* (1950), plate 15.

178 LF: lbdan.

179 See note 51.

180 A. Hussey. Ford manor house and lands in 1647. *Arch. Cantiana* (1904), **26**, 119–32; C. E. Woodruff. A seventeenth century survey of the estates of the Dean and chapter of Canterbury in east Kent. *Arch. Cantiana* (1926), **38**, 29–44.

181 R. H. Goodsall. *Whitstable, Seasalter and Swalecliffe* (1938), pp. 59, 185. This book also reproduces a print (p. 163) entitled *Opening of the first passenger railway in the south of England, 3rd May, 1830.*

Returning from Whitstable. Elm is conspicuous in the landscape depicted.

182 See note 175. There are many references to elm at Chislet, and one, p. 11a, for Nonington.

183 LF: English Elm DWn; Narrow-leaved Elm lbdan.

184 The elms of Leicestershire have been previously described by Horwood and Gainsborough (1933) and Messenger (1971).

185 G. F. Farnham. *Leicestershire medieval village notes* (1929–33), vol. 4, p. 167.

186 A. H. Thompson. The building accounts of Kirby Muxloe castle, 1480–1484. *Trans. Leic. Arch. Soc.* (1913–20), **11**, 193–345.

187 LF: LbwpaN.

188 LF: DWn.

189 M. Bateson. *Records of the borough of Leicester* (1901), vol. 2, p. 305.

190 H. Stocks. *Records of the borough of Leicester, 1603–88* (1923), pp. 72, 311.

191 S. H. Skillington. Medieval Cossington. *Trans. Leic. Arch. Soc.* (1935–7), **19**, 331.

192 LF: BDPn.

193 LF: Wych Elm LbDwpaN; Narrow-leaved Elm a.

194 G. F. Farnham. The Charnwood manors. *Trans. Leic. Arch. Soc.* (1927–8), **15**, p. 163.

195 LF: Narrow-leaved Elm LBDPn; English Elm DW.

196 See note 185, vol. 1, p. 74.

197 M. W. Beresford. Glebe terriers and open field Leicestershire. *Trans. Leic. Arch. Soc.* (1948), **24**, 112.

198 LF: Wych Elm LBDpaN; Narrow-leaved Elm BPn; English Elm DW.

199 W. G. Hoskins. A history of the Humberstone family. *Trans. Leic. Arch. Soc.* (1938–9), **20**, 257.

200 The elms of Norfolk have been previously described by Petch and Swann (1968).

201 Fig. 152 is Henry Ninham. *Elm at Haveringland Park*, etching. In J. Grigor. *The eastern arboretum* (1841), p. 151. Other etchings by Ninham: *English elm, St Peter at Hungate, Norwich*; *English elm, Cossey* [Costessey] *Park*; *Elm, Salhouse.* Ibid., pp. 16, 34, 305.

202 A. G. Legge. *Ancient churchwardens' accounts in the parish of North Elmham* (1891).

203 W. Hudson. On an ancient timber roadway across the river Wensum at Eye Bridge, Norwich. *Norfolk Arch.* (1895–7), **13**, 217–32.

204 W. H. St John Hope and W. T. Bensly. Recent discoveries in the cathedral church of Norwich. *Norfolk Arch.* (1898–1900), **14**, 105–27.

205 J. Kirkpatrick. The streets and lanes of the city of Norwich (ed. W. Hudson, 1889), p. 60.

206 W. Rye. An old cannon at the great hospital, Norwich. *Norfolk Arch.* (1905–7), **16**, 85–90.

207 P. Millican. The rebuilding of Wroxham Bridge in 1576. *Norfolk Arch.* (1935–7), **26**, 281–95.

208 G. A. Carthew. Extracts from papers in the church chest of Wymondham. *Norfolk Arch.* (1894), **9**, 121–52.

209 LF: LDwPa.

210 Place names and surnames not otherwise attributed are from Ekwall (1960).

211 See note 201, p. 361.

212 LF: LBDwPAn.

213 H. Clarke and A. Carter. Excavations in Kings Lynn,

1963–70. *Monogr. Ser. Soc. Med. Arch.* (1977), no. 7.

214 A. D. Stallard. *A transcript of the churchwardens' accounts of the parish of Tilney All Saints, Norfolk. 443–1589.* (1922), p. 215.

215 LF: LBPAn.

216 LF: LPA.

217 LF: LDwaN.

218 LF: LP.

219 LF: Dan.

220 LF: LDA.

221 LF: LDwa.

222 LF: Narrow-leaved Elm LBDwan; English Elm BDWn.

223 The elms of Northamptonshire have been previously described by Druce (1930).

224 Place names and surnames not otherwise attributed are from J. E. B. Gover, A. Mawer and F. M. Stenton. *The place-names of Northamptonshire* (1933), Cambridge University Press.

225 S. T. Winckley. Royalist papers relating to the sequestration of the estates of Sir Lewis Watson. *Rep. Pap. Assoc. Architect. Soc.* (1899), **25**, 371–404.

226 J. Cornforth. The making of Boughton landscape. Ph. repr. *Country Life* (1971), **149**, 536–9.

227 Boughton estate accounts, 1707–8. Northampton, Northamptonshire Record Office. Boughton estate accounts (1690–1708), nos. 9 and 10.

228 Ph. repr. *Country Life* (1954), **116**, 1758–61.

229 LF: DW.

230 J. C. Cox. The records of the borough of Northampton (1898), vol. 2, pp. 153, 154; O. F. Brown and G. J. Roberts. *Passenham* (1973), Phillimore, p. 66.

231 Badby manorial court rolls transcript (1616). Northampton, Northamptonshire Record Office. ZA 913.

232 Account of timber sales (1634). Aylesbury, Buckinghamshire Record Office. D13/12.

233 R. G. Scriven. 'Walks and avenues'. *J. Northants. Nat. Hist. Soc.* (1889), **5**, 277–84.

234 LF: Lbwa.

235 LF: Narrow-leaved Elm LDwPa; English Elm DW.

236 Lease (1603). Northampton, Northamptonshire Record Office. SS 1702.

237 Rhyming public-house signs in Northamptonshire. *Northants. NQ* (1890), **3**, 62–3.

238 The elms of western Suffolk have been described previously by Richens (1955).

239 Place names and surnames not otherwise attributed are from Ekwall (1960).

240 A. W. Clapham and W. H. Godfrey. The Saxon cathedral of Elmham. *Antiq. J.* (1926), **6**, 402–9; A. W. Clapham and W. H. Godfrey. The Saxon cathedral of Elmham. *Norfolk Arch.* (1927–9), **23**, 56–67; R. Howlett. The ancient see of Elmham. *Norfolk Arch.* (1911–4), **18**, 105–28; S. E. Rigold. The Anglian cathedral of North Elmham, Norfolk. *Med. Arch.* (1962–3), **6–7**, 67–108.

241 H. Harrod. On the site of the bishopric of Elmham. *Proc. Suffolk Arch. Inst.* (1874), **4**, 7–13; J. T. Micklethwaite. The old minster of South Elmham. *Proc. Suffolk Arch. Inst.* (1918), **16**, 30–5; J. J. Raven. The 'Old Minster' of South Elmham. *Proc. Suffolk Arch. Inst.* (1900), **10**, 1–6; V. B. Redstone. South Elmham deanery. *Proc. Suffolk Inst. Arch. Nat. Hist.* (1912), **14**, 323–31. F. S. Stevenson. The present state of the Elmham controversy. *Proc. Suffolk Arch. Inst.* (1927), **19**, 110–6; B. B. Woodward. The old minster, South Elmham. *Proc. Suffolk. Arch. Inst.* (1874), **4**, 1–7.

242 N. Scarfe. *The Suffolk landscape* (1972); P. Wade-Martins. Excavations at North Elmham, 1967–8. *Norfolk Arch.* (1969), **34**, 352–97; P. Wade-Martins. Excavations at North Elmham, 1969. *Norfolk Arch.* (1970), **35**, 25–78.

243 N. Scarfe, see note 242, p. 215.

244 J. T. Munday. *Lakenheath records. No. 6. Parliament's plunder* (1970), p. 19; J. T. Munday. *Lakenheath records. No. 7. Man's manor* (1970), p. 4.

245 *Ickworth survey boocke, 1665* (1893), pp. 38–41.

246 Account book of the Guildhall Feoffment Trust (1603–22). Bury St Edmunds, Suffolk Record Office, Bury St Edmunds Branch. HD1150.

247 LF: LBDwPa.

248 F. Warner and A. Warner. Book of remembrance (1659–77). Bury St Edmunds, Suffolk Record Office, Bury St Edmunds Branch. 1341/5/1.

249 W. Addison. *Suffolk* (1950), p. 23.

250 LF: LDwan.

251 Cartulary of SS Peter and Paul, Ipswich. Ipswich, Suffolk Record Office, Ipswich Branch. HD226/1, f.15v.

252 Walsham le Willows field book (ed. K. M. Dodd). [*Publ.*] *Suffolk Rec. Soc.* (1974), **17**, 52, 132.

253 Transcript of the court rolls of Cardon's Hall manor, Witnesham. Ipswich, Suffolk Record Office, Ipswich Branch. S. Witnesham 333.32, p. 71.

254 Indenture (1617). Norwich, Norfolk Record Office. WLS XXXII/13/10.

255 LF: Dan.

256 A. H. Denny. The Sibton Abbey estates. [*Publ.*] *Suffolk Rec. Soc.* (1960), **2**, 107.

257 Extracts from Hinton court rolls. Ipswich, Suffolk Record Office, Ipswich Branch. HA30/369/379.

258 Survey and valuation of demesne lands of the manor of Virleys in Sternfield (1601). Ipswich, Suffolk Record Office, Ipswich Branch. HA18/DF5/2.

259 LF: Narrow-leaved Elm BDwan; English Elm BDWn.

260 LF: Narrow-leaved Elm LDan; English Elm DWAn.

261 J. Constable. Correspondence (ed. R. B. Beckett). [*Publ.*] *Suffolk Rec. Soc.* (1962), **4**, 217.

15: Conservation

1 The conservation policy outlined here is an elaboration of that proposed by Richens (1977c, 1978).

2 The effects of Dutch elm disease on other organisms have been collated by Archibald and Stubbs (1980).

3 Booth (1977); Greig (1979); Holmes (1977).

General bibliography

Afanasijev, D. (1957). Plutasti brest. *Shumarstvo*, **10**, 100–4.

Aiton, W. (1789). *Hortus kewensis*, vol. 1, pp. 319–20.

Allen, A. J. W. (1975). *Aulonium trisulcum* Fourc. (Col. Colydiidae) in Gloucestershire. *Ent. Month. Mag.* **111**, 1328–30.

Anderson, E. (1933). The mucilage from slippery elm bark. *J. Biol. Chem.* **104**, 163–70.

Andronov, N. M. (1955). Novyĭ vid vyaza. *Bot. Material. Gerbar. Bot. Inst. V. L. Komarov.* **17**, 106–9.

Andronov, N. M. (1969). K systematike nekotorykh vidov roda. *Ulmus. Nauch. Trud. Leningrad. Lesotekh. Akad.*, no. 128, 66–71.

Anon. (1852). The Wellington tree on the field of Waterloo. *Illustr. London News*, **21**, 469.

Anon. (1906). The great elm between Towcester and Northampton. *Proc. Wesley Hist. Soc.* **5**, 63.

Anon. (1908). Seeding of English elm. *Gardeners' Chron.* **43**, 10.

Anon. (1911). European elms. *Gardeners' Chron.* **50**, 202–3.

Anon. (1912). *Ulmus plotii. J. Bot.* **50**, 96–7.

Anon. (1916). The elm and the blizzard. *Country Life*, **39**, 458–60.

Anon. (1926). Marshall's elm. *Proc. Somerset. Arch. Nat. Hist. Soc.* **72**, 88.

Arcangeli, G. (1882). *Compendio della flora italiana*, p. 601.

Archibald, J. F. and Stubbs, A. E. (1980). The effects of Dutch elm disease on wildlife. *Quart. J. For.* **74**, 30–7.

Armfield, M. (1939). In praise of elms. *Tree Lover*, **3**, 76–7.

Armitage, E. (1917). Fruiting of the English elm. *J. Bot.* **55**, 162–3.

Armitage, E. (1922). Further notes on elm flowering. *J. Bot.* **60**, 141–2.

B., A. (1855). Nottingham elms. *Gardeners' Chron.*, 206.

Backhouse, W. O. (1934). Classification of elms. *Gardeners' Chron.* **96**, 180.

Baker, R. St B. (1934). Historic trees of England. II. The elm tree – philosopher and friend. *My Garden*, **3**, 415–20.

Baldacci, A. (1891). Nel Montenegro. Una parte delle mie raccolte. *Malpighia*, **5**, 62–82.

Bancroft, H. (1934). Notes on the status and nomenclature of the British elms. *Gardeners' Chron.* **96**, 122–3, 138–40, 208–9, 244–5, 298–9, 334–5, 372–3.

Bancroft, H. (1935). The elm problem. *Quart. J. For.* **29**, 102–6.

Bancroft, H. (1936). Elm notes for 1935. *Gardeners' Chron.* **99**, 104, 268; **100**, 127, 392–3, 429–30, 445–7.

Bancroft, H. (1937). The British elms. *J. Bot.* **75**, 337–46.

Barker, E. A. (1916). The elms of College Green, Worcester. *Country Life* (1916), **39**, 576.

Barnes, W. (1962). The elm in home-ground. In *Poems* (ed. B. Jones), Centaur, pp. 653–4.

Bate-Smith, E. C. and Richens, R. H. (1973). Flavonoid chemistry and taxonomy in *Ulmus*. *Biochem. Systematics*, **1**, 141–6.

Bauhin, C. (1623). *Pinax theatri botanici*, pp. 426–7.

Bean, W. J. (1907). Fastigiate trees. *Gardeners' Chron.* **41**, 149–52.

Bean, W. J. (1914). *Trees and shrubs hardy in the British Isles*, vol. 2, p. 620.

B[ean], W. J. (1931). The English elm. *Country Life*, **69**. 95–7.

Beaver, R. A. (1966a). Notes on the biology of bark beetles attacking elm in Wytham Wood, Berks. *Ent. Month. Mag.* **102**, 156–62.

Beaver, R. A. (1966b). Notes on the fauna associated with elm bark beetle in Wytham Wood, Berks. – I. Coleoptera. *Ent. Month. Mag.* **102**, 163–70.

Benson, R. B. (1936). Some more new or little known British sawflies (Hymenoptera Symphyta) *Ent. Month. Mag.* **72**, 203–7.

Bentham, G. (1865). Handbook of the British flora, vol. 2, pp. 746–7.

Berger, W. (1957). Untersuchungen an der obermiozänen (sarmatischen) Flora von Gabbro (Monti Livornesi) in der Toskana. *Palaeontographica ital.* **51**, 1–96.

Blight, M. M., Mellon, F. A., Wadhams, L. J. and Wenham, M. J. (1977). Volatiles associated with *Scolytus scolytus* beetles on English elm. *Experientia* **33**, 845–7.

Blight, M. M., Wadhams, L. J. and Wenham, M. J. (1978). Volatiles associated with unmated *Scolytus scolytus* beetles on English elm: differential production of α-multistriatin and 4-methyl-3-heptanol, and their activities in a laboratory bioassay. *Insect Biochem.* **8**, 135–42.

Blight, M. M., Wadhams, L. J., Wenham, M. J. and King, C. J. (1979). Field attraction of *Scolytus scolytus* (F.) to the enantiomers of 4-methyl-3-heptanol, the major component of the aggregation pheromone. *Forestry*, **52**, 83–90.

Bock, H. (Tragus, H.) (1539). *De stirpium, maxime eorum, quae in Germania nostra nascuntur, 1086–1088*.

Booth, A. G. (1977). When the elms have gone. *Times*, 18 June.

Borbás, V. (1881). Békésvármegye flórája. *Értekezéseb a Természettudományok Köreből*, **11**, 55.

Borbás, V. (1891). Közlemények Békés- és Bihar-vármegyék flórájából. In *Nagyváradon tartott XXV. vándorgyülésének történeti vázlata és munkálatai*, pp. 479–504.

Borkhausen, M. B. (1800). *Theoretisch-praktisches Handbuch der Forstbotanik und Forstechnologie*, pp. 625–46.

Boulger, G. S. (1879). British elms. *Gardeners' Chron.* **12**, 298.

Boulger, G. S. (1912). *Ulmus Plotii*, Druce. *Gardeners' Chron.* **51**, 35.

Brasier, C. M. (1979). Dual origin of recent Dutch elm disease outbreaks in Europe. *Nature, London* **281**, 78–80.

Brasier, C. M. and Gibbs, J. N. (1973). Origin of Dutch elm disease epidemic in Britain. *Nature, London* **242**, 607–9.

Brasier, C. M. and Gibbs, J. N. (1975). Variation in *Ceratocystis ulmi*: significance of the aggressive and nonaggressive strains. In D. A. Burdekin and H. M. Heybroek (eds.) *Dutch elm disease*. Proceedings of IUFRO Conference, Minneapolis–St Paul, USA, September 1973.

Braun, A. F. (1963). The genus *Bucculatrix* in America north of Mexico (Microlepidoptera). *Mem. Amer. Ent. Soc.*, no. 18.

Brett, D. W. (1978*a*). Dendrochronology of elm in London. *Tree-ring Bull.* **38**, 35–44.

Brett, D. W. (1978*b*). Elm tree rings as a rainfall record. *Weather*, **33**, 87–94.

Bridges, R. (1936). The great elm. In Poetical works, pp. 537–9.

Brockmann-Jerosch, H. (1936). Futterläubbaume und Speiselbäume. *Ber. Schweiz. Bot. Ges.* **46**, 594–613.

Browicz, K. and Zieliński, J. (1977). Polish herbaria collection of trees and shrubs from southwest Asia made in the years 1972–1975. *Arboretum Kórnickie*, **22**, 285–321.

Burbidge, F. W. (1896). The winged elm. *Gardeners' Chron.* **19**, 527.

Buszewicz, G. M. and Holmes, G. D. (1961–2). Forest tree seed investigations: seed storage. *Rep. For. Res. For. Commission*, 17–18.

Butcher, R. W. (1930). *Further illustrations of British plants*, pp. 310–13.

Butcher, R. W. (1961). *A new illustrated British flora*, vol. 1, 948–54.

Byron, Lord (1970). Lines written beneath an elm in the churchyard of Harrow. In *Poetical works*, Oxford University Press, p. 47.

Cadbury, D. A., Hawkes, J. G. and Readett, R. C. (1971). *A computer-mapped flora. A study of the county of Warwickshire*, Academic Press, pp. 180–1.

Carson, H. L. (1958). The population genetics of *Drosophila robusta*. *Advances in Genetics*, **9**, 1–40.

Carter, J. C. (1946). Inheritance of foliage variegation in variegated English elm. *Trans. Illinois Acad. Sci.* **39**, 43–6.

Cejp, K. (1967). New or rare species of the genus *Phyllosticta* Pers. in Czechoslovakia. *Nova Hedwigia*, **13**, 183–97.

Cejp, K. (1969). Miscellaneous notes on the *Phyllosticta* Pers., *Septoria* Fr. and *Ascochyta* Lib. from Czechoslovakia. *Nova Hedwigia*, **18**, 557–76.

Cheng, W. C. (1936). In S. S. Chien, W. C. Cheng and C. P'ei. An enumeration of the vascular plants from Chekiang. IV. *Contrib. Biol. Lab. Sci. Soc. China, Bot. Ser.* **10**, 93–155.

Cheng, W. C. (1939). Une espèce nouvelle d'*Ulmus* chinois, *U. gaussenii*. *Trav. Lab. For. Toulouse*, **3**, Art. III.

Cheng, W. C. (1963). In W. C. Cheng *et al.* Species novae et nomina emendata arborum utilium Chinae. *Scientia Silvae, China*, **8**, 1–14.

Chernyavskaya, T. A. (1978). Protsess obrazovaniye pyl'tsy u razlichnykh form i gibridov vyaza prizemistogo. *Bjull. VNII Agrolesomelior.* no. 1, 26–9.

Chernyavskaya, T. A. (1979). O mikrosporogeneze u raznykh form vyaza prizemistogo (*Ulmus pumila* L.). *Nauch. Dokl. Vyssh. Shkol. Biol. Nauk*, no. 5, 77–9.

Christy, M. (1909–10). On the abnormal fruiting of the common elm in 1909. *Essex Nat.* **16**, 73–81.

Christy, M. (1922). The flowering times of some British elms. *J. Bot.* **60**, 36–41.

Clapham, A. R. (1969). *Flora of Derbyshire*, pp. 229–30.

Clapham, A. R., Tutin, T. G. and Warburg, E. F. (1952). Flora of the British Isles, 1st edn., Cambridge University Press, pp. 715–24.

Clapham, A. R., Tutin, T. G. and Warburg, E. F. (1962). Flora of the British Isles, 2nd edn., Cambridge University Press, pp. 562–6.

Clare, J. (1935*a*). The fallen elm. In *Poems* (ed. J. W. Tibble) vol. 2, pp. 18–20.

Clare, J. (1935*b*). The shepherd's tree. In *Poems* (ed. J. W. Tibble) vol. 2, pp. 18–20.

Clinton, G. P. and McCormick, F. A. (1936). Dutch elm disease–*Graphium ulmi*. *Bull. Comm. Agric. Exp. Sta.*, no. 389, 701–52.

Codrington, R. (1814). Of the height and hollowness of the Great Elm at Hampstead. In J. J. Park (ed.) *The topography and natural history of Hampstead*, pp. 35–7.

Columella (1941–68). Loeb edn., Heinemann, vol. 2, pp. 42–71.

Crabbe, G. (1907). On the drawing of the elm tree, under which the Duke of Wellington stood several times during the Battle of Waterloo. In *Poems* (ed. A. W. Ward), vol. 3, p. 431.

Crampon, M. (1936). Le culte de l'arbre et de la forêt en Picardie. *Mem. Soc. Antiq. Picardie*, **46**.

Crété, P., Guignard, J.-L. and Mestre, J.-S. (1966). Embryogénie des Ulmacées. Développement de l'embryon chez l'*Ulmus campestris* L. *CR Acad. Sci. Paris, D* (1966) **262**, 986–8.

Cullum, T. G. (1774). *Florae anglicae specimen*, p. 96.

Dalechamps, J. (1586). *Historia generalis plantarum*, pp. 80–2.

D'Amato, F. (1940). Embriologia di *Ulmus campestris* L. *Nuovo G. Bot. Ital.* **47**, 247–63.

Danielsson, R. (1979). The genus *Eriosoma* Leach in Sweden, with descriptions of two new species. *Ent. Scandinavica*, **10**, 193–208.

Darby, S. (1908). A historic tree. *Berks. Bucks. Oxon. Arch. J.* **14**, 30.

Daumann, E. (1975). Ein rudimentäres Blutennektarium und unterschiedliche Pollenkittreste bei der Rüster (*Ulmus*). *Preslia*, **47**, 14–21.

Davey, F. H. (1909). *Flora of Cornwall*, pp. 401–2.

[Defoe, D.] (1704). *The storm*.

Demetz, P. (1958). The elms and the vine: notes towards the history of a marriage topos. *Publ. Modern Lang. Assoc. Amer.* **73**, 521–32.

Denson, J. (1831). Notes on *S. destructor* Ol. *Ann. Mag. Nat. Hist.* **4**, 152–7.

Dodoens, R. [Dodonaeus, R.] (1554). *Cruyde-boek*, pp. ccccccccc–ccccccccciii.

Dodoens, R. [Dodonaeus, R.] (1557). *Histoire des plantes*, pp. 527–30.

Dodoens, R. [Dodonaeus, R.] (1583). *Stirpium historiae pemptades sex*, pp. 815–27.

Donisthorpe, H. (1941). *Aulonium trisulcatum*, Fourc., extending its range in England. *Ent. Rec. J. Var.* **53**, 18–19.

Donn, J. (1826). Hortus Cantabrigiensis, 11th edn., p. 96.

Dony, J. G. (1953). *Flora of Bedfordshire*, Luton Museum and Art Gallery, pp. 307–9.

Doskotch, R. W., Chatterji, S. K. and Peacock, J. W. (1970). Elm bark derived feeding stimulants for the smaller European elm bark beetle. *Science*, **167**, 380–2.

Doskotch, R. W., Mikhail, A. A. and Chatterji, S. K. (1973). Structure of the water-soluble feeding stimulant for *Scolytus multistriatus*: a revision. *Phytochemistry*, **12**, 1153–5.

Draper, A. T. (1869). Elm timber. *Gardeners' Chron. Agric. Gaz.*, p. 194.

Drobov, V. P. (1953). *Ulmus*. In *Flora Uzbekistana*, vol. 2, p. 77.

Druce, G. C. (1897). *The flora of Berkshire*, pp. 440–2.

Druce, G. C. (1907). *Ulmus sativa*, Mill. var. *Lockii*. *Rep. Bot. Exchange Club*, 258.

Druce, G. C. (1908). *List of British plants*, p. 63.

Druce, G. C. (1910). Northamptonshire plant notes. *J. Northampton Nat. Hist. Soc.* **15**, 281–96.

Druce, G. C. (1911a). *Ulmus Plotii*, sp. nov. *J. Northampton Nat. Hist. Soc.* **16**, 88.

Druce, G. C. (1911b). *Ulmus Plotii* Druce. *J. Northampton Nat. Hist. Soc.* **16**, 107–8.

Druce, G. C. (1911c). *Ulmus Plotii* Druce sp. nov. *Gardeners' Chron.* **50**, 408.

Druce, G. C. (1911d). *Ulmus Plotii* Druce. *Proc. Rep. Ashmolean Nat. Hist. Soc. Oxford*, 33–5.

Druce, G. C. (1912). *Ulmus Plotii* Druce. *Gardeners' Chron.* **51**, 35.

Druce, G. C. (1925). Huntingdonshire plants. *Rep. Bot. Exchange Club*, **7**, 949–57.

Druce, G. C. (1926). *The flora of Buckinghamshire*, pp. 298–304.

Druce, G. C. (1927). *The flora of Oxfordshire*, pp. 375–9.

Druce, G. C. (1928). *British plant list*, 2nd edn., pp. 102–3.

Druce, G. C. (1930). *The flora of Northamptonshire*, pp. 204–6.

Druce, G. C. (1932). *Comital flora*, p. 266.

Ducros, H. (1935–6). Note sur le derdar. *Bull. Inst. Égypte*, **18**, 117–21.

Duhamel du Monceau, H. L. (1755). *Traité des arbres et arbustes qui se cultivent en France en pleine terre*, vol. 2, pp. 367–70.

du Mont de Coursey, G. L. M. (1811). Le botaniste cultivateur, 2nd edn., vol. 6, p. 384.

Dumortier, B.-C. (1827). *Florula Belgica*, p. 25.

Edees, H. L. (1972). *Flora of Staffordshire*, pp. 108–9.

Edlin, H. L. (1965). A modern sylva or a discourse of forest trees. 12: elms – *Ulmus*. *Quart. J. For.* **59**, 41–51.

Ehrenberg, C. E. (1949). Studies on asynapsis in the elm, *Ulmus glabra* Huds. *Hereditas*, **35**, 1–26.

Eilers, W. and Mayrhofer, M. (1961). Kurdisch *būz* und die indogermanische 'Buchen'-sippe. *Mitt. Anthropol. Ges. Wien*, **91**, 61–92.

Ekdahl, J. (1941). Die Entwicklung von Embryosack und Embryo bei *Ulmus glabra* Huds. *Svensk. Bot. Tid.* **35**, 143–56.

Ekwall, E. (1960). The concise Oxford dictionary of English place-names, 4th edn., Oxford University Press.

Elton, C. S. (1970). *Aulonium trisulcum* Fourc. (Col. Colydiidae) in Wytham Wood, Berkshire; with remarks on its status as an invader. *Ent. Month. Mag.* **106**, 190–2.

Elwes, H. J. (1907). *Ulmus alata* Mich. *Gardeners' Chron.* **42**, 143.

Elwes, H. J. and Henry, A. (1913). *The trees of Great Britain and Ireland*, vol. 7, pp. 1847–1929.

Emmett, A. M. (1973–4). Notes of some of the British Nepiculidae, II. *Ent. Rec.* **85**, 77–80, 176–80, 278–83; **86**, 75–80, 103–8, 147–53.

Endtmann, J. (1967). Zur Taxonomie der mitteleuropaischen Sippen der Gattung *Ulmus*. *Arch. Forstwes.* **16**, 667–72.

Enescu, V. and Enescu, V. (1957). Cîteva cercetări asupra germinaţiei seminţelor de ulm. *Rev. Pădurilor*, no. 4. 230–5.

Evelyn, J. (1664). *Sylva*, pp. 16–20.

Evelyn, J. (1959). Diary (ed. E. S. de Beer).

Farkas, H. K. (1960). Über die Eriophyiden (Acarina) Ungarns. I. *Acta Zool. Acad. Sci. Hung.* **6**, 315–39.

Fisher, R. C. (1928). The relation of the elm bark-beetles to their host trees. *Forestry*, **2**, 53–61.

Fisher, R. C. (1931). Notes on the biology of the large elm bark-beetle, *Scolytus destrictor* Ol. *Forestry*, **5**, 120–31.

Fisher, R. C. (1937). The genus *Scolytus* in Great Britain, with notes on the structure of *S. destructor* Ol. *Ann. Appl. Biol.* **24**, 110–30.

Forsyth, A. (1877). The elm tree. *Gardeners' Chron.* **8**, 229–30.

Franchet, A. (1884). Plantae Davidianae ex Sinarum imperio. *Nouv. Archiv. Muséum Hist. Nat. Paris*, **7**, 77

Fransen, J. J. (1939). Iepenziekte, iepenspintkevers en beider bestrijding. 'Thesis Wageningen Agric. Coll.'

Friedrich, P. (1970a). Proto-Indo-European trees. In G. Cardona, H. M. Hoenigswald and A. Senn, (eds.) *Indo-European and the Indo-Europeans*, Pennsylvania University Press, pp. 11–34.

Friedrich, P. (1970b). *Proto-Indo-European trees*, Chicago University Press.

Gamble, J. S. (1902). *A manual of Indian timbers*, pp. 627–8.

Garry, F. N. A. (1904). Notes on the drawings for 'English Botany'. *J. Bot.* **42**, suppl., 121–2.

Genner, V. and Bonnevie, P. (1938). Eczematous eruptions produced by leaves of trees and bushes. *Archiv. Dermatol. Syphilology*, **37**, 583–9.

Gerard, J. (1597). *Herball*, 1st. edn., pp. 1295–6.

Gerard, J. (1633). *Herball*, 2nd. edn., pp. 1479–82.

Giannasi, D. E. and Niklas, K. J. (1977). Flavonoid and other constituents of fossil Miocene *Celtis* and *Ulmus* (Succor Creek Flora). *Science*, **197**, 765–7.

Gibbs, J. N. (1978a). Development of the Dutch elm disease epidemic in southern England. *Ann. Appl. Biol.* **88**, 219–28.

Gibbs, J. N. (1978b). Intercontinental epidemiology of Dutch elm disease. *Ann. Rev. Phytopath.* **16**, 287–307.

Gibbs, J. N. and Brasier, C. M. (1973). Correlation between

cultural characters and pathogenicity in *Ceratocystis ulmi* from Britain, Europe and America. *Nature, Lond.,* **241**, 381–3.

Gibbs, J. N. and Howell, R. S. (1972). *Dutch elm disease survey 1971.* For. Comm. For. Rec., no. 82.

Gibson, W. (1926). The elm. In *Collected poems, 1905–1925,* pp. 403–6.

Gilibert, J. E. (1792). *Exercita phytologica quibus omnes plantae europeae,* vol. 2, pp. 395–6.

Gill, R. E., Hirst, E. L. and Jones, J. K. N. (1939). Constitution of the mucilage from the bark of *Ulmus fulva* (slippery elm mucilage). Part I. The aldobionic acid obtained by hydrolysis of the mucilage. *J. Chem. Soc.* 1469–71.

Gill, R. E., Hirst, E. L. and Jones, J. K. N. (1946). Constitution of the mucilage from the bark of *Ulmus fulva* (slippery elm mucilage). Part II. The sugars found in the hydrolysis of the methylated mucilage. *J. Chem. Soc.* 1025–9.

Gleditsch, J. (1773). *Pflanzenverzeichniss,* pp. 353–4.

Goeppert, H. R. (1855). *Die tertiäre Flora von Schossnitz in Schlesien,* p. 28.

Gordon, C. (1943). Natural breeding sites of *Drosophila obscura. Nature, London,* **149**, 499–500.

Green, P. S. (1964). Registration of cultivar names in *Ulmus. Arnoldia,* **24**, 41–80.

Greguss, L. (1966). Vplyv podmienok prostredia na kľčenie peľu brestov. *Biológia, Czechoslovakia,* **21**, 813–20.

Greguss, L. (1972). Výsledky vrúbľovania niektorých druhov brestov na podpníky bresta horského. *Lesn. Čas.* **18**, 33–47.

Greig, B. J. W. (1979). *English elm regeneration.* For. Comm. Res. Inf. Note., no. 46.

Greville, M. (1949). Trees of a royal forest. *Country Life,* **105**, 1317–18.

Grigson, G. (1963). Elms under cloud. In *Collected poems,* p. 103.

Groschke, F. (1944). Neues über Minierer aus dem Mittelmeergebiet. *Mitt. münch. ent. Ges.* **34**, 115–21.

Grose, D. (1957). *The flora of Wiltshire,* Wiltshire Archaeological and Natural History Society, pp. 496–9.

Grosset, G. É. (1967). Puti i vremya migratsii lesnykh krymsko-kavkazskykh vidov na territoriju russkoĭ ravniny i posledushchie izmeneniya ikh arealov svyazi s ėvolyutsieĭ landshaftov. *Bjull. Moskov. Obshchestva Ispytaleleĭ Prirod.* **47**(5), 47–76.

Grudzinskaya, I. A. (1956). K systematike nekhotorykh vidov *Ulmus. Bot. Zh. USSR,* **41**, 97–104.

Grudzinskaya, I. A. (1957). Chto takoe *Ulmus carpinifolia* Gleditsch? *Bot. Mater. Gerbar. Bot. Inst. Nauk SSSR,* **18**, 48–50.

Grudzinskaya, I. A. (1966). Sotsvetiya vidov *Ulmus* L. *Bot. Zh. USSR,* **51**, 15–27.

Grudzinskaya, I. A. (1968). Kharakteristiki semeĭstv Ulmaceae Mirb. i Celtidaceae Link. *Novosti Sistematiki Vyssh. Rasten.* 95–8.

Grudzinskaya, I. A. (1971a). O zabytom gimalaĭskom vide *Ulmus* L. *Novosti Sistematiki Vyssh. Rasten.* **8**, 131–4.

Grudzinskaya, I. A. (1971b). O nazvanii vyaza polevogo (*Ulmus campestris* L.). *Novosti Sistematiki Vyssh. Rasten.* **8**, 135–9.

Grudzinskaya, I. A. (1977). Novyĭ vid il'ma – *Ulmus boissieri* (Boiss.) Grudz. (Ulmaceae) iz Irana. *Bot. Zh. USSR,* **62**, 856.

Grudzinskaya, I. A. and Chernik, V. V. (1976). Periginnaya trubka ('gipantiĭ') i sochlenennaya tsvetonozhka u *Ulmus* L.). *Bot. Zh., USSR,* **61**, 25–31.

Grudzinskaya, I. A. and Zahar'eva, O. I. (1967). O diagnosticheskom znachenii tsitologicheskikh priznakov u taksonomii nekotorikh drevesnykh porod (na primere roda *Ulmus* L.). *Bot. Zh., USSR,* **52**, 641–50.

Guignard, J.-L. and Mestre, J.-C. (1966). Sur le développement d'embryons à partir des antipodes chez l'*Ulmus campestris* L. *Bull. Soc. Bot. France,* **113**, 227–8.

Gürke, M. (1897). In K. Richter, *Plantae europaeae,* vol. 2, pp. 72–3.

Gysseling, M. (1960). *Toponymisch woordenboek van België, Nederland, Luxemburg, Noord-Frankrijk en West-Duitsland (vóór 1226),* Belgisch Inter-Universitair Centrum voor Neerlandistiek.

Hadfield, M. (1960). Elms of the English hedgerow. *Country Life,* **128**, 1302–4.

Hanbury, S. J. and Marshall, E. S. (1899). *Flora of Kent,* pp. 310–11.

Hance, H. F. (1868). Sertulum chinense tertium. *J. Bot.* **6**, 328–33.

Hans, A. S. (1981). Compatibility and crossability studies in *Ulmus. Silvae Genetica,* **30**, 149–52.

Hartig, T. (1851). *Vollständige Naturgeschichte der forstlichen Culturpflanzen Deutschlands,* pp. 452–67.

Hassall, C. (1957). The unsafe elm. In *The red leaf,* pp. 3–4.

Hayata, B. (1913). *Icones plantarum Formosanarum,* vol. 3, pp. 174–5.

Heer, O. (1856). *Flora tertiaria Helvetiae,* vol. 2, p. 59.

Hemsley, W. B. (1889–1902). In F. B. Forbes and W. B. Hemsley. An enumeration of all the plants known from China Proper, Formosa, Hainan, Corea, the Lucha Archipelago, and the Islands of Hongkong, together with their distribution and synonymy. *J. Linn. Soc. London (Bot.),* **26**, 1–592.

Henry, A. (1910). On elm-seedlings showing Mendelian results. *J. Linn. Soc. London (Bot.),* **39**, 290–300.

Henry, A. (1913). In H. J. Elwes and A. Henry. *The trees of Great Britain and Ireland,* vol. 7, pp. 1847–1929.

Hering, E. M. (1931). Minenstudien 12. *Z. Pflanzenkrank.* **41**, 530–51.

Hering, E. M. (1932). Minenstudien 13. *Z. Pflanzenkrank.* **42**, 567–79.

Heslop Harrison, J. W. (1920). New and rare British Aleurodidae. *Entomologist,* **53**, 255–7.

Heybroek, H. M. (1963). Diseases and lopping for fodder as possible causes of a prehistoric decline of *Ulmus. Acta Bot. Neerlandica,* II, 1–11.

Heybroek, H. M. (1965). Iep en onderstamm. *Korte Meded. Bosbouwproefsta. Dorschkamp,* no. 74, 3–4.

Hille Ris Lambers, D. (1968–9). A study of *Tetraneura* Hartig, 1841 (Homoptera, Aphididae), with descriptions of a new subgenus and new species. *Boll. Zool. Agrar. Bachicoltura,* II, **9**, 21–101.

Hillier, E. L. (1936). The elm problem. *Gardeners' Chron.* **99**, 102.

Hirst, E. L., Hough, L. and Jones, J. K. N. (1951). Constitution of the mucilage from the bark of *Ulmus fulva* (slippery elm mucilage). Part III. The isolation of 3-monomethyl D-galactose from the product of hydrolysis. *J. Chem. Soc.* 323–5.

Hjelmqvist, H. and Grazi, F. (1965). Studies on the

variation in embryo sac development. Second Part. *Bot. Not.* 118, 329–60.

Holden, M. (1980). Elm disease. In *The country heart*, p. 16.

Holmes, G. (1977). When the elms have gone. *Times*, 9 July.

Hood, T. (1906). The elm tree. In *Complete poetical works* (ed. W. Jerrold), pp. 616–22.

Hooker, W. J. and Arnott, G. A. W. (1855). *British flora*, 7th edn., pp. 386–8.

Hoppe, B. (1969). *Das Kräuterbuch des Hieronymus Bock*, pp. 370–1.

Horne, E. (1938). The unknown decorator. *Country Life*, 83, 206.

Horwood, A. R. (1933). *Ulmus elegantissima* mihi. The Midland elm. *In* Horwood, A. R. and Gainsborough, Earl of. The flora of Leicestershire and Rutland, 482–4.

Horwood, A. R. and Gainsborough, Earl of (1933). *The flora of Leicestershire and Rutland*, pp. 478–85.

Host, N. T. (1827). *Flora austriaca*, vol. 1, pp. 327–30.

Hough, L., Jones, J. K. N. and Hirst, E. L. (1950). Constitution of slippery elm mucilage: isolation of 3-methyl D-galactose from the hydrolysis products. *Nature, Lond.* 165, 34–5.

Howitt, R. C. L. and Howitt, B. M. (1963). *A flora of Nottinghamshire*, pp. 170–2.

Huberty, J. (1904). Étude forestière et botanique sur les ormes. *Bull. Soc. Cent. For. Belg.* 11, 408–27.

Huddesford, G. (1804). On a threat to destroy the tree at Winchester. In *The Wiccemical chaplet*, pp. 198–9.

Hudson, W. (1762). *Flora anglica*, 1st edn, pp. 94–5.

Hudson, W. (1778). *Flora anglica*, 2nd edn, p. 109.

Ingram, J. (1847). The Huntingdon elm – Bates v. Rivers. *Gardeners' Chron. Agric. Gaz.* 526.

Iversen, J. (1941). Landnam i Danmarks Stenalder. En pollenanalytisk Undersøgelse over det første Landbrugs Indvirkning paa Vegetationsudviklingen. *Danm. Geol. Unders.* Række 2, no. 66.

Jacobs, S. N. A. (1962). *Stigmella ulmifoliae* Hering, a species new to Britain. *Ent. Rec.* 74, 122–3.

Jacquin, N. J. (1798). *Plantorum rariorum horti caesarei Schoenbrunnensis descriptiones et icones*, vol. 3, pp. 6–7, plate 262.

Janjić, N. (1975). Sistematika poljskih brestova u Bosni i Hercegovini. *Godišnjaka Biol. Inst. Univ. Sarajevu*, 28, 117–24.

Jeffers, J. N. R. (1972). Dutch elm disease and botanical variation in English elm. *Nature, Lond.* 236, 407–8.

Jeffers, J. N. R. and Richens, R. H. (1970). Multivariate analysis of the English elm population. *Silvae Genetica*, 19, 31–8.

Johansson, K. (1921). Bidrag til kännedom om Gottlands *Ulmus*-former. *Svensk. Bot. Tid.* 15, 1–19.

Johansson, K. (1922). *Ulmus*-studier på Öland. *Bot. Not.* 197–202.

Johns, C. A. (1849). *The forest trees of Britain*, vol. 2, pp. 99–124.

Juba (1919). The elm tree. *Country Life*, 46, 816.

Kalm, J. (1966). Resejournal (ed. M. Kerkkonen), vol. 1.

Keate, E. M. (1912). The season of elm blossom. *Country Life*, 31, 343–5.

Keifer, H. H. (1939). Eriophyid studies VI. *Bull. Dep. Agric. California*, 28, 416–26.

Keifer, H. H. (1940). Eriophyid studies IX. *Bull. Dep. Agric. California*, 29, 112–7.

Keifer, H. H. (1959). Eriophyid studies XXVIII. *Occ. Pap. Bur. Ent. California Dep. Agric.* no. 2.

Keifer, H. H. (1962). Eriophyid studies B-6.

Keifer, H. H. (1965). Eriophyid studies B-13.

Keifer, H. H. (1971). Eriophyid studies C-5.

Keifer, H. H. (1975). Eriophyid studies C-11.

Kent, D. H. (1975). *The historical flora of Middlesex*, pp. 358–9.

Kirchner, G. (1864). In E. C. A. Petzold and G. Kirchner. *Arboretum muscaviense*, p. 560.

Kittel, M. B. (1844). *Taschenbuch der Flora Deutchlands*, vol. 2, pp. 224–6, 1198–9.

Klimesch, J. (1941). *Nepticula preissekeri* spec. nov. *Z. Wien. Ent. Ver.* 26, 162.

Klimesch, J. (1975). Die an Ulmen lebenden europaischen Nepticuliden-Arten (Lepidoptera: Nepticulidae). *Opuscula Zool.*, no. 135.

Klokov, M. V. (1963). *Materialy po istorii flory i rastitel'nosti SSSR*, no. 4.

Koch, K. (1849). Beiträge zu einer Flora des Orientes. *Linnaea*, 22, 599.

Kotov, M. I. (1940). Novii vid v'yaza – *Ulmus Wyssotzky* Kotov. sp. nov. *Bot. Zh., Ukrainian SSR*, 1, 333–4.

Krstić, M. (1950). Doprinos proučavanju konzerviranja brestovog remena (*Ulmus campestris* L.). *Rad. Inst. Nauč. Istraž. Šum. Beograd*, no. 1, 53–61.

Kumata, T. (1959). Redescriptions of the species of the genus *Lithocolletis* described by Prof. Dr S. Matsumura. *Insecta Matsumurana*, 22, 71–81.

Kumata, T. (1963). Taxonomic studies on the Lithocolletinae of Japan (Lepidoptera: Gracillariidae). II. *Insecta Matsumurana*, 25, 53–90; 26, 1–48, 69–88.

Kumata, T. (1967). New or little-known species of the genus *Lithocolltis* occurring in Japan (Lepidoptera: Gracillariidae). *Insecta Matsumurana*, 29, 59–72.

Kuznetsov, V. I. (1962). Il'movaya krivoucaya mol' – *Bucculatrix ulmicola* Kuznetz. sp. n. (Lepidoptera: Bucculatrigidae) – vreditel' il'mov v Zakavkaz'e i Srednei Azii. *Dokl. Akad. Nauk. Armyan. SSR*, 35, 81–3.

Laidlaw, W. B. R. (1932). The enemies of elm bark beetle. (*Scolytus destructor* Oliv.). *Scot. For. J.* 46, 117–29.

Laut, J. G., Schomaker, M. E., Stieger, T. M. and Metzler, J. (1979). *Dutch elm disease – a bibliography* (revised).

Lees, E. (1853–70). The forest and chace of Malvern. *Trans. Malvern Nat. Field Club*, 57–118.

Lees, E. (1874a). Notes on remarkable elms. *Gardeners' Chron.* 1, 790–1.

Lees, E. (1874b). Notes on old and curious wych elms. *Gardeners' Chron.* 2, 102–3.

Leliveld, J. A. (1933). Cytological studies in the genus *Ulmus* L. I. The supposed hybrid nature of the common Dutch elm. *Genetica*, 15, 425–32.

Le Sueur, A. D. C. (1941). Elms have a bad name. *Country Life*, 89, 304–5.

Ley, A. (1909). Fructification of elms. *J. Bot.* 47, 355.

Ley, A. (1910a). Notes on British elms. *J. Bot.* 48, 65–72.

Ley, A. (1910b). Notes on synonymy in *Ulmus*. *J. Bot.* 48, 130–2.

Liebmann, F. (1850). *Chaetoptelea*, en ny Slægt af Elmenes Familie. *Vidensk. Meddel. Naturhist. Foren. Kjöbenhavn*, 74–8.

Lindley, J. (1829). *A synopsis of the British flora*, pp. 226–7.

Lindquist, B. (1929–31). Two varieties of north-west

European *Ulmus glabra* Hudson. *Rep. Bot. Exchange Club*, **9**, 785.

Lindquist, B. (1930). Nya skånska växtlokaler. *Bot. Not.* 214–32.

Lindquist, B. (1932). Om den vildväxande skogsalmens raser och deras utbredning i Nordvästeuropa. *Acta phytogeograph. suecica*, **4**.

Linnaeus, C. (1737). *Hortus cliffortianus*, p. 83.

Linnaeus, C. (1753). *Species plantarum*, vol. 1, pp. 225–6.

Litchfield, F. (1889). In M. H. Bloxham, *Rugby*, p. 93.

Little, J. E. (1923). The Huntingdon elm in Sloane Herb. *J. Bot.* **61**, 201.

Litvinov, D. J. (1922). *Ulmus androssowi* sp. nova *Sched. Herbar. Flor. Ross.* **8**.

Lobel, M. de (1576). Plantarum seu stirpium historia, pp. 606–7.

Lobel, M. de (1581). *Kruydtboeck*, pp. 221–2.

Lobel, M. de and Pena, P. (1570). *Stirpium adversaria nova*, p. 440.

Loddiges, C. and Sons (1836). *Catalogue of plants*.

Loudon, J. C. (1838). *Arboretum et fruticetum britannicum*, vol. 3, pp. 1371–409.

Lousley, J. E. (1976). *Flora of Surrey*, pp. 234–7.

Lowbury, E. (1980). Elms in winter. In *Poems from the medical world* (ed. H. Sergeant), MTP Press, p. 152.

Lupe, I. Z. (1956). Contribuţii la cunoaşterea influenţei luminii asupra germinaţiei seminţelor de ulm. *Rev. Pădurilor*, **71**, 506–8.

Lyte, H. (1578). *A niewe herball*, pp. 751–4.

MacDougall, R. S. (1900). The biology and forest importance of *Scolytus* (*Eccoptogaster*) *multistriatus* (Marsh.). *Proc. Roy. Soc. Edinburgh*, **23**, 359–64.

Mädler, K. (1939). Die pliozäne Flora von Frankfurt am Main. *Abhand. senckenberg. naturforsch. Ges.*, no. 446.

Manaresi, A. (1935). Sulla germinazione delle samare de olmo campestre. *Ital. Agric.* **72**, 733–6.

Marshall, J. E. (1978). The larva of *Aulonium trisulcum* (Fourcroy) (Coleoptera: Colydiidae) and its association with elm bark beetles (*Scolytus* spp.). *Ent. Gaz.* **29**, 59–69.

Martin, W. K. and Fraser, G. T. (1939). Flora of Devon, 577–8.

Mayr, H. (1906). *Fremdländische Wald- und Parkbaume für Europa*, pp. 523.

Melville, J. C. (1909). *Ulmus glabra* Huds. *J. Bot.* **47**, 324.

Melville, R. (1937). The accurate definition of leaf shapes by rectangular coordinates. *Ann. Bot.* **1**, 673–80.

Melville, R. (1938a). Contributions to the study of British elms. I. What is Goodyer's elm. *J. Bot.* **76**, 185–92.

Melville, R. (1938b). Is *Ulmus campestris* L. a nomen ambiguum? *J. Bot.* **76**, 261–5.

Melville, M. (1939a). Contributions to the study of British elms. II. The East Anglian elm. *J. Bot.* **77**, 138–45.

Melville, R. (1939b). Ambiguous elm names. I. *Ulmus sativa* Mill. *J. Bot.* **77**, 244–8.

Melville, R. (1939c). Ambiguous elm names. II. *Ulmus minor* Mill. *J. Bot.* **77**, 266–70.

Melville, R. (1939d). The application of biometrical methods to the study of elms. *Proc. Linn. Soc. London (Bot.)* **151**, 152–9.

Melville, R. (1939e). On elm seedlings. *Quart. J. For.* **31**, 258–66.

Melville, R. (1940). Contributions to the study of British elms. III. The Plot elm, *Ulmus Plotii* Druce. *J. Bot.* **78**, 181–92.

Melville, R. (1946). Typification and variation in the smooth-leaved elm, *Ulmus carpinifolia* Gleditsch. *J. Linn. Soc. London (Bot.)*, **53**, 83–90.

Melville, R. (1949). The Coritanian elm. *J. Linn. Soc. London (Bot.)*, **53**, 263–71.

Melville, R. (1951). The elms of the Dumortier herbarium. *Bull. Jardin Bot. État Bruxelles*, **21**, 347–51.

Melville, R. (1955). Morphological characters in the discrimination of species and hybrids. In J. E. Lousley (ed.). *Species studies in the British flora*, 55–64.

Melville, R. (1956). An early specimen of *Ulmus carpinifolia* Gleditsch. *Kew Bull.* 179–81.

Melville, R. (1957). *Ulmus canescens*: an eastern Mediterranean elm. *Kew Bull.* 499–502.

Melville, R. (1960). The names of the Cornish and the Jersey elm. *Kew Bull.* **14**, 216–8.

Melville, R. (1975). *Ulmus.* In C. A. Stace (ed.), *Hybridization and the flora of the British Isles*, Academic Press, pp. 292–9.

Melville, R. (1978). On the discrimination of species in hybrid swarms with special reference to *Ulmus* and the nomenclature of *U. minor* Mill. and *U. carpinifolia* Gled. *Taxon*, **27**, 345–51.

Melville, R. and Heybroek, H. M. (1972). The elms of the Himalaya. *Kew Bull.* **26**, 5–28.

Messenger, G. (1971). *Flora of Rutland*, pp. 69–70.

Meyer, H. J. and Norris, D. M. (1967). Behavioural response by *Scolytus multistriatus* (Coleoptera: Scolytidae) to host- (*Ulmus*) and beetle-associated chemotactic stimuli. *Ann. Ent. Soc. Amer.* **60**, 642–7.

Meyerhof, M. (1935–6). Sur le nom dardär (orme et frêne) chez les arabes. *Bull. Inst. Égypte*, **18**, 137–49.

Michalski, J. (1973). *Revision of the palearctic species of the genus* Scolytus *Geoffroy (Coleoptera, Scolytidae)*.

Miles, P. M. (1978). Blomer's rivulet, *Discoloxia blomeri* (Curtis) (Lepidoptera, Geometridae) and Dutch elm disease, *Ceratocystis ulmi* Buism. *Ent. Gaz.* **29**, 43–6.

Miller, P. (1731). *Gardeners dictionary*, 1st edn.

Miller, P. (1739). *Gardeners dictionary*, 2nd edn.

Miller, P. (1748). *Gardeners dictionary*, 3rd edn.

Miller, P. (1759). *Gardeners dictionary*, 7th edn.

Miller, P. (1768). *Gardeners dictionary*, 8th edn.

Moench, C. (1785). *Verzeichniss*, pp. 136–7.

Moench, C. (1794). *Methodus plantas horti botanici et agri marburgensis*, p. 333.

Moggridge, E. (1912). Elm-blossom. *Country Life*, **31**, 341.

Mordvilko, A. K. (1929). Anolocyclic elm aphids, Eriosomea, and the distribution of elms during the Tertiary and Glacial Periods. *Dokl. Akad. Nauk. USSR (A)* **8**, 197–202.

Morrison, M. E. S. (1959). Evidence and interpretation of 'Landnam' in the north-east of Ireland. *Bot. Not.* **112**, 185–204.

Moss, C. E. (1912). British elms. *Gardeners' Chron.* **51**, 199, 216–7, 234–6.

Moss, C. E. (1914). *The Cambridge British flora*, vol. 2, pp. 88–96.

Mueller, C. H. (1936). New and noteworthy trees in Texas and Mexico. *Bull. Torrey Bot. Club*, **63**, 147–55.

Muhlenberg, H. (1793). Index florae Lancastriensis. *Trans. Amer. Phil. Soc.* **3**, 157–84.

Munro, J. W. (1926). British bark-beetles. *Bull. For. Comm.*, no. 8.

Nicholson, N. (1974). The elm decline. In *A local habitation*, pp. 32–4.

Nicoll, Lady R. (1944). A tale of a tree. *Country Life*, **95**, 693.

Nordhagen, R. (1954). Om barkebrød og treslaget alm i kulturhistorisk belysning. *Danm. Geol. Unders.* Række 2, no. 80.

Nuttall, T. (1837). Collections towards a flora of the territory of Arkansas. *Trans. Amer. Phil. Soc.* **5**, 139–203.

P., J. (1766). Some account of the Watch Elm, at Stoke Gifford in Gloucestershire. *Gentleman's Mag.* **36**, 504.

Pallas, P. S. (1784). *Flora rossica*, vol. 1, p. 75.

Parkinson, J. (1640). *Theatrum botanicum*, pp. 1403–8.

Peace, T. R. (1960). The status and development of elm disease in Britain. *For. Comm. Bull.*, no. 33.

Peace, T. R. (1962). *Pathology of trees and shrubs*, pp. 265–6, 424–5.

Peacock, J. W., Lincoln, A. C., Simeone, J. B. and Silverstein, R. M. (1971). Attraction of *Scolytus multistriatus* (Coleoptera: Scolytidae) to a virgin-female-produced pheromone in the field. *Ann. Ent. Soc. Amer.* **64**, 1143–9.

Pearce, G. T., Gore, W. E., Silverstein, R. M., Peacock, J. W., Cuthbert, R. A., Lanier, G. N. and Simeone, J. B. (1975). Chemical attractants for smaller European elm bark beetle, *Scolytus multistriatus* (Coleoptera: Scolytidae). *J. Chem. Ecol.* **1**, 115–24.

Pearce, N. J. and Richens, R. H. (1977). Peroxidase isozymes in some elms (*Ulmus* L.) of eastern England. *Watsonia*, **11**, 382–3.

Petch, C. P. and Swann, E. L. (1968). Flora of Norfolk, 169–70.

Pflug, H. (1953). In P. W. Thomson and H. Pflug, Pollen und Sporen des mitteleuropäischen Tertiärs. *Palaeontographica*, **94B**, 90.

Pilcher, J. R. (1969). Archaeology, palaeoecology and radiocarbon dating of the Beaghmore Stone Circle site. *Ulster J. Arch.* **32**, 73–91.

Planchon, J. E. (1848). Sur les Ulmacées. *Ann. Sci. Nat., Bot.* **10**, 244.

Planchon, J. E. (1873). Ulmaceae. In A. de Candolle, *Prodromus systematis naturalis regni vegetabilis*. vol. **17**, 151–60.

Plath, S. (1965). Elm. In *Ariel*, pp. 25–6.

Pliny (1938–62). *Naturalis historia*, Loeb edn, vol. 4, pp. 434–5.

Plot, R. (1677). *The natural history of Oxfordshire*, pp. 158–69.

Plukenet, L. C. (1696). *Almagestum botanicum*, p. 393.

Poederlé, Baron de (1792). *Manuel de l'arboriste et du forestier belgiques*, 3rd edn, vol. 2, pp. 114–9.

Pool, C. J. C. (1904). *Aulonium sulcatum*, Oliv. (trisulcum, Fourc.) a species of colydiid Coleoptera new to Great Britain. *Ent. Rec. J. Var.* **16**, 310.

Popovski, P. (1970a). Morfologiškkite odliki na nizinskite brestovi vo Srednoto Povardarje i Ovče Pole. *Godišen Zborn. Zemjod.-Šumarsk. Fak. Univ. Skopje, Šumarstvo*, **23**, 87–116.

Popovski, P. (1970b). Nizinskite brestovi vo okolinata na Ulcinj (Crna Gora). *Godišen Zborn. Zemjod.-Šumarsk. Fak. Univ. Skopje, Šumarstvo*, **23**, 219–27.

Popovski, P. (1972). Komparativni proučavanja na 'rtlivosta na semeto od nizinski brest (*U. campestris* L., *U. minor* Mill.) što e čuvano vo različni uslovi. *Godišen Zborn. Zemjod.-Šumarsk. Fak. Univ. Skopje, Šumarstvo*, **24**, 17–31.

Pospichal, E. (1897). *Flora des Oesterreichischen Küstenlandes.*

Preissecker, F. (1942). Zwei neue *Nepticula*-Arten aus dem Gebiet des heutigen Reichsgaues Wien. *Z. Wien. Ent. Ver.* **27**, 208.

Preston, T. A. (1888). *The flowering plants of Wilts*, pp. 268–9.

Rawnsley, H. D. (1877). The knotted elm, at Abbots Leigh. In *A book of Bristol sonnets*, p. 65.

Redenbaugh, M. K., Westfall, R. D. and Karnosky, D. F. (1981). Dihaploid callus production from *Ulmus americana* anthers. *Bot. Gaz.* **142**, 19–26.

Rehder, A. (1917). *Ulmus*. In L. H. Bailey, *The standard cyclopedia of horticulture*, vol. 6, pp. 3408–13.

Rehder, A. (1938). New species, varieties and combinations from the collection of the Arnold Arboretum. *J. Arnold Arboretum* **19**, 264–78.

Rehder, A. (1940). Manual of cultivated trees and shrubs, 2nd edn, pp. 174–82.

Ribezzo, F. (1950). Di quattro nuove voci mediterranee gia credute celtiche. *Rev. Internat. Onomastique*, **2**, 13–25.

Richards, W. R. (1967). A review of the *Tinocallis* of the World (Homoptera: Aphididae). *Canadian Ent.* **99**, 536–53.

Richardson, A. D. (1911). The weeping varieties: *Ulmus montana*. *Gardeners' Chron.* **50**, 221.

Richens, R. H. (1955). Studies on *Ulmus*. I. The range of variation of East Anglian elms. *Watsonia*, **3**, 138–53.

Richens, R. H. (1956). Elms. *New Biol.* **20**, 7–29.

Richens, R. H. (1958). Studies on *Ulmus*. II. The village elms of southern Cambridgeshire. *Forestry*, **31**, 132–46.

Richens, R. H. (1959). Studies on *Ulmus*. III. The village elms of Hertfordshire. *Forestry*, **32**, 138–54.

Richens, R. H. (1960). Cambridgeshire elms. *Nature in Cambridgeshire*, 18–22.

Richens, R. H. (1961a). Studies on *Ulmus*. IV. The village elms of Huntingdonshire and a new method for exploring taxonomic discontinuity. *Forestry*, **34**, 47–64.

Richens, R. H. (1961b). Studies on *Ulmus*. V. The village elms of Bedfordshire. *Forestry*, **34**, 181–200.

Richens, R. H. (1961c). Vagrant eriophyid mites on elm. *Ent. Month Mag.* **97**, 264.

Richens, R. H. (1962). Western Himalayan elm galls and leaf mines. *Agra Univ. J. Res. (Science)*, **11**, 167–72.

Richens, R. H. (1963a). Monophage analysis of elm populations. World Consultation on Forest Genetics and Tree Improvement. FAO/FORGEN 63-66/4.

Richens, R. H. (1963b). Four new stigmellid elm-leaf mines. *Ent. Gaz.* **14**, 36–8.

Richens, R. H. (1965). Studies on *Ulmus*. VI. Fenland elms. *Forestry*, **38**, 225–35.

Richens, R. H. (1967). Studies on *Ulmus*. VII. Essex elms. *Forestry*, **40**, 185–206.

Richens, R. H. (1968). The correct designation of the European field elm. *Feddes Repertorium*, **79**, 1–2.

Richens, R. H. (1972). The stately elms of England. *Times*, 9 Dec.

Richens, R. H. (1976). Variation, cytogenetics and breeding of the European field elm (*Ulmus minor* Miller sensu latissimo = *U. carpinifolia* Suckow). *Ann. For., Zagreb*, **7**, 107–45.

Richens, R. H. (1977a). New designations in *Ulmus minor* Mill. *Taxon*, **26**, 583–4.

Richens, R. H. (1977b). Elms and motorways. In R. K. Tabor, *Motorways and the biologist*, pp. 58–65.

Richens, R. H. (1977c). When the elms have gone. *Times*, 23 June.

Richens, R. H. (1978). The tougher elm. *Times*, 15 May.

Richens, R. H. (1980). On fine distinctions in *Ulmus* L. *Taxon*, **29**, 305–12.

Richens, R. H. (1981). Elms (genus *Ulmus*). In B. Hora, *The Oxford encyclopedia of trees of the world*, pp. 150–2.

Richens, R. H. and Jeffers, J. N. R. (1975). Multivariate analysis of the elms of northern France. *Silvae Genetica*, **25**, 141–50.

Richens, R. H. and Jeffers, J. N. R. (1978). Multivariate analysis of the elms of northern France. II. Pooled analysis of the elm populations of northern France and England. *Silvae Genetica*, **27**, 85–95.

Riddelsdell, H. J., Hedley, G. W. and Price, W. R. (1948). *Flora of Gloucestershire*, pp. 423–8.

Rivers, T. (1855). Cork-barked elms. *Gardeners' Chron. Agric. Gaz.* 790.

R[ivers], T. (1861). A golden-leaved elm. *Gardeners' Chron. Agric. Gaz.* 718.

Rivers, T. (1869). Weeping elms. *Gardeners' Chron. Agric. Gaz.* 840–1.

Rodgers, F. (1955). The Marriage Elms. *Country Life*, **118**, 1033.

Rogers, W. H. (1869). Elms. *Gardeners' Chron. Agri. Gaz.* 54.

Ross-Craig, S. (1970). Drawings of British plants, part 27, plates 1–6.

Rowe, J. W., Seikel, M. K., Roy, D. N. and Jorgensen, E. (1972). Chemotaxonomy of *Ulmus*. *Phytochemistry*, **11**, 2513–7.

Ruellius (1536). *De natura stirpium*, p. 370.

Ruppius, H. B. (1745). *Flora ienensis*, p. 330.

Salisbury, R. A. (1796). *Prodromus stirpium in horto ad Chapel Allerton vigentium*, p. 391.

Salmon, C. E. (1931). *Flora of Surrey*, pp. 580–1.

Santamour, F. S., Jr. (1972a). Flavonoid distribution in *Ulmus*. *Bull. Torrey Bot. Club*, **99**, 127–31.

Santamour, F. S., Jr. (1972b). Interspecific hybridization with fall- and spring-flowered elms. *For. Sci.* **18**, 283–289.

Săvulescu, O. and Eliade, E. (1962). Contribuţie la cunoaşterea micromicetelor din Republica Populară Romînă. *Stud. Cerc. Biol., Biol. Veg.* **14**, 9–27.

Sax, K. (1933). Chromosome numbers in *Ulmus* and related genera. *J. Arnold Arboretum* **14**, 82–4.

Schkuhr, C. (1791). *Botanisches Handbuch*, vol. 1, pp. 178–9.

Schneider, C. K. (1906–12). *Illustriertes Handbuch der Laubholzkunde*, vol. 1, p. 220; vol. 2, pp. 901–4.

Schneider, C. K. (1916a). Beiträge zur Kenntnis der Gättung *Ulmus*. I. Gliederung der Gattung und Übersicht der Arten. *Öst. bot Z.* **66**, 21–34.

Schneider, C. K. (1916b). Beiträge zur Kenntnis der Gattung *Ulmus*. II. Über die richtige Benennung der europäischen *Ulmen-Arten. Öst. bot. Z.* **66**, 65–81.

Scott-Snell, E. (1940). Elms at Kelmscott. *Country Life*, **88**, 241.

Scovell, E. J. (1958). Leaves of elm. In *Poems of the mid-century* (ed. J. Holloway), p. 60.

Serra, G. (1951). Tracce del culto dell' olmo e del tiglio nella

toponomastica e negli usi civili dell' Italia medioevale. *Proc. Trans. 3rd Internat. Congr. Toponymy Anthroponymy*, **3**, 546–63.

Seymour, W. (1965). A plea for the disappearing elm. *Country Life*, **137**, 1429.

Shkhiyan, A. S. (1953). Novyĭ vid roda *Ulmus* L. iz Vostochnoĭ Gruzii. *Zametki Sistematike Geograf. Rasten., Tbilisi*, no. 17, 78.

Simonkai, L. (1890). Nagyváradnak és vidékének növényvilága. In *Nagyvárad termeszetrajza*, p. 124.

Simon-Louis (1869). *Catalogue*, p. 98.

Skala, H. (1933). Neue Neptikel. *Z. öst. Ent. Ver.* **18**, 30–2.

Skala, H. (1934). Neptic. ulmi spec. nov. und andere wenig bekannte Neptikel. *Z. öst. Ent. Ver.* **19**, 51.

Skala, H. (1936). Minen aus Mittel- und Südeuropa. *Z. öst. Ent. Ver.* **21**, 78–9.

Skala, H. (1939). Miner in deutschen Landen. *Z. öst. Ent. Ver.* **24**, 27–30.

Skala, H. (1941). Neues uber Miner. *Z. Wien. Ent. Ver.* **26**, 55.

Skala, H. (1942). Falter von Haid und Anderes. *Z. Wien. Ent. Ver.* **27**, 5.

Smith, A. G. and Willis, E. H. (1961–2). Radiocarbon dating of the Fallahogy Landnam Phase. *Ulster J. Arch.* **24–25**, 16–24.

Smith, J. E. (1790–1814). In J. Sowerby and J. E. Smith, *English botany*.

Smith, J. E. (1800). *Flora britannica*, vol. 1, pp. 281–2.

Sørensen, I. and Søltoft, P. (1958). Gas-liquid chromatographic determination of the fatty acid composition in oil from seeds of *Ulmus*. *Acta Chem. Scandinavica*, **12**, 814–22.

Sowerby, J. and Smith, J. E. (1790–1814). *English botany*.

Spach, E. (1841a). Note sur les Planera. *Ann. Sci. Nat., Bot.* **15**, 349–59.

Spach, E. (1841b). Revisio Ulmorum europaeorum et boreali-americanum. *Ann. Sci. Nat., Bot.* **15**, 359–65.

Stearn, W. T. and Gilmour, J. S. L. (1932). Notes from the University Herbarium, Cambridge, **70**, suppl. 1–29.

Stearn, W. T. and Gilmour, J. S. L. (1933). *Ulmus campestris* Linn. *Kew. Bull.* 503–4.

Stockmarr, J. (1970). Species identification of *Ulmus* pollen. *Danm. Geol. Unders.* Række 4, **4**(11).

Stockmarr, J. (1974). SEM studies on pollen grains of north European *Ulmus* species. *Grana*, **14**, 103–7.

Stokes, J. (1787). In W. Withering, *A botanical arrangement of British plants*, vol. 1, pp. 259–60.

Stokes, J. (1812). *A botanical materia medica*, vol. 2, pp. 34–9.

Strode, W. (1907). On a great hollow tree. In *Poetical works* (ed. B. Dobell), pp. 21–5.

Strutt, J. (1822). *Sylva britannica*.

Suckow, G. (1777). *Oekonomische botanik*, pp. 39–41.

Syme, J. T. B. (1868). In J. Sowerby, *English botany*, 3rd edn, vol. 8, pp. 137–43.

Szafer, W. (1961). Mioceńska flora ze Starych Gliwic na Ślasku. *Prace Inst. Geol.* **33**, 1–203.

T., J. L. (1944). A tree's perserverance. *Country Life*, **95**, 648.

Tabernaemontanus, J. T. (1588). *Neuw Kreuterbuch*, vol. 3, pp. 689–91.

Takai, S. (1974). Pathogenicity and cerato-ulmin production in *Ceratocystis ulmi*. *Nature, Lond.* **252**, 124–6.

Takai, S. (1980). Relationship of the production of the toxin, cerato-ulmin, to synnemata formation, pathogenicity, mycelial habit, and growth of *Ceratocystis ulmi* isolates. *Canadian J. Bot.* **58**, 658–62.

Takhtadzhyan, A. L. (1945). Zametka o dvukh novykh vidakh roda *Ulmus* iz Yuzhnogo Zakavkaz'ya. *Dokl. Akad. Nauk. Armyan. SSR*, **2**, 57–8.

Tanai, T. and Wolfe, J. A. (1977). Revisions of *Ulmus* and *Zelkova* in the Middle and Late Tertiary of western North America. *Prof. Pap. US Geol. Survey*, no. 1026.

ten Hove, H. A. (1968). The *Ulmus* fall at the transition Atlanticum – Subboreal pollen diagrams. *Palaeogeogr. Palaeoclimat. Palaeoecol.* **5**, 359–69.

Theophrastus (1644). *De historia plantarum.*

Theophrastus (1916). *De historia plantarum*, Loeb edn, Heinemann, vol. 1, pp. 248–51.

Thornton, R. J. (1810). *A new family herbal*, p. 241.

Thurston, E. (1930). *British and foreign trees and shrubs in Cornwall.*

Toll, S. (1934). Jeszcze o krajowych gatunkach rodzaja *Nepticula* Zell. *Polskie Pismo Ent.* **8**, 77.

Tournefort, J. P. de (1700). *Institutiones rei herbariae*, vol. 1, p. 601.

Townsend, A. M. (1975). Crossability patterns and morphological variation among elm species and hybrids. *Silvae Genetica*, **24**, 18–23.

Townsend, A. M. (1979). Influence of specific combining ability and sex of gametes on transmission of *Ceratocystis ulmi* resistance in *Ulmus*. *Phytopathology*, **69**, 643–5.

Townsend, F. (1904). *Flora of Hampshire*, pp. 371–2.

Trautvetter, E. R. de (1859). In C. J. Maximowicz. Primitiae florae amurensis. *Mém. Savants Étrangers Acad. Imp. Sci. St Petersbourg*, **9**, 246–9.

Trimen, H. and Thistleton Dyer, W. T. (1869). *Flora of Middlesex*, 254–5.

Troels-Smith, J. (1959). En Elmetræs-Bue fra Aamosen og andre Træsager fra tidlig-neolitisk Tid. *Aarbøg. Nord. Oldkyndighed Hist.* 91–145.

Troels-Smith, J. (1960). Ivy, mistletoe and elm. Climate indicators – fodder plants. *Danm. Geol. Unders.* Række 4, **4**(4).

Turk, S. M. (1981). Biggest landscape disaster. *West Briton*, 2 April.

Unger, F. (1847). *Chloris protogaea*, pp. 91–101.

Verdcourt, B. (1947). The increase of *Aulonium trisulcum*

Fourc. (Col., Colydiidae) in Britain. *Ent. Month. Mag.* **83**, 185–6.

Verzhutskiĭ, B. N., Gronina, L. M. and Naĭmushi, E. P. (1976). Miniruyushchie pilil'shchiki Primor'ya. *Zaschita Rasten.*, no. 2, 45.

Vincent, G. (1958). Die Lagerung von Saatgut in geschlossen Gefässen. *Cbl. ges. Forstw.* **75**, 257–67.

Vinje, J. M. and Vinje, M. M. (1955). A note on the origin of 'blue rain'. *Proc. Iowa Acad. Sci.* **62**, 92.

W. (1839). The Kingsborough elm, Isle of Sheppey. *Gentleman's Mag.* **11**, 28–9.

W., J. (1968). The Fyfield elm. *Country Life*, **103**, 187.

Wahlenberg, G. (1814). *Flora Carpetorum principalium*, p. 71.

Walker, H. (1927). An immemorial elm. *Country Life*, **61**, 219.

Wallich, N. (1831). *Plantae asiaticae rariores*, vol. 2, p. 200.

Webster, A. D. (1926). Our vanishing elm trees. *Gardeners' Chron*, **79**, 216.

Wells, R. (1967). Whipping trees. *Daily Telegraph*, 15 March.

Went, J. C. (1954). The Dutch elm disease. Summary of fifteen years' hybridisation and selection work (1937–1952). *Tijd. Plantenziekt.* **60**, 109–27.

Went, J. C. (1955). Verlsag van de onderzoekingen over de iepenziekte en andere boomziekten, uitgevoerd op het Phytopathologisch Laboratorium 'Willie Commelin Scholten' te Baarn gedurende 1952 en 1953. *Meded. Com. Ziekt. Boomsorten*, no. 48.

Westerdijk, J. and Buisman, C. (1929). *De iepenziekte. Rapport over het onderzoek verricht op versoek van de Nederlandsche Heidemaatschappij.*

Weston, R. (1770). *Botanicus universalis*, pp. 314–5, 352.

Weston, R. (1775). *English flora*, p. 46.

White, J. W. (1912). *The flora of Bristol*, pp. 532–4.

Wilczinski, T. (1921). *Flora polska*, vol. 2.

Wilkinson, G. (1978). *Epitaph for the elm*, Hutchinson.

Wilmott, A. J. (1922). In C. C. Babington, *Manual of British botany*, 10th edn, 591–3.

Winieski, J. A. (1959). Artificial hybridization and grafting methods with *Ulmus americana*. *Proc. 7th NE For. Tree Improvement Conf.* 48–51.

Wolff, H. (1934). Mikrofossilien des pliozänen Humodils. *Arb. Inst. Paläobot. Petrogr. Brennsteine*, **5**, 55–86.

Woodman (1957). Two notable elms. *Country Life*, **122**, 1391.

Wooley-Dod, A. H. (1937). *Flora of Sussex*, pp. 400–1.

Young, A. (1974). The elm beetle. In *Complete poems* (ed. L. Clark), Secker & Warburg, pp. 128–9.

Index

Numbers in bold type refer to the serial numbers of the illustrations.
Scientific Latin names and vernacular names are in italics.